Hazardous Waste Measurements

Edited by

Milagros S. Simmons

Library of Congress Cataloging-in-Publication Data

Hazardous waste measurements / edited by Milagros S. Simmons.
p. cm.
Includes bibliographical references and index.
ISBN 0-87371-171-8
1. Hazardous wastes—Analysis. 2. Hazardous wastes—Toxicology—Measurement. I. Simmons, Mila S.
TD1030.H39 1990
628.4′2—dc20 90-41121
CIP

LEWIS PUBLISHERS, INC.
121 South Main Street, P.O. Drawer 519, Chelsea, Michigan 48118

PRINTED IN THE UNITED STATES OF AMERICA

Preface

An essential component of all programs relating to waste management is the ability to perform measurements on site for safe handling and disposal of hazardous wastes. This volume is designed to focus on recent developments in field testing methods and quality assurance, of importance to both RCRA and CERCLA hazardous waste management programs. The book highlights sampling strategies, field measurements and toxicity screening of complex waste matrices and describes requirements for quality assurance intended to be used in hazardous wastes remediation, management, and control. The chapters are organized into five main sections: (1) an introductory chapter describing regulatory requirements for hazardous wastes measurements; (2) a section on sampling, with emphasis on sampling designs, methods, and equipment; (3) a section on state-of-the-art field techniques and instrumentation; (4) a section on toxicity screening using short-term assays and plant test systems; (5) a section on quality assurance issues in hazardous wastes. The chapters describe work in progress at each contributor's institution.

Today's rapidly developing and changing technologies, industrial products, and practices result in increased generation of solid and hazardous wastes. The importance of hazardous wastes measurements in dealing with the management and control of these materials cannot be overemphasized.

Milagros S. Simmons
The University of Michigan
Ann Arbor, Michigan

Milagros S. Simmons is an Associate Professor in the Department of Environmental and Industrial Health, School of Public Health, The University of Michigan in Ann Arbor. She is an advisor to students in the Environmental Chemistry and Hazardous Wastes Programs in that department. Dr. Simmons received her B.S. in Chemistry from the University of Santo Tomas in Manila, Philippines and her M.S. and Ph.D. in Chemistry from Wayne State University in Detroit. She has also taught several short courses in the area of environmental pollutants and hazardous wastes during the last ten years. She is likewise affiliated with the Michigan Universities Consortium on Hazardous Substance Research and Training, the NIOSH-Educational Resource Center on Hazardous Waste Training and is chair of a subcommittee on Hazardous Wastes of the Pilot Secretariat on the Measurement of Pollution of the International Organization of Legal Metrology.

Contents

INTRODUCTION

SAMPLING

FIELD TECHNIQUES AND INSTRUMENTATION

TOXICITY SCREENING METHODS FOR HAZARDOUS WASTE

QUALITY ASSURANCE/QUALITY CONTROL

SECTION I

INTRODUCTION

CHAPTER 1

Regulatory Requirements for Hazardous Waste Measurements

James E. Martin

INTRODUCTION

The key perspective pertaining to waste materials is perhaps best described in the National Environmental Policy Act of 1969, which states that each generation has a responsibility "as trustee of the environment for succeeding generations." Waste products represent an intergenerational linkage because many are not only toxic, but very long-lived, thereby posing risks to both present and future generations. This long-term element is basic to most policy issues pertaining to wastes; it is now being formally recognized in stated levels of constituents leached from waste forms requiring measurements to determine if wastes are to be controlled as hazardous wastes.

Waste problems are due to rapid industrial change in the past few decades; most of the sophisticated chemicals that pose environmental contamination were not in existence 50 years ago. The newness and large quantities of such substances are significant to society's challenge in dealing with them and in incorporating societal values on potential risks of waste constituents in control decisions. The amount, the extent, and the apparent seriousness of solid and hazardous wastes were markedly below those of today.

Perhaps the most valuable lesson from past management of wastes is that casual handling can result in costly future endeavors in location, evaluation, and remediation. Waste disposal practices have left a legacy of environmental contamination levels that are essentially irretrievable because of the extreme technical requirements and costs of remediating them. Methane generation at old landfills, the presence of acids and other contaminants that have destroyed habitats in numerous streams, the discovery of old radium dumps, the presence of toxic chemicals in ground and surface waters near old industrial sites, and improperly managed landfills are all attributable to improper disposal of such wastes or improper closure and management following closure. Such practices have produced serious health impacts; others are of concern as potential threats of environmental contamination. Intensive federal efforts are underway to identify wastes from current industrial activities, to locate and investigate old sites, to prescribe the necessary controls for each, and to detail measurement programs to decide which wastes require control.

The Hazardous Waste Problem

Hazardous wastes, when disposed, represent a complex system that poses a risk to human health through all elements of the environment. After generation, a decision must be made about their eventual management based on measurements of important characteristics which determine their hazard. In the process of becoming wastes, the constituents may be reused, modified by treatment, or disposed as produced or as modified by treatment. These processes can impact air, land, surface water, and groundwater, each of which can be in contact with human populations. Improper management during each of these processes may create risks to people. Such risks may involve long-term risks due to environmental contamination if improperly managed including effects on health. Many hazardous waste materials could produce cancer or birth defects if ingested by humans in contact with contaminated air, land, or water. The key mechanism for preventing such risks is isolation: it is essential for the short term while wastes are handled, but critical for long-term protection since many such wastes are long-lived. Many chemical wastes contain metals and stable chemical compounds that are extremely long-lived (some forever) and are toxic over those lifetimes.

The potential risks posed by chemical wastes require such risks to be either avoided or mitigated during their management. This can be done by restricting storage, treatment, and disposal to proper sites; by encouraging maximum reuse and treatment of wastes before disposal; and remediating past problems to assure protection of public health and the environment. Since ultimate disposal of either such waste depends to a large degree on internment in the ground, it is particularly important in this context to assure protection of the long-term reservoirs of clean groundwater which are more and more scarce.

Conceptual Basis of the Hazardous Waste Control System

Control of hazardous chemical wastes is mandated by Subtitle C of the Resource Conservation and Recovery Act (RCRA) of 1976, which was substantially amended in 1984. The system for control of hazardous chemical wastes is shown in Figure 1. In concept, the RCRA control system is based on producers of waste being classified as generators if they generate any listed waste above cutoff quantities or if a waste meets criteria for corrosivity, ignitability, reactivity, or toxicity based on an extraction procedure test. The bulk of the responsibility for assuring proper management of such wastes is on generators who must identify wastes and manifest them to transporters (subject to EPA and DOT regulations) for transfer to permitted treatment, storage, or disposal

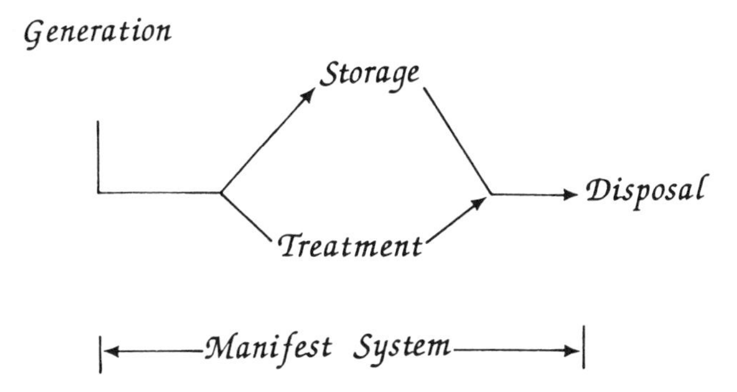

Figure 1. RCRA "Cradle to Grave" hazardous waste control.

facilities. Conceptually, proper management is brought about by (1) providing generators with requirements for identifying wastes and requiring them to enter their waste into the system and (2) controlling the balance of the system by permit conditions to assure that such wastes are properly managed or disposed once they leave the generator's control. The manifest system provides the documentation that this systematic approach has been followed for any waste determined to be hazardous. The control system does not require generators to be licensed (i.e., chemical waste generators don't require a permit to generate waste; these are required only to store, treat, or dispose).

MEASUREMENTS OF HAZARDOUS WASTES

Deciding what constitutes a hazardous chemical waste is many times difficult, but as a general rule, measurements are made of those materials that meet the statutory definition of first being a solid waste. This has been a larger problem in the chemical industry because so many materials could be recycled; if not now, at some future point. This is a key issue in implementation of the 1984 RCRA amendments. The Environmental Protection Agency has dealt with this issue for chemical wastes by specifying criteria that satisfy the definition of hazardous waste given by Congress and then listing specific wastes that are considered hazardous or extremely hazardous. In addition to those listed wastes, other waste materials are encompassed in the hazardous waste control system if they meet one or more of four characteristics: ignitability, corrosivity, reactivity, and a toxicity characteristic based on a leaching test. Tests for these characteristics are detailed in 40 CFR 261.20.

The leachability test for extractable toxic constituents (the EP toxic test) from a generated waste has been problematic. The congress in the 1984 RCRA amendments required that the EP toxicity test (40 CFR 261.23) be updated. On June 29, 1990, EPA issued a final rule (55 FR 26986) to replace the existing extraction procedure test (EP) contained in 40 CFR 261.24 with a new toxicity characteristic leaching procedure (TCLP), which can be used to test the leach-

Table 1. Maximum Concentration of Contaminants for EP Toxicity Characteristic

EPA No.	Contaminant	Maximum Concentration (mg/L)
D004	Arsenic	5.0
D005	Barium	100.0
D006	Cadmium	1.0
D007	Chromium	5.0
D008	Lead	5.0
D009	Mercury	0.2
D010	Selenium	1.0
D011	Silver	5.0
D012	Endrin	0.02
D013	Lindane	0.4
D014	Methoxychlor	10.0
D015	Toxaphene	0.5
D016	2,4-D	10.0
D017	2,4,5-TP (Silvex)	1.0

ing potential of both organic and inorganic compounds. The new TCLP test impacts identification and management of wastes in several areas: (1) response to current control requirements under RCRA, (2) land disposal restriction of wastes, (3) handling of waste waters, and (4) evaluation and remediation of old or existing problem sites under CERCLA. We examine each of these in turn to present the basic rationale upon which each is based and to appreciate the significance of the changes introduced by the new TCLP test.

The EP Toxicity Characteristic (EP Test)

The basis for the EP toxicity characteristic test as well as the TCLP is to identify wastes that have the potential to leach significant concentrations of hazardous constituents into ground water. The EP toxicity characteristic test entails the use of a leaching medium containing acetic acid designed to simulate leaching that results when an industrial solid waste is codisposed with municipal wastes in a sanitary landfill. This represents a worst-case scenario for the mismanagement of hazardous wastes. Both the EP test and the TCLP test require a representative waste sample to be mixed with an aqueous leaching medium containing acetic acid, which is one of the more dominant carboxylic acids present in municipal waste leachate.

After a specified extraction period, the liquid extract is analyzed and if the extract contains a contaminant which exceeds a concentration in Table 1, the waste is considered to be hazardous. The maximum concentrations in Table 1 were derived by multiplying National Interim Primary Drinking Water Standards by a factor of 100 to account for groundwater dilution prior to reaching a receptor.

Three principal difficulties with the EP test have been identified: (1) toxicity standards were only available for 14 contaminants of concern to drinking water; thus the listed constituents did not reflect many hazardous materials

Table 2. Comparison of the Extraction Procedure (EP) and the Toxicity Characteristic Leaching Procedure (TCLP)[a]

Item	Extraction Procedure (EP)	Toxicity Characteristic Leaching Procedure (TCLP)
Leaching media	0.5 *N* acetic acid added to distilled deionized water to a pH of 5 with 400 ml maximum addition; continual pH adjustment	0.1 *N* acetic acid solution (pH 2.9) for moderate to high alkaline wastes and 0.1 *N* acetate buffer (pH 4.9) for other wastes
Liquid/solid separation	0.45-μm filtration to 75 psi in 10-psi increments; no specific filter type	0.6–0.8-μm glass-fiber filter filtration to 50 psi
Monolithic material/particle size reduction	Use of Structural Integrity Procedure, or grinding and milling	Grinding or milling only; Structural Integrity Procedure.
Extraction vessels	Unspecified design; blade/stirrer vessel acceptable	Zero-headspace vessel required for volatiles; bottles used for nonvolatiles; blade stirrer vessel not used
Agitation	Broad requirement for agitation	Rotary agitation only in an end-over-end fashion at ± 2 rpm
Extraction time	24 hr	18 hr
Quality control requirements	Standard additions required; one blank per sample batch	Standard additions required in some cases; one blank per 10 extractions and every new batch of extract; analysis specific to analyte

Source: 51 FR 21678; 55 FR 26986.
[a]All other attributes between the two tests are generally the same although there are some minor differences.

likely to be in industrial wastes; (2) although acetic acid approximated the leaching of metals from an industrial waste, it did not test the leaching potential of organic compounds; and (3) there was no clear risk basis for selected contaminants based on groundwater modeling and toxicity considerations.

The Toxicity Characteristic Leaching Procedure (TCLP)

As required under the Hazardous and Solid Waste Amendments of 1984 (HSWA), the EPA proposed significant changes to the toxicity leaching test to overcome the deficiencies in the EP test. These changes principally involve adding additional compounds to the toxicity characteristic test, specification of regulatory levels based on risk, and redesign of the leaching procedure. The main features of the TCLP test compared to the EP test are shown in Table 2. The major changes in the test procedure are to realistically extract constituents typical of hazardous wastes, including volatile organics.

The general conceptual framework for deriving the regulatory levels for passing or failing the TCLP test represents a significant evolution in the approach to determining controls for the disposal of hazardous wastes or limiting the concentrations that may be in hazardous waste forms. The objective of the procedure is to limit disposal of wastes where leachability would present hazardous conditions under a postulated mismanagement scenario. The municipal landfill scenario where hazardous wastes are codisposed with municipal solid wastes was selected by the EPA as a worst-case condition for regulating industrial solid wastes. This may be quite conservative since industrial wastes are rarely disposed in municipal landfills, but instead are usually disposed in industrial landfills or managed in other ways (e.g., incineration, recycling, treatment on the land, or treatment in a surface impoundment).

The municipal landfill scenario is a reasonable mismanagement scenario for industrial wastes because municipal landfills have traditionally accepted nonhazardous industrial wastes. Although fewer industrial wastes are being disposed in municipal landfills now as compared to a few years ago, a substantial quantity of industrial wastes continue to be sent to such facilities, and even though it may not be the most likely scenario, it still represents a reasonable worst-case management scenario.

A management-contingent approach using alternate characteristics in the TCLP was considered but not adopted because it could substantially complicate effective implementation of the RCRA regulations because it is not always possible to determine—at the point of generation, during transport, or even at a treatment, storage, or disposal facility—how a solid waste will ultimately be managed and thus to ensure that a waste is in fact properly managed. For this reason, management-contingent characteristics could encourage "scenario shopping," wherein a handler of solid waste, in good faith or bad, claims that a waste actually destined for one type of management is destined for another. Therefore, wastes are regulated not as they are actually managed, but according to the ways in which they could plausibly be mismanaged.

The general approach used for derivation of the regulatory levels allowed for extracted constituents in the leaching procedure is shown in Figure 2. Fundamentally, the model in Figure 2 sets a framework for three key determinations: (1) the allowable, or acceptable, level at the ground-water consumption point based on risk; (2) the modeled dilution/attenuation factor (DAF) between the disposal unit and the receptor; and (3) the leachate concentration (from the regulatory level) from the waste form that would be permitted without exceeding the allowed contamination level.

The key factor in deriving the extraction concentration for constituents leached from wastes in the TCLP test involves the explicit determination of allowed concentrations from risks of exposure to the leached constituents. The risks are based on the following where applicable:

- Risk-Specific Doses (RSDs) for carcinogenic compounds that will result in an incidence of cancer equal to or less than 10^{-5} (i.e., the probability of one person in 100,000 contracting some form of cancer in his or her lifetime).
- Reference Doses (RfDs) for noncarcinogenic constituents based on an estimate

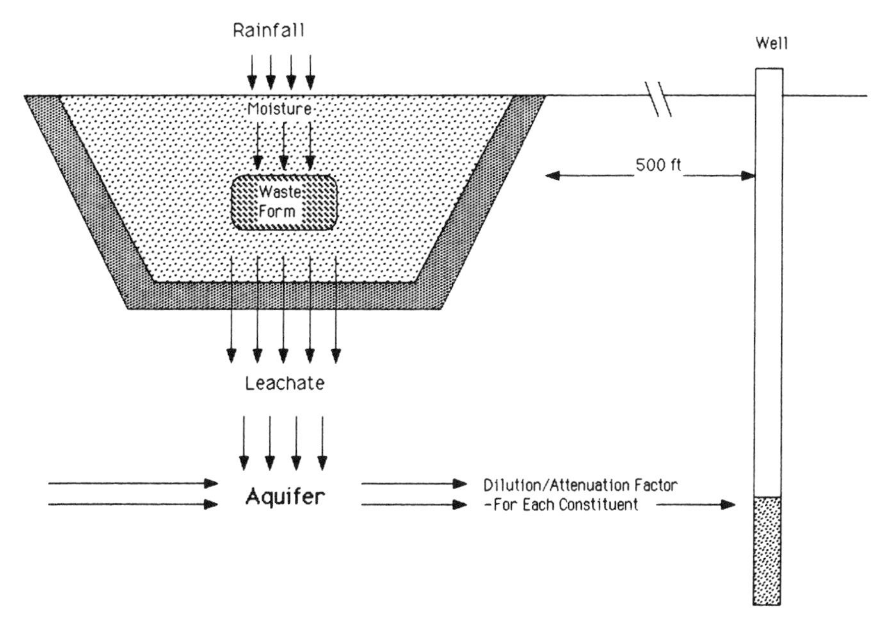

Figure 2. Conceptual model for establishing allowable leachability of water constituents that would not exceed regulated contamination levels at ground water use points.

of the daily dose of a substance that will result in no adverse effect even after a lifetime of such exposure. It is assumed that a 70-kg person ingests 2 L of contaminated drinking water per day.

- Proposed Maximum Contaminant Levels (PMCLs) in drinking water. Eight PMCLs for synthetic volatile organic chemicals were issued on November 13, 1985 (50 FR 46880); the finalized MCLs take precedence in developing regulatory limits for the toxicity leaching characteristic.

After chronic toxicity levels were identified, they were multiplied by a compound-specific dilution/attenuation factor (DAF) derived via a groundwater transport model. The model estimates the reduction in contaminant concentration that occurs as the contaminant migrates to an underground drinking water well located 500 ft down gradient. The model used compound-specific hydrolysis and soil absorption data and generic values for the groundwater flow rate, pH, soil porosity, and other relevant factors. DAFs are generally in the range of a 100-fold dilution, which has been adopted as a generic factor. The allowed chronic toxicity levels were thus multiplied by 100 to determine the TCLP extract concentration limits. These values are shown in Table 3 for the new leachate constituents required by the TCLP along with the basis for their determination. In some cases, the calculated regulatory thresholds were below the measurable analytical levels using current methods; therefore, in these instances the regulatory limit for these contaminants is the quantitation limit, which is the lowest level that can be reliably achieved using routine laboratory procedures, or five times the detection limit.

Table 3. Toxicity Characteristic Contaminants and Regulatory Levels

EPA Hazardous Waste Number	Contaminant	Chronic Toxicity Level (mg/L)	Basis	Calculated Level (mg/L)[a]	Quantitation Limit (mg/L)[b]	Regulatory Level (mg/L)[c]
D004	Arsenic	0.05	DWS[d]	5.0	0.025	5.0
D005	Barium	1.0	DWS	100	0.05	100
D019	Benzene	0.005	MCL[e]	0.5	0.5	0.5
D006	Cadmium	0.01	DWS	1.0	0.05	1.0
D019	Carbon tetrachloride	0.005	MCL	0.5	0.05	0.5
D020	Chlordane	0.0003	RSD[g]	0.029	0.0025	0.03
D021	Chlorobenzene	1.0	RfD[f]		0.05	100
D022	Chloroform	0.06	RSD		0.05	6.0
D007	Chromium	0.05	DWS	5.0	0.1	5.0
D023	o-Cresol	2.0	RfD	10.0	0.5	200
D024	m-Cresol	2.0	RfD	10.0	0.5	200
D025	p-Cresol	2.0	RfD	10.0	0.5	200
D026	Cresol	2.0	RfD	10.0	0.5	200
D016	2,4-D	0.1	DWS	10.0	0.025	10
D027	1,4-Dichlorobenzene	0.075	MCL	10.8	0.125	7.5
D028	1,2-Dichloroethane	0.005	MCL	0.5	0.05	0.5
D029	1,1-Dichloroethylene	0.007	MCL	0.7	0.05	0.7
D030	2,4-Dinitrotoluene	0.0005	RSD	0.1	0.125	0.13[c]
D012	Endrin	0.0002	DWS	0.02	0.0005	0.02
D031	Heptachlor (and its hydroxide)	0.00008	RSD	0.008	0.0005	0.008
D032	Hexachlorobenzene	0.0002	RSD	0.02	0.125	0.13[c]
D033	Hexachlorobutadiene	0.005	RSD	0.5	0.125	0.5
D034	Hexachloroethane	0.03	RSD	3.0	0.125	3.0
D008	Lead	0.05	DWS	5.0	0.4	5.0
D013	Lindane	0.004	DWS	0.4	0.0005	0.4
D009	Mercury	0.002	DWS	0.2	0.002	0.2
D014	Methoxychlor	0.1	DWS	10.0	0.0025	10.0
D035	Methyl ethyl ketone	2.0	RfD	200	0.05	200
D036	Nitrobenzene	0.02	RfD	2.0	0.125	2.0
D037	Pentachlorophenol	1.0	RfD	1.0	0.05	100
D038	Pyridine	0.04	RfD	4.0	4.0	5.0
D010	Selenium	0.01	DWS	1.0	0.05	1.0
D011	Silver	0.05	DWS	5.0	0.05	5.0
D039	Tetrachloroethylene	0.007	RSD	0.7	0.05	0.7
D015	Toxaphene	0.005	DWS	0.5	0.025	0.5
D040	Trichloroethylene	0.005	MCL	0.5	0.05	0.5
D041	2,4,5-Trichlorophenol	4.0	RfD	400.0	0.25	400
D042	2,4,6-Trichlorophenol	0.02	RSD	2.0	0.25	2.0
D017	2,4,5-TP (Silvex)	0.01	DWS	1.0	0.025	1.0
D043	Vinyl chloride	0.002	MCL	0.2	0.05	0.2

Note: Chronic toxicity levels were originally proposed in 51 FR 21652–75.

[a]Chronic toxicity level multiplied by a dilution/attenuation factor, usually 100.

[b]Quantitation limit equals five times the detection limit.

[c]If the quantitation limit is greater than the calculated level, the quantitation limit becomes the regulatory level (indicated by level in parenthesis).

[d]DWS = National Interim Primary Drinking Water Standard.

[e]MCL = Maximum Contaminant Level (originally proposed in 50 FR 46880).

[f]RfD = Reference Dose.

[g]RSD = Risk Specific Dose.

In the EPA's original proposal of June 13, 1986, the Agency included seven organic compounds which are no longer in the TCLP test: acrylonitrile, bis(2-chloroethyl) ether, methylene chloride, 1,1,1,2-tetrachloroethane, 1,1,2,2-trichloroethane, 1,1,1-trichloroethane, and 1,1,2-trichloroethane. The Agency is considering adding these seven compounds to a future revision of the testing requirement when better data are available on risks and environmental behavior.

IMPACT OF THE TCLP TEST ON WASTE MANAGEMENT

The new TCLP toxicity leaching procedure will have a significant impact on management of hazardous wastes under RCRA land disposal restrictions for hazardous wastes, CERCLA wastes, various solid wastes, and wastewaters. The largest impacts are not caused by the analytical changes in the test procedure itself, but are due to the inclusion of the regulatory levels for the additional constituents in the leachate extracted from waste samples. These additional levels mean that a broader universe of wastes will be required to be controlled as hazardous, especially under the RCRA Subtitle C program, which is what Congress intended in requiring the test to be improved.

Land Disposal Restrictions

The TCLP test imposes significant changes on land disposal wastes since, under RCRA 3004(g)(4), treatment standards must be established for all such wastes within 6 months after the TCLP rule is promulgated. Some treatment standards for TCLP wastes might be below the regulatory levels, which will require the treatment of solid wastes which are nonhazardous via the TCLP test.

Soil and debris from corrective actions and closures appear to be subject to regulation as hazardous wastes if they fail the TCLP test. On the other hand, the EPA does not intend for the TCLP test to be used as a cleanup standard for corrective action or clean closure.

Other Regulatory Programs

Wastes that are currently excluded from the definition of hazardous waste are not subject to the TCLP test since such exclusion has been broadly established by regulatory action. Such wastes are household wastes, special study wastes, and arsenical-treated wood. Alternatively, the current exemption for arsenical-treated wood has not been expanded to include creosote-and pentachlorophenol-treated wood since such wood is expected to rarely fail the TCLP test.

The TCLP test is also not appropriate for some special study wastes such as mining wastes; however, decisions on the operation of the TCLP test to such wastes will be made when the determination is made to regulate the particular waste under RCRA Subtitle C.

Other Waste Materials

The TCLP test characteristic will apply to ash from municipal waste incinerators if those units manage any nonhousehold solid wastes; however, it is not expected that the TCLP test will affect the regulatory status of ash. This requirement may be a significant burden to municipalities which plan to use incineration as a basic strategy for relieving the burdens of solid waste landfills which are increasingly more difficult to site and operate. The TCLP test will also apply to landfill gas condensate and leachate from municipal solid waste landfills.

The TCLP test will also apply to sludges from publicly owned treatment works (POTWs) despite the fact that these sludges may soon be regulated under Section 405(d) of the Clean Water Act. It is not expected that many POTW sludges will fail the TCLP test. The toxicity characteristics as determined by the TCLP test will also be applied in CERCLA remedial actions on a case-by-case basis when it is found to be applicable or relevant.

Wastewaters

Some wastes because of their volume and physical properties cannot reasonably be placed in a municipal landfill (e.g., mining wastes or ash from municipal solid waste incinerators). With respect to wastewaters in surface impoundments, the EPA performed an extensive analysis of the waste management techniques, attenuative mechanisms, and geohydrologic processes that govern constituent transport from surface impoundments and concluded that the DAFs for nondegrading constituents managed in surface impoundments were similar to those for the same constituents managed in landfills.

The TCLP test will be applied to wastewaters in the same way that it applies to other solid wastes. The applicable regulatory levels are not expected to change substantially for a surface impoundment mismanagement scenario rather than a landfill mismanagement scenario. Compliance with the TCLP procedures will be at the point of generation rather than in any wastewater treatment impoundments. Treatment by a series of surface impoundments will be regulated to the extent that each individual surface impoundment receives wastewaters that exhibit the toxicity characteristic as determined by the TCLP test.

SUMMARY

Significant changes in testing for and managing hazardous wastes are represented in the toxicity characteristic leaching procedure being required for determining one of the four characteristics of hazardous waste. The TCLP test itself is significantly different from the previous EP test in attempting to more realistically represent landfill leaching conditions for disposed wastes. Of greater significance is the addition of 31 new extractant concentrations in the TCLP test which will result in more wastes being identified as hazardous. This will correct a major deficiency of the EP test.

The TCLP testing concept also represents a significant step in the conceptual basis for establishing which wastes are hazardous. The derivation of the levels of TCLP extractants (i.e, landfill leachability potential) for wastes is based on chronic toxicity levels determined on the basis of the health risks they pose and a subsurface/groundwater transport model to establish DAFs between the disposal unit and a point of groundwater use. Previous decisions on which levels of solid wastes were hazardous did not use this explicit risk basis; thus, the new TCLP test represents a significant step toward controlling hazardous wastes on the basis of the risk they may represent when disposed.

SECTION II

SAMPLING

CHAPTER 2

Sampling and Analysis of Hazardous Wastes for Toxic Pollutants

Aida Fuentes and Milagros S. Simmons

INTRODUCTION: NATURE OF HAZARDOUS WASTES

The Resource Conservation and Recovery Act (RCRA) was enacted by the Congress of the United States in 1976 and amended in 1984 with the primary goal of controlling the problems derived from the generation, transportation, and disposal of hazardous wastes. This act establishes cradle-to-grave tracking and permit systems for the regulation of all hazardous wastes.

RCRA defines a hazardous waste as:

> "a solid waste, a combination of solid wastes, which because of its quantity, concentration, or physical, chemical or infectious characteristics may
>
> a. cause or significantly contribute to an increase in serious irreversible, or incapacitating reversible illness; or
> b. pose a substantial present or potential hazard to human health or the environment when improperly treated, stored, transported, or disposed of, or otherwise managed."[1]

In this definition of hazardous wastes, the term solid waste includes any solid, semisolid, liquid, or contained gaseous material present in the wastes. The U.S. Environmental Protection Agency (EPA) to whom the administration of the RCRA was delegated utilizes two mechanisms in order to determine if a waste is hazardous or not. The mechanisms are

1. identification of a hazardous waste by source or chemical name as classified and listed by EPA, and
2. description of waste properties that may be harmful to human health or the environment. These hazardous properties include ignitability, corrosivity, reactivity, toxicity (determined by the extraction procedure), radioactivity, infectiousness, phytotoxicity, teratogenicity, and mutagenicity.[2]

Since there are no validated test protocols for some of the properties, i.e., infectiousness, phytotoxicity, teratogenicity, and mutagenicity, and because radioactive wastes are separately regulated, EPA only examines a solid waste for ignitability, corrosivity, reactivity, and toxicity in order to determine whether it is hazardous or not.

A waste is considered ignitable if it consists of liquids with flash points of less than 140°F (60°C) or nonliquids that may cause fires through friction, spontaneous chemical change, etc. or if the waste consists of ignitable compressed air or oxidizers.[1,4] The EPA describes two methods from which ignitability of a waste is determined. These methods are the Pensky-Martens closed-cup method and the Seta flash close-cup method.[8] The former is utilized to determine flash points of oils, solids suspensions, liquids that form surface film, and other liquids, whereas the Seta flash close-cup method is used to determine flash points of paints, enamels, lacquers, varnishes, related products, and some of their components.

Corrosive wastes are those that have the ability to corrode containers' standard materials or those that have the ability to dissolve toxic contaminants. Aqueous wastes with pHs of less than or equal to 2 or greater than or equal to 12.5 are considered corrosive. Also, liquid wastes that can corrode steel at a rate greater than 0.250 in./year are classified as corrosive wastes by the EPA.[1,3,4] In order to classify a waste as corrosive, the EPA describes a method that measures the corrosivity toward steel of aqueous and nonaqueous liquid wastes by exposing the waste to coupons of SAE type 1020 steel.[8]

Hazardous wastes that exhibit the property of reactivity include those wastes that may react spontaneously, those that react violently with water, or wastes that could generate toxic fumes when mixed with water or exposed to basic or mild acidic conditions. It also includes wastes that may explode at normal temperatures and pressures or during their management.[1,3,4] Available methods for testing liquids and solids to determine their reactivity are described by the EPA in *Test Methods for Evaluating Solid Waste.*[8]

Toxicity of a solid waste is determined through the extraction procedure (EP) developed by the EPA to simulate waste leachate in sanitary landfills. This procedure involves the extraction of a waste sample with 0.5 *N* acetic acid at pH 5 for 24 hr. The resulting extract is analyzed for the presence of eight specific elements and six specific organic compounds (four pesticides and two herbicides) to determine whether or not their established thresholds are exceeded. The EP extract should not contain elements or compounds levels equal to or greater than the following levels:[4,8]

Compound	**Maximum Concentration (mg/L)**
Arsenic	5.0
Barium	100.0
Cadmium	1.0
Chromium	5.0
Lead	5.0

Mercury	0.2
Selenium	1.0
Silver	5.0
Endrin	0.02
Lindane	0.4
Methoxychlor	10.0
Toxaphene	0.5
2,4-D (2,4-Dichloro-phenoxyacetic acid)	10.0
2,4,5-TP Silvex (2,4,5-Trichlorophenoxypropionic acid)	1.0

If any one of these limits are exceeded then the waste is said to have the characteristic of EP toxicity and is classified as a hazardous waste. In 1986, EPA issued a proposed rule to replace EP toxicity with a new toxicity characteristic leaching procedure (TCLP). TCLP replaced EP in September, 1990. A description of the basic differences of these tests is provided in Chapter 1.

Besides description of waste characteristics for classification as hazardous waste, the EPA lists wastes resulting from either nonspecific or specific sources and classifies them as hazardous wastes. Examples of listed hazardous wastes from nonspecific sources are wastewater treatment sludges from the chemical conversion coating of aluminum and spent halogenated solvents used in degreasing, such as tetrachloroethylene, trichloroethylene, methylene chloride, 1,1,1-trichloroethane, carbon tetrachloride, and chlorinated fluorocarbons.

Listed hazardous wastes from specific sources include wastes generated during the manufacturing processes of the following industries: wood preservation, inorganic pigments, organic chemicals, inorganic chemicals, pesticides, explosives, petroleum refining, iron and steel, secondary lead smelting, veterinary pharmaceuticals, ink formulation, and coking.[4]

Different methods are utilized for disposal of hazardous wastes among which surface impoundments and landfills are the most common methods used by industries. Surface impoundments are "natural depression, manmade excavations, or dike-in areas used to contain liquids or wastes containing free liquids," whereas landfills are "disposal facilities in which hazardous wastes are placed directly in or on the ground."[4]

Surface impoundments for hazardous wastes include pits, ponds, and lagoons. These types of impoundments mostly present two potential environmental problems: leakage of hazardous substances to groundwater and air emissions of volatile materials. Environmental problems presented by landfills include contamination of surface and groundwaters and their potential human hazard from fires, explosions, and generation of toxic fumes among others.[4,5]

To minimize these harmful environmental effects that such disposal sites may cause, the EPA suggests the use of safeguards such as synthetic liners, leachate detection systems, runoff controls, and groundwater monitoring systems.[4,6]

Other methods that were, are, or will be utilized for the disposal of hazardous wastes are underground storage tanks, deep-well injection, and the so-called "midnight dumping," which is an illegal and unsafe method of disposal. Also the application of hazardous waste onto the soil known as land treatment and the recycling and reuse of hazardous waste known as resource recovery are methods available for handling hazardous wastes.

Besides groundwater or water supply contamination, air pollution, and fire or explosion waste disposal sites. have been associated with other adverse effects to the environment and to public health. As an EPA survey of 350 hazardous waste disposal sites reveals, incidents such as destruction of indigenous species' habitat, human health damages, soil contamination, fish kills, livestock losses, damages to crops or wildlife, and sewer systems' contamination have been attributed to contaminant produced by hazardous wastes.[7]

ENVIRONMENTAL SAMPLES

In order to assess whether a solid waste is a hazardous waste or not, a sampling plan should be developed from which the waste's characteristics and constituents can be determined. Samples should be representative of the waste and should be taken considering both sampling accuracy and precision to increase the validity of the waste analysis.

Four sampling strategies are recommended for solid wastes depending upon the degree of information available on the nature of the wastes. The first strategy is known as simple random sampling which is used when wastes are heterogenous in nature and no information on chemical properties is available. This sampling strategy first identifies all locations or points from which a waste sample could be collected and then selects at random a number of samples from the population. If more information about the waste is available so that it can be determined that the waste is not randomly heterogeneous in its chemical properties, the stratified random sampling is used. Stratified random sampling consists of the stratification of the population in order to isolate the sources of nonrandom chemical heterogeneity into strata which are identified by numbers. After stratification is done, a simple random sample is taken from each stratum.[8]

Another sampling strategy utilized for solid wastes is known as systemic random sampling. This sampling strategy is used when the population from which samples will be collected is almost random. The first sample is chosen at random from the population and the following samples are taken at fixed space or time intervals. When a nonrandomized sample is collected from a solid waste by a person who knows its nature, then the sampling strategy is known as authoritative sampling. This type of sampling is not recommended when one intends to use the samples for the chemical characterization of the waste.[8]

Given the diversity of the nature of wastes and waste disposal sites, different equipment and procedures exist for the sampling of waste that might serve as a guide for those performing solid waste analyses. However, each situation should be examined for its particularities and the most convenient sampling procedure and equipment should be employed so that reliable results be obtained. Depending upon the ulterior objective of the study, environmental samples could be drawn from wastewaters (industrial effluents), soil and sediments at landfills and impoundments, air, biological tissues (fish, plants, etc.), groundwaters, drums, and labpacks.

Whether a simple, a stratified, or a systemic random sampling is used as a sampling strategy, composite sampling or subsampling techniques might be used to ensure a representative sample of the hazardous waste. Composite sampling consists of obtaining random samples from a waste and combining them into a single sample. For the analysis of hazardous waste, the use of a large number of composite samples is recommended, which can be determined as discussed in *Test Methods for Evaluating Solid Waste*. Subsampling consists of splitting a composite sample or random samples into a number of subsamples in order to obtain replicate measurements during the waste analysis.

Different types of samplers are recommended for sampling of hazardous wastes. Some of the sampling methods or equipment utilized for liquid wastes are the Coliwasa (composite liquid waste sampler), the dipper or pond sampler, the weighted bottle sampler, and tap sampling. Sampling tools available for the sampling of solid wastes are the grain sampler (thief), the sampler corer (trier), the Auger, and the trowel (scoop) and shovel. Sampling tools utilized for liquid wastes sampling can be used for free-flowing sludges and slurries, whereas those tools used for solid wastes can be used for nonfree-flowing or compacted sludges.[8,9] The sampling tools mentioned can be used for the withdrawal of samples from drums, storage tanks, ponds, lagoons and pits, and soils among others.

In order to identify any hazardous substance in groundwaters, samples should be collected from leachate seeps, groundwater monitoring wells, or leachate collection systems. Leachate, which is the liquid that forms as wastes decompose in a landfill and mix with rainwater, can enter groundwater and either remain below the ground surface or emerge on the ground surface as a leachate seep. Leachate samples may be collected by using either of two sampling techniques: grab sampling and space composite sampling. Grab sampling consists of taking grab samples over a short period of time, whereas space composite sampling consists of combining samples taken at different locations into one sample.[10] For sampling of leachate collection systems, the same sampling techniques as for leachate seeps are used. Sampling monitoring wells can be done by utilizing a cleaned bailer to collect the sample from the well after which the water sample is transferred to a sample container.[10]

Examples of equipment utilized for biological sampling are provided by

Janisz and Butterfield[11] in their assessment of the effects of waste sites on aquatic organisms. Their primary goal was to determine potential hazards to human health due to the presence of inorganic and organic materials which could concentrate in aquatic organisms and reach toxic levels for humans. They designed site-specific biological sampling programs in order to obtain sufficient tissues for analyses and utilized established methodologies for determining impacts on aquatic ecosystems (e.g., diversity index comparison and bioassays) to complement the results of tissue analyses.

Five different sites were studied where contamination with hazardous substances from chemical and municipal landfill leachates were suspected. Janisz and Butterfield selected most of the sampling stations upstream, adjacent to, and downstream of known or suspected discharges. They utilized a portable backpack electrofishing to sample freshwater streams that had firm bottom and shallow depths. To sample large open-water habitats such as ponds or lagoons, the investigators utilized gill, seine, or throw nets. For narrow, soft-bottomed, brackish tidal creeks, they utilized minnow traps to sample small fish and a petite ponar dredge to sample benthic invertebrates. Crabs and oysters were collected by hand or purchased.

An example of methods for the sampling of gases from landfills is obtained from the work of Young and Parker.[12] They describe a modified drive-in piezometer they used for sampling landfill gases in order to determine trace constituents. These gas sampling probes were inserted into the studied landfills to take samples of the pure landfill gas as it was generated. Three different trapping methods were used for obtaining representative gas samples for analysis by either GC/MS or emission spectroscopy. Tenax absorption tubes were used when trapped samples would be analyzed by GC/MS for nonpolar species and general "fingerprinting," whereas Porapak Q absorption tubes were used when the analysis objective was determination of polar and low molecular weight species. Condensate samples were taken at –60°C for analysis of water-soluble bases or acids and neutral components by emission spectroscopy.

Once appropriated environmental samples are collected, samples should be preserved and prepared for their subsequent analysis. Preservation, storage, and preparation procedures of collected waste samples depend upon the type of analysis that will be performed and the parameters to be measured. The next section presents available methods for sample preparation and analysis that are utilized for the characterization of wastes.

ANALYTICAL METHODS

Waste characterization involves the use of test procedures to determine either the characteristics of the waste or its composition. Available methods for the determination of waste characteristics (ignitability, corrosivity, reactivity, and toxicity characteristic leaching procedure) were mentioned early in this paper. Test procedures for the analyses of waste constituents are classified into three types of analyses: proximate analyses, survey analyses, and directed analyses.[9] Proximate analyses are directed toward the determination of the waste's physical form and approximate mass balance including determination of "moisture (volatile), solid, and ash content; elemental composition (carbon, hydrogen, nitrogen, sulfur, phosphorus, fluorine, chlorine, bromine, iodine); heating value of the waste; and viscosity or physical form."[9]

Survey analyses involve analytical methods for obtaining qualitative descriptions of the waste in terms of major types of organic compounds and inorganic elements present. The analytical methods include the use of chromatographic and gravimetric procedures, infrared and probe mass spectrometric procedures, gas chromatographic/mass spectrometric (GC/MS) or high performance liquid chromatographic/infrared (HPLC/IR) or high performance liquid chromatographic/mass spectrometric (HPLC/MS) procedures, and coupled argon plasma emission spectroscopic (ICAP) and atomic absorption spectroscopic (AAS) procedures.[9]

Directed analyses provide a means by which specific compounds are identified and their concentrations are quantified. This type of analysis is more useful in determining specific toxic pollutants that might be present in waste samples. In directed analyses, instruments such as GC, GC/MS, and HPLC are utilized for the analysis of organic compounds and instruments such as an atomic absorption spectrophotometer and ICAP are utilized for the analysis of inorganic elements. Atomic absorption spectrophotometers may differ in terms of sample introduction methods, atomizers, and lamps. GC methods may differ in terms of sample introduction techniques and detectors; for example, electron-captured detectors (ECDs), flame ionization detectors (FIDs), and halide-specific detectors (HSDs).

In order to accomplish directed analyses of waste samples, different procedures prior to the analyses should be followed for preparation of the samples. Samples that will be used for inorganic analyses may be prepared following EPA specifications which include acid digestion of samples, dissolution procedures, and alkaline digestion.[3,8] Samples that are intended to be used for organic analyses can be prepared by different extraction methods or headspace and purge-and-trap methods both used for volatile organic compound determination.[8,14,15]

The EPA describes in its *Test Methods for Evaluating Solid Waste*[8] two acid digestion methods for the preparation of aqueous samples, EP, now modified to TCLP, and mobility extracts, and some nonaqueous wastes that are to be analyzed for certain inorganic elements by flame atomic absorption spectroscopy (AAS) or by furnace atomic absorption spectroscopy. The methods consist of heating a mixture of sample and nitric acid to near dryness, repeating this step until the digestate stabilizes or becomes light in color after which dilute hydrochloric acid is added to the digestate. Examples of inorganic elements that can be determined in the subsequent analysis of the digestate include aluminum, barium, chromium, lead, and vanadium among others. Microwave digestion is also being evaluated for the preparation of samples.

The EPA describes separated methods of acid digestion for samples that contain great amounts of oils, greases, or waxes and for sludge-type and soil samples. When samples that contain oil, greases, or waxes are to be analyzed for barium, cadmium, chromium, lead, and silver, the samples can also be prepared by a dissolution procedure which consists of dissolving the samples in solvents such as xylene or methyl isobutyl ketone and preparing organometallic standards with the same solvents. Both dissolved samples and standards are subsequently analyzed by either atomic absorption spectroscopy or inductively coupled argon plasma emission spectroscopy.

For the determination of hexavalent chromium in solid wastes, an alkaline digestion of the waste samples is specified by the EPA. It utilizes a 3% sodium carbonate-2% sodium hydroxide solution to dissolve all Cr(VI) and to prevent its reduction to trivalent chromium. Hexavalent chromium can also be determined by three other methods, namely the coprecipitation method to analyze samples containing more than 5 μg of Cr(VI) per liter, the colorimetric method to analyze samples containing from 0.5 to 50 mg of Cr(VI) per liter, and the chelation/extraction method to analyze samples containing from 1.0 to 25 μg of Cr(VI) per liter. The coprecipitation method utilized either flame or furnace atomic absorption spectroscopy and the chelation/extraction method utilized flame atomic absorption spectrophotometer (direct aspiration technique) for the analysis of Cr(VI).

For the determination of mercury in liquid, solid, or semisolid wastes, the EPA describes a cold-vapor atomic absorption procedure based upon the absorption of radiation at 253.7 nm by mercury vapors.

Preparation of waste samples for determination of nonvolatile or semivolatile organic compounds includes extraction of the compounds by different procedures. Separatory funnel liquid-liquid extraction utilizes standard separatory funnel techniques employing methylene chloride as a solvent when no solvent is specified. The extracts are dried with anhydrous sodium sulfate and concentrated using Kuderna-Danish evaporation apparatus. Alternatively, continuous liquid-liquid extraction can be used which extract samples continuously over a period of 16 hr. Its advantages over the standard separatory method is that continuous extraction minimizes emulsion formation. Both techniques are used for extraction of organics from liquid wastes.

Three additional extraction procedures are described in Reference 8 for the preparation of nonvolatile or semivolatile organic compounds. These are acid-base cleanup extraction to remove interferences that prevent direct chromatographic measurements, and Soxhlet extraction and sonication extraction, both for extraction of nonvolatile or semivolatile organic compounds from solid wastes.

Warner et al.[16] describe a solvent extraction method developed for subsequent determination of semivolatile organic compounds in solid wastes. They refer to it as the dry neutral extraction procedure which consists of a single extraction with anhydrous sodium sulfate used to remove water. To ensure interaction between the solvent and the waste, they used sonification with a high-intensity sonic probe. Extracts obtained are suitable for GC/MS analyses and GC or HPLC analyses.

Preparation of waste samples for determining volatile organic compounds involves the use of two methods: headspace method and purge-and-trap method. The headspace method consists of collecting a waste sample in a sealed glass container and allowing it to equilibrate at 90°C after which a sample of the headspace gas is obtained using a gas-tight syringe for subsequent GC analysis. The purge-and-trap method consists of transferring the purgeables from prepared solutions to the vapor phase, which is trapped into a sorbent column. This column is then heated and backflushed with inert gas to desorb the volatile compounds onto a GC column from which the compounds are eluted and detected by the appropriate method. New technology incorporates the use of silicone rubber permeators coupled with gas sensors.

Since most of the described techniques and analytical methods are time-consuming and laboratory-oriented, there is an increased interest in the development of rapid-onsite methods of chemical analysis for the screening, monitoring, and assessment of potential contamination from hazardous wastes. In the next section, some of the rapid monitoring and screening techniques are presented along with other methods.

MONITORING AND SCREENING FOR HAZARDOUS WASTES

Monitoring contamination from hazardous wastes embraces all environmental media, air, soil, sediment, water as well as biological systems. Toxicity tests for assessment of contamination in biological systems will be presented in the fifth part of this paper. Methods for air, soil/sediment, and water monitoring are presented next.

Water Monitoring

Hinton et al.[17] describe an on-line, computer-controlled monitor for mercury in streams and aqueous discharges that has some advantages over grab sampling methods. The prototype instrument consists of a Spectro Products HG-3 mercury analyzer which is a dual-wavelength spectrophotometer, a modified Perkin-Elmer MHS-20 mercury/hidrite generator for sample preparation, and a Hewlett-Packard HP 9825 desktop computer. The instrument calibrates itself every 6 hr and analyzes samples every hour. It sounds an alarm in a central monitoring facility if downstream samples exceed some preset level. The system can detect small fluctuations in the streams within the range of 0.5 to 10 ppb of mercury.

Crathorne et al.[18] utilized coupled-column high-performance liquid chromatography and mass spectrometry with field desorption (FD) and fast atom bombardment (FAB) ionization for identification of nonvolatile organics in water. The coupled-column system was shown to increase chromatographic resolution allowing detection and identification of more components of the HPLC fractions by mass spectrometric methods.

A system that permits in situ analysis of volatile chloroorganics in groundwater is presented by Milanovich.[19] The technique is known as remote fiber fluorimetry (RFF). It consists of fiber-optic chemical sensors (FOCSs) which are sensitive and specific to a compound or class of compounds, a spectrometer which measures fluorescence of target molecules, and an optical coupler used to separate the excitation light from the returning fluorescent light.

Several field test instruments and kits exist for onsite analysis of water that provides rapid qualitative and semiquantitative measurements of water samples.[20,21] Among the field test instruments are the Spectronix Mini-20 which is a compact battery-operated spectrophotometer with a wavelength range of 400 to 800 nm, the Model DP/1 Portable Colorimeter which measures pH and performs 50 different water and wastewater tests, and the Model OR-EL/4 Portable Laboratory consisting of spectrophotometer, titrator, and cartridges.[20] Examples of field analysis kits for water monitoring include the Model CYN-2 Cyanide test kit, which is a colorimetric test determining 0 to 2.6 ppm of free cyanide, and the Hazardous Materials Detector kit that tests for pH, heavy metals, benzene, phenol, cyanide, conductivity, turbidity, nitrate, nitrogen, color, sulphate, phosphate, NH_3, and fluoride.[20]

Other methods employed for water monitoring include remote sensing and the EPA mobile spill alarm system.[10,21] Remote sensing consists of "infrared photographic detection of changes in the radiant energy status of a discharge area."[10] The EPA mobile spill alarm system can detect organic and inorganic hazardous materials in water. It has several detectors to sense wide classes of these materials, a pH, oxidation-reduction potential and conductivity sensor package, and an ultraviolet absorptionmeter and a total organic carbon analyzer integrated into a detection package.[21]

Soil/Sediment Monitoring

Spittler[22] used a portable gas chromatograph for rapid field determination of polychlorinated biphenyls (PCBs) in soil and sediments. The gas chromatograph was used along with a linearized electron capture detector (GC/EC). Preparation of field samples for the analyses was done by weighing out 400 mg of soil into a 2-cc septum vial and adding 100 μg/L of water, 40 μL of methanol, and 500 μL of technical-grade hexane to the soil. One to three μL of the sample was injected into the GC for the analysis. The oven of the instrument can be preheated using AC power, and the temperature maintained for 8 hr on battery operation. The method provides rapid assessment of soil contamination.

Another portable instrument available for field use is the portable X-ray fluorescence unit.[20] This instrument might be used to detect elements in concentrations as low as 10 ppb and to detect PCBs and organo-metallic compounds in soil samples.

Few field kits exist for the analysis of contaminated soil. The Soil Extraction kit is used to perform rapid aqueous extraction of the soil that can be analyzed using available water test kits.[20]

Other testing approaches that might be employed for screening of contaminants in soils include the use of chemical surrogates. Thielen et al.[23] define past migration of chemicals from landfills into groundwater or surface water by analyzing the soil for determination of two chemical surrogates, monochlorobenzene (MCB), and monochlorotoluene (MCT). Pojasek and Scott[24] describe a surrogate screening test for determination of locations containing volatile organic solvents that might require further analyses. The test involves homogenize-preserved specimens with organic-free water in a vial from which a portion is withdrawn and placed into another vial. The second vial is sealed with mercury and equilibrated at 70°C (158°F). A headspace sample is obtained and analyzed on a GC using a fast elute isothermal or temperature program. Results are compared to standards. The test is inexpensive and allows screening of large numbers of samples.

Air Monitoring

There are several instruments and methods available for the monitoring of air that can be applied at waste disposal sites. One of them, the Model 128 Century Organic Vapor Analyser (OVA), can be employed to monitor for volatile organic compounds.[20] It consists of a lightweight shoulder-borne gas chromatograph (GC) with a flame ionization detector (FID) and an externally mounted GC column. Different column lengths (225 mm, 300 mm, 600 mm) and packing materials can be used. The gas used as carrier gas is hydrogen gas which also served as fuel for detector combustion. The OVA uses either a battery or a 120-V AC power supply for its operation. The detection limit of the instrument is 0.2 ppm, although air temperature fluctuations may affect its functioning.

Another portable instrument capable of detecting organic and inorganic vapors is the Model PI 101 Hazardous Waste Detector.[20] It is a gas analyzer which works by photoionization. This process consists of introducing the gases into a sample probe and then into a lamp/ion chamber where molecules are ionized after absorption of ultraviolet light. Different sample probes can be employed, expanding the instrument capabilities of detection. Its sensitivity is between 0.1 and 2000 ppm with an accuracy of 96% at the 5-ppm concentration level. However, one of its limitations is that the Model PI 101 cannot detect hydrogen, cyanide, or methane gases.

There are other portable air monitoring instruments, such as the Photovac, Inc. 10A10 Portable Photoionization Gas Chromatograph which is mostly used to monitor low concentrations (in micrograms per liter) of volatile compounds, a portable gas chromatograph with thermal conductivity detector, and a portable infrared analyzer based upon the absorption of infrared radiation by gases and vapors.[20,21] Other portable GCs with ECD, FID, and argon ionization detector (AID) are also commercially available.

Kerkhoff et al.[25] describe a simple screening technique for airborne polycyclic aromatic hydrocarbons (PAHs). They used a high-volume sampling technique along with constant energy synchronous luminescence spectroscopy (CESLS). The sampler uses polyurethane foam adsorbent and glass-fiber filters for gas and particulate-phase PAH collection. CESLS is based upon the absorption-fluorescence transition molecules may undergo. It eliminates Rayleigh and Raman solvent scatter interferences producing simple, reliable, and reproducible spectra.

For screening of polynuclear aromatic hydrocarbons (PAH) Vo-Dinh et al.[26] employed two simple and rapid analytical methods, synchronous luminescence (SL), and room temperature phosphorescence (RTP). During synchronous excitation, both emission and excitation monochromators are scanned at the same time and the luminescence signal recorded. This technique only required a switch to be installed in available spectrometers. RTP is based on the phosphorescence from inorganic compounds adsorbed on solid substrates at room temperature. The preferred solid substrate is filter paper, although silica, alumina, paper, and asbestos can be used. The screened samples are ranked in terms of an SL and RTP ranking index which provides a good estimate of the contents of the major PAH species in the samples.

A mobile mass spectrometry technique exists which monitors air for the analysis of trace levels of organic and inorganic pollutants. The TAGA™ (Trace Atmospheric Gas Analyzer) 6000 Mobile Mass Spectrometer System has been employed for sampling and analysis of volatile organic compounds emanating from waste disposal sites.[20,27,28] It is a mobile laboratory equipped with a mass spectrometer that can operate in the APCI (atmospheric pressure chemical ionization) mode or in the DASCI (direct air sampling chemical ionization) mode. The system is capable of continuous air monitoring of the contaminated area as the laboratory moves. The use of special probes expands its use for soil screening.

Other Methods

For rapid onsite chemical analysis at waste disposal sites, the EPA's OHMSB (Oil and Hazardous Materials Spills Branch) counts on a mobile laboratory equipped with various analytical instruments. It includes two computerized GCs with flame ionization, electron capture; a nitrogen-phosphorus detector; a computerized GC/MS; an emission spectrometer; infrared and fluorescence spectrophotometers; and a total organic carbon analyzer.[20] Different procedures for preparation and pretreatment of the samples have been developed to be used in this mobile laboratory. This system allows for the analysis of most organic and inorganic substances often found at hazardous waste sites.

Fourier-transform infrared (FTIR) is another analytical technique that is available for screening of organic hazardous waste. Puskar et al.[29] employed this technique to identify major components of hazardous waste drum mixture samples. They used two automated infrared analysis techniques: the forward searching and PAWMI (Program for Automated Waste Mixture Interpretation). The first technique involves the interpretation of spectra by forward searching the unknown spectrum agains a library of known spectra of pure compounds, polymers, and commercial mixtures. PAWMI is an automated artificial intelligence program that has been developed to identify major components in a complex mixture spectrum.

An airborne imager has been developed that can be utilized for detecting and monitoring stress to vegetation. The Programmable Multispectral Imager (PMI) detects damage by monitoring subtle changes in the vegetation color.[35]

TOXICITY BIODIRECTED ANALYSIS

Analyses of hazardous wastes should include tests for detection of potential toxic pollutants, especially of those known to be mutagens, carcinogens, or teratogens. Environmental samples can be evaluated for these toxic characteristics by means of short-term tests.

Numerous short-term tests exist that can be employed for screening of mutagens in environmental samples. Among the most widely used tests are the Salmonella/mammalian microsome test, cultured Chinese hamster or mouse lymphoma cell gene mutations tests, and the sex-linked recessive lethal test in Drosophila among others.[31] Some of these tests have been applied to a variety of situations, such as mutagenicity studies of industrial effluents, ambient air samples, and municipal sewage sludge.[31] Short-term tests provide an inexpensive means for testing large numbers of environmental samples requiring small volumes of samples.

Donnelly et al.[32] employed a battery of short-term bioassays along with GC/MS/DS chemical analysis to evaluate the hazardous characteristics of two waste streams from the petroleum industry. Among the bioassays were the Salmonella/microsome assay using *S. typhimurium* strains, the haploid eukaryotic bioassay using the meth G1 bi A1 Glasgow strain of *A. nidulans*, and a diploid bioassay using *A. nidulans*, all of them used to determine frequency of mutation. Also five different strains of *B. subtilis* were used in order to test for lethal DNA-damage. The bioassays served to determine the potential genotoxicity of the waste samples, whereas the GC/MS/DS technique served to define the type and quantity of the genotoxic compounds.

Samolioff et al.[33] developed a protocol for the chemical fractionation of sediments and biological testing of these fractions in order to establish which components of the contaminated samples produce the greatest toxic effects. They used two bioassay methods to test the different fractions: the *S. typhimurium* test as an indicator of mutagenesis and a developmental assay using the nematode *Pangrellus redivivus* to detect lethal, semilethal, developmental, and mutagenic effects.

Barfknecht and Naismith[30] suggest the use of a battery of genotoxicity bioassays with different endpoints to evaluate hazardous complex mixtures. They proposed the use of a two-tiered testing matrix where the first tier is intended to use for detection of genotoxic agents and the second one is designed to confirm the results of the tier one tests. Tier one includes well validated and established in vitro assays, such as the Ames/Samonella histidine reversion assay or the Salmonella/Forward Mutation, the Chinese Hamster V79 and Chinese Hamster Ovary (CHO) Cell Forward Mutation Assays or the L5178Y Mouse Lymphoma Cell Mutation Assay, Cytogenetics Chromosome Aberrations and Sister Chromatid Exchange (SCES) assays, and Primary Hepatocyte DNA Repair assay.

Tier two consists of a battery of short-term in vivo bioassays and in vitro cellular malignant transformation assays. Both tiers allow for the evaluation of complex environmental samples.

Other tests are used for in situ monitoring of mutagens in the environment. These tests measure mutagenic effects in test organisms that grow in the studied environment.[31] For example, the *Tradescantia stamen* hair test which is sensitive to gaseous and airborne mutagens is used for in situ monitoring of mutagenic air pollutants. Its endpoint is a change in the pigmentation of stamen hair cells from blue to pink that arise from a somatic gene mutation.[31]

Another example of in situ monitoring for mutagens is the waxy locus test in corn used for monitoring air and soil.[31] The mutations at the waxy locus are detected by stain changes of pollen grains.

There are also a series of portable instruments for assessing toxicity of hazardous materials. These are

1. The Cholinesterase Antagonist Monitor (CAM-4) – a battery-powered (12 V dc) portable instrument which can determine toxic or subtoxic levels of cholinesterase-antagonistic substances such as organophosphate and carbamate insecticide in water.
2. The Beckman Microtox System – used to evaluate the acute toxicity of aqueous samples by measuring bioluminescence. It might be used as an alternative to bioassays such as the 96-hr fish toxicity test.
3. The Portable Fish and Macroinvertebrate Toxicity-Monitoring System – a gravity-feed proportional diluter system for small-volume exposures.[20,21]

Dorward and Barisas[34] describe a rapid, inexpensive, and quantitative instrumental bioassay for screening the acute toxicity of substances in water. It consists of an *Escherichia coli* electrode which functions by measuring potentiometrically the CO_2 produced by *E. coli* cells attached to the surface of a CO_2-sensing electrode. Inhibition of CO_2 production by any pollutant is measured as a decrease in CO_2 production by the bacteria from which dose-effect curves can be obtained. By using this method, the acute toxicity of numerous substances can be determined including metals, anions, gases, and organic compounds.

CONCLUSION

Disposal of hazardous wastes should follow established technologies in order to make the process safe. Waste disposal sites are monitored and screened for determination of potential hazardous and toxic pollutants.

Traditional sampling and analytical methods as well as new and rapid techniques were presented in this paper. It also included biodirected analyses for toxicity assessment of wastes.

REFERENCES

1. Greenberg, M. R. and Anderson, R. F. *Hazardous Waste Sites*, The Center for Urban Policy Research, New Brunswick, NJ, 1984.
2. Municipal Environmental Research Laboratory. *Project Summary, Evaluation of the RCRA Extraction Procedure: Lysimeter Studies with Municipal/Industrial Wastes*, EPA-600/52-84-022, U.S. EPA, Cincinnati, OH, March 1984.
3. Princeton University Water Resources Program. *Groundwater Contamination from Hazardous Wastes*, Prentice-Hall, Inc., Englewood Cliffs, NJ, 1984.
4. Quarles, J., *Federal Regulation of Hazardous Wastes*. The Environmental Law Institute, N.W., Washington, D.C., 1982.
5. Guiney, P. D. "Use of Predictive Toxicology Methods to Estimate Relative Risk of Complex Chemical Waste Mixtures," *Hazardous Waste Hazardous Mater.* 2:177–189 (1985).
6. Office of Public Affairs. *Region 5 Hazardous Waste and Toxic Substances*, U.S. EPA Region 5, Chicago, IL, 1985.
7. Epstein, S. S., Brown, L. O., and Pope, C. *Hazardous Waste in America*, Sierra Club Books, San Francisco, CA, 1982.
8. Office of Solid Waste and Emergency Response. *Test Methods for Evaluating Solid Waste–Physical/Chemical Methods*, 2nd ed., U.S. EPA, Washington, D.C., 1982.
9. Harris, J. C., Larsen, D. J., Rechsteiner, C. E., and Thrun, K. E. *Combustion of Hazardous Wastes–Sampling and Analysis Methods*, Noyes Publications, Park Ridge, NJ, 1985.
10. Cheremisinoff, P. N. and Gigliello, K. A. *Leachate from Wastes Sites*, Technomic Publishing Co., Lancaster, PA, 1963.
11. Janisz, A. J. and Butterfield, W. S. "Biological Sampling Methods and Effects of Exposure to Municipal and Chemical Landfill Leachate on Aquatic Organisms," in *Hazardous and Industrial Waste Management and Testing: Third Symposium*, American Society for Testing and Materials, Philadelphia, PA, 1984.
12. Young, P. and Parker, A. "Vapors, Odors, and Toxic Gases from Landfills," in *Hazardous and Industrial Waste Management and Testing: Third Symposium*, American Society for Testing and Materials, Philadelphia, PA, 1984.
13. White, D. K. "Inorganic Analytical Methods–General Description and Quality Control Considerations," in *Quality Control in Remedial Site Investigation: Haz-*

ardous and Industrial Solid Waste Testing, Vol. 5, ASTM STP 925, American Society for Testing and Materials, Philadelphia, PA, 1986.
14. Fisk, J. F. "Semi-Volatile Organic Analytical Methods—General Description and Quality Control Considerations," in *Quality Control in Remedial Site Investigation: Hazardous and Industrial Solid Waste Testing, Vol. 5, ASTM STP 925*, American Society for Testing and Materials, Philadelphia, PA, 1986.
15. Fisk, J. F. "Volatile Organic Analytical Methods—General Description and Quality Control Considerations," *Quality Control in Remedial Site Investigation: Hazardous and Industrial Solid Waste Testing, Vol. 5, ASTM STP 925*, American Society for Testing and Materials, Philadelphia, PA, 1986.
16. Warner, J. S., Landes, M. C., and Slivon, L. E. "Development of a Solvent Extraction Method for Determining Semivolatile Organic Compounds in Solid Wastes," in *Hazardous and Industrial Solid Waste Testing: Second Symposium, ASTM STP 885*, American Society for Testing and Materials, Philadelphia, PA, pp. 203–213.
17. Hinton, E. A., Rawlins, L. K., and Flanagan, E. B. "Development of an On-line Mercury Stream Monitor," *Environ. Sci. Technol.* 21:198–202 (1987).
18. Crathorne, B., Fielding, M., Steel, C. P., and Watts, C. D. "Organic Compounds in Water: Analysis using Coupled-Column High-Performance Liquid Chromatography and Soft-Ionization Mass Spectrometry," *Environ. Sci. Technol.* 18:797–802 (1984).
19. Milanovich, F. P. "Detecting Chloroorganics in Groundwater," *Environ. Sci. Technol.*, 20:441–442 (1986).
20. Smith, M. A. *Contaminated Land, Reclamation and Treatment*, Plenum Press, New York, NY, 1985.
21. Bennett, G. F., Feates, F. S., and Wilder, I. *Hazardous Materials Spills Handbook*, McGraw-Hill, New York, 1982.
22. Spittler, T. M. "Field Measurement of Polychlorinated Biphenyls in Soil and Sediment Using a Portable Gas Chromatograph," in *Environmental Sampling for Hazardous Wastes*, American Chemical Society, Washington, D.C., 1984, pp. 37–42.
23. Thielen, D. R., Foreman, P. S., Davis, A., and Wyeth, R. "Delineation of Landfill Migration Boundaries Using Chemical Surrogates," *Environ. Sci. Technol.* 21:145–148 (1987).
24. Pojasek, R. B. and Scott, M. F., "Surrogate Screening for Volatile Organics in Contaminated Media," in *Hazardous Solid Waste Testing: First Conference, ASTM STP 760*, American Society for Testing and Materials, Philadelphia, PA, pp. 217–224.
25. Kerkhoff, M. J., Lee, T. M., Allen, E. R., Lundgren, D. A., and Winefordner, J. D. "Spectral Fingerprinting of Polycyclic Aromatic Hydrocarbons in High-Volume Ambient Air Samples by Constant Energy Synchronous Luminescence Spectroscopy," *Environ. Sci. Technol.* 19:695–699 (1985).
26. Vo-Dinh, T., Bruewer, T. J., Colovos, G. C., Wagner, T. J., and Jungers, R. H. "Field Evaluation of a Cost Effective Screening Procedure for Polynuclear Aromatic Pollutants in Ambient Air Samples," *Environ. Sci. Technol.* 18:447–482 (1984).
27. Miller, S. "Recent Advances in Monitoring Air Pollution, a Report on the Fourth EPA Symposium," *Environ. Sci. Technol.*, 18:253A–254A (1984).
28. Lane, D. A. "Mobile Mass Spectrometry, a New Technique for Rapid Environmental Analysis," *Environ. Sci. Technol.* 16:38A–46A (1982).

29. Puskar, M. A., Levine, S. P., and Lowry, S. R. "Qualitative Screening of Hazardous Waste Drum Mixtures," *Environ. Sci. Technol.* 21:90–96 (1987).
30. Barfknecht, T. R. and Naismith, R. W. "Methodology for Evaluating the Genotoxicity of Hazardous Environmental Samples," *Hazardous Waste* 1:93–109 (1984).
31. Hoffman, G. R. "Mutagenicity Testing in Environmental Toxicology," *Environ. Sci. Technol.* 560A–573A (1982).
32. Donnely, K. C., Brown, K. W., Thomas, J. C., Davol, P., Scott, B. R., and Kampbell, D. "Evaluation of the Hazardous Characteristics of Two Petroleum Wastes," *Hazardous Waste Hazardous Mater.* 2: 191–208 (1985).
33. Samolioff, M. R., Bell, J., Birkholz, D. A., Webster, G. R., Arnott, E., Pulak, R., and Madrid, A. "Combined Bioassay-Chemical Fractionation Scheme for the Determination and Ranking of Toxic Chemicals in Sediments," *Environ. Sci. Technol.* 17:329–334 (1983).
34. Dorward, E. J. and Barisas, B. G. "Acute Toxicity Screening of Water Pollutants using a Bacterial Electrode," *Environ. Sci. Technol.* 18:967–972 (1984).
35. Reid, N. J. "Remote Sensing and Forest Damage," *Environ. Sci. Technol.* 21:428–429 (1984).

CHAPTER 3

The Homogenization of Environmental Soil Samples in Bulk

G. A. Raab, M. H. Bartling, M. A. Stapanian, W. H. Cole, III, R. L. Tidwell, and K. A. Cappo

INTRODUCTION

As technology for measuring hazardous waste rapidly progresses in the areas of instrumentation and technique, there is an urgent need for standardization of quality control procedures. Crucial to this standardization is the availability of uniform referee material. This need is especially critical in the area of soil remediation, due to the dynamic variability found in soil matrices. The ability to prepare a homogeneous, representative soil sample is essential to ensure the quality of any soil survey, feasibility study, or remediation project.

Much research is being conducted in an effort to improve the control of soil quality assurance samples. Site comparison samples (Barich et al., 1988), spiked reference materials, natural quality assurance samples, and synthetic quality assurance samples are currently being employed or investigated as quality assurance controls. For these sample types to be effective as quality assurance checks, they must meet certain criteria. They must be representative of the routine sample in physical appearance, physical composition, and chemical composition. They must also maintain a high degree of analytical reproducibility. To maintain reproducibility in soils, the soil material must remain in a homogeneous state throughout the study.

Several methods have been developed for homogenizing small amounts of soil material. For long-term surveys or large-scale remediation programs, however, it is often necessary to deal with large quantities of soils. Our investigation focused in developing a standard homogenization procedure for large bulk samples. The generalized procedure is shown in Figures 1 and 2 and is summarized below.

A 150-kg soil sample was collected from each of five soil horizons (A, Bs, Bt2, Bw, and C). A variation on the long-standing cone-and-quartering tech-

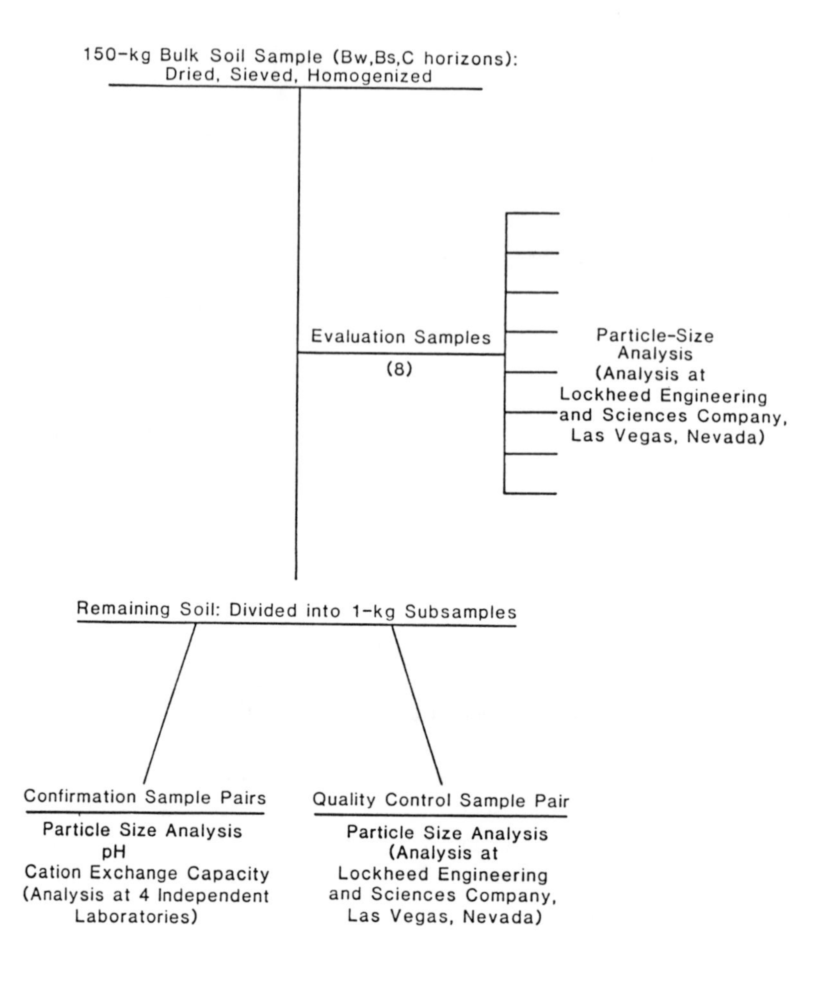

Figure 1. Flow diagram outlining the analysis of the samples taken from the B_w, B_s, and C horizons. (Graphics by Steve Garcia, LESC.)

nique provided homogeneous 1-kg subsamples from the bulk quantities. Sieve analysis data showed that there were no significant variations in particle-size distribution between replicate evaluation samples collected from arbitrary locations within the 150-kg bulk samples or between the 1-kg confirmation samples.

To further support the validity of the homogenization technique, the 1-kg subsamples were analyzed for pH and cation exchange capacity (CEC) as a means of identifying any chemical heterogeneities in the bulk (150-kg) samples. The pH and CEC results confirmed that there was no significant heterogeneity in the original bulk samples or in the respective subsamples.

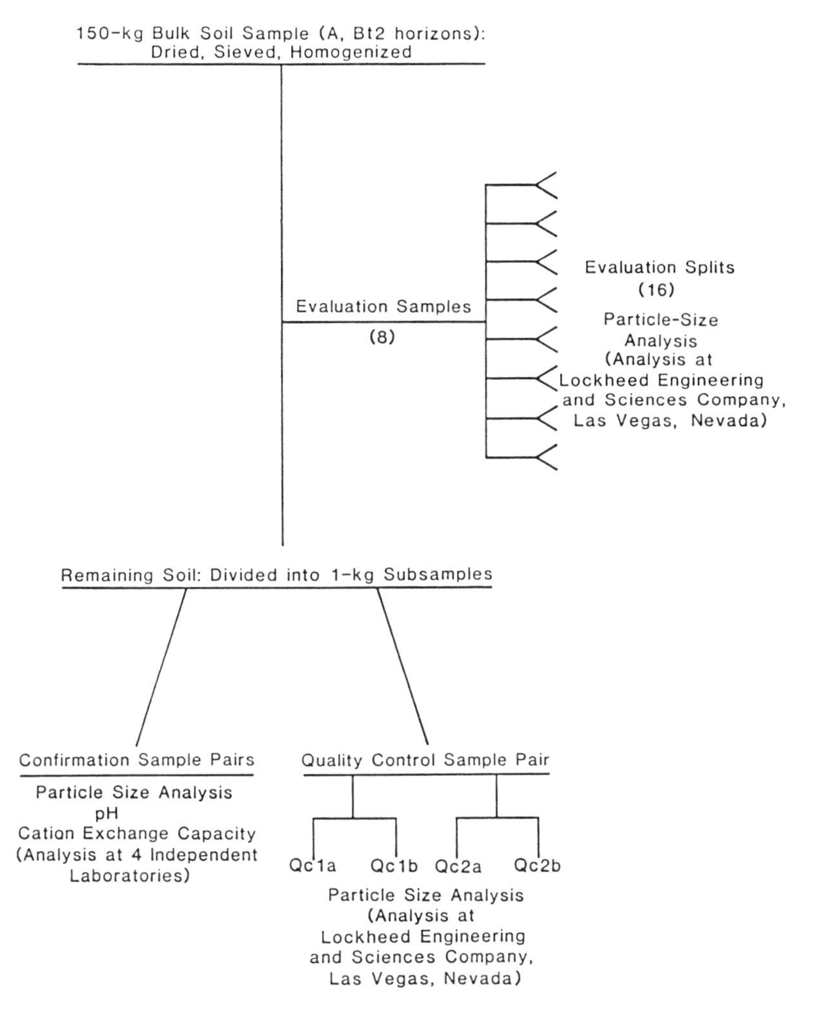

Figure 2. Flow diagram outlining the analysis of the samples taken from the A and Bt2 horizons. (Graphics by Steve Garcia, LESC.)

OBTAINING HOMOGENEOUS SAMPLES

Sample Preparation

The 150-kg bulk samples were processed at a preparatory laboratory operated by Lockheed Engineering & Sciences Company for the U.S. Environmental Protection Agency in Las Vegas, NV. Ambient temperatures in the laboratory varied from 100 to 120°F and relative humidity varied from 5 to 10%.

The bulk samples contained varying amounts of moisture and had to be dried before they could be homogenized. The samples were spread out on 8- x 4-ft trays and were turned with a shovel each day to speed the drying process, which lasted 14 to 20 days. The soil moisture content was measured each day by using the oven-drying method (Soil Conservation Service, 1984). The bulk samples were considered to be air dry when the soil moisture content did not vary more than 2.5% in 24 hr. Clods that remained in the air-dried samples were broken up by rotating the bulk samples for 6 hr in a 55-gal drum containing 4-in. diameter stainless steel balls.

After clod disaggregation, the bulk samples were passed through a standard 2-mm soil sieve. Material that did not pass through the sieve was saved for future study. Material that did pass through the sieve was collected on the drying tray and was piled into a cone.

Homogenization Procedure

The procedure used to homogenize the sample cone was a variation of the cone-and-quartering reduction method described in ASTM C 702–80 (1984). Instead of using the method for directly reducing a large volume sample, we modified the technique by homogenizing the entire 150-kg sample before preparing the 1-kg subsamples. The cone was divided into four approximately equal "quarters," which were numbered clockwise from 1 through 4 (see Figure 3). The first quarter was removed with a shovel and was placed away from the old cone to make a new cone; the third, second, and fourth quarters were piled

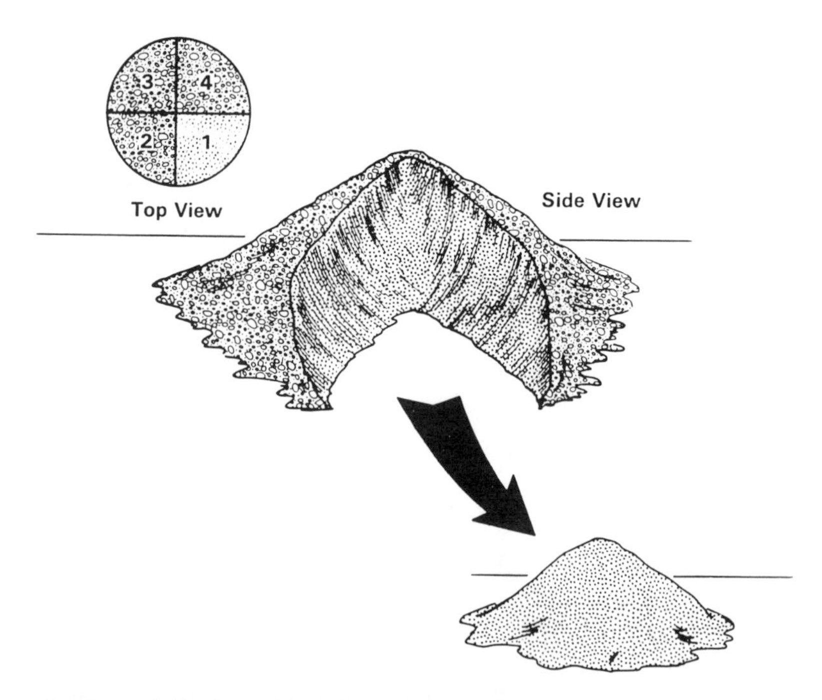

Figure 3. Top and side views of the bulk sample cone. The top view shows the sequence for removal of the quarters. The side view illustrates the original cone with the first quarter removed to form the new cone. (Graphics by Paula Fry, Pipkins Group.)

sequentially onto the first quarter. This homogenization process was performed six times to ensure that all particle-size fractions were distributed evenly throughout the bulk sample. Under the assumption that uniform particle-size distribution represents uniform mineralogical and chemical distribution, we hypothesized that the modified cone-and-quartering technique could be used to produce aliquots of uniform composition from large-volume samples and that these aliquots could be used with confidence as quality assurance samples or performance evaluation samples.

Subsampling of the Bulk Sample (Evaluation Samples)

Eight evaluation samples were taken from the middle third region of each homogenized bulk sample cone (see Figure 4). The selection of this region was arbitrary. To delineate the sampling region, we assumed that it is bounded by two imaginary horizontal planes that pass through the cone; the higher plane intersects the cone one third of the distance from the peak of the cone. The lower horizontal plane intersects the cone one third of the distance from the base of the cone. Evaluation samples 1 through 5 were collected at equidistant points just inside the perimeter of the sampling region, and samples 6 through 8 were collected at equidistant points near and radial to the center of the sampling region (see Figure 5). To collect each evaluation sample, a 2-ft-long plastic tube (2-in. inside diameter) was inserted into the sampling region. The tube was removed and the soil collected in it was transferred to a plastic bottle. The eight evaluation samples collected from each bulk sample ranged in weight from 100 to 350 g. The weights were varied intentionally to determine if a variation in sample size would affect homogeneity.

To accurately evaluate the thoroughness of the cone-and-quartering homogenization process, we also had to analyze the particle-size distribution within a single evaluation sample. Consequently, each of the evaluation samples collected from the Bt2 and A horizons was split in half by using a 0.75-in. riffle

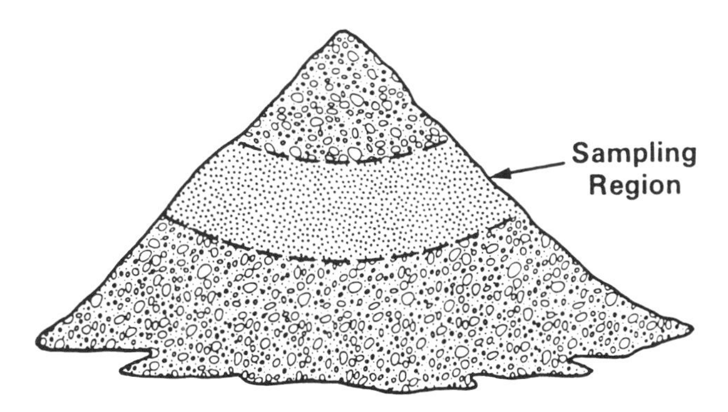

Figure 4. Side view of the homogenized cone showing sample region. (Graphics by Paula Fry, Pipkins Group.)

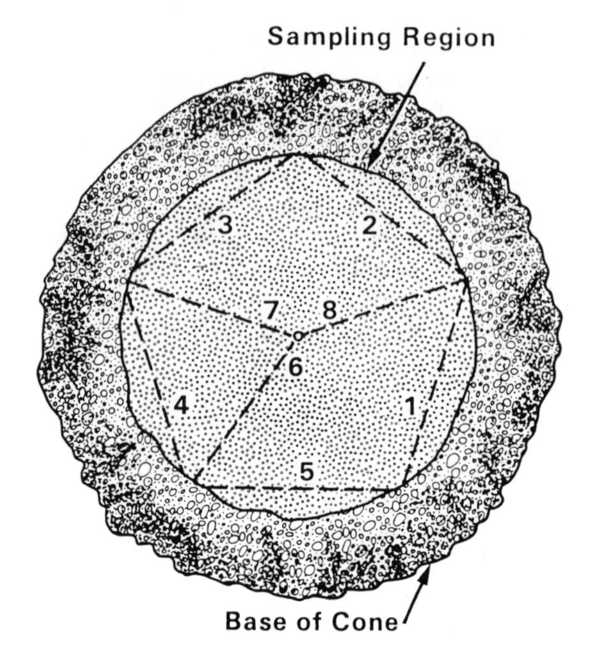

Figure 5. Horizontal section through sampling region of the homogenized cone. Numbers represent samples and the sequence in which they were taken. (Graphics by Paula Fry, Pipkins Group.)

splitter (see Figures 2 and 6). The splits were sieved separately to check variance among the evaluation samples.

Subsampling: Confirmation Samples

After homogenizing each bulk sample completely and collecting the evaluation samples, we collected samples of approximately 1 kg each. To obtain a representative sample of the bulk material, we filled a 1-kg sampling container by making one pass along the exterior of the cone from bottom to top. Each 1-kg subsample was placed in a plastic bag and set aside for sealing and cold storage. This procedure was performed as many as 110 times for each of the five homogenized bulk samples. The subsamples were placed in cold storage at 4°C to reduce the possibility of chemical change with time.

Two of the 1-kg subsamples were arbitrarily selected from each horizon and were used as quality control checks on the homogenization of the other 1-kg subsamples. These quality control samples were analyzed in Las Vegas along with the evaluation samples. As with the evaluation samples, the quality control samples taken from the A and Bt2 horizons were split by using a Jones-type riffle splitter (see Figures 2, 3, and 6).

Other 1-kg subsamples were sent to four independent laboratories for analyses; these samples are referred to here as confirmation samples.

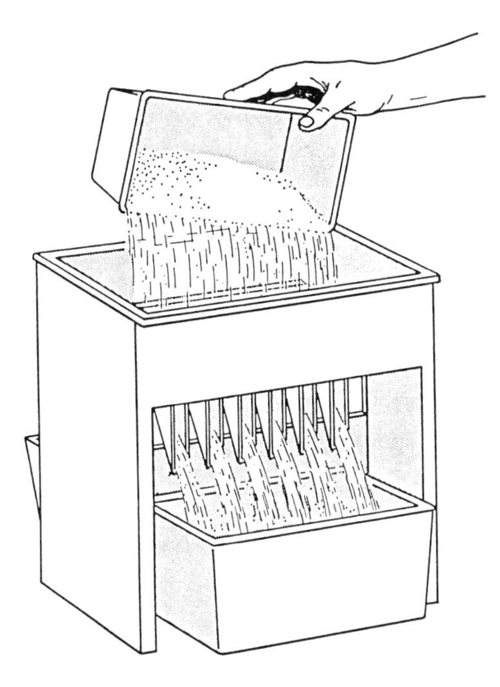

Figure 6. Drawing of a Jones Riffle Splitter. (Graphics by Steve Garcia, LESC.)

EVALUATION OF HOMOGENIZATION

Overview

Sieve analyses were conducted on all evaluation samples to assess the homogeneity achieved by using the modified cone-and-quartering procedure. We applied particle-size analysis as a single-parameter test, for which we assumed that the soil was mineralogically and chemically homogeneous if the individual particle sizes were randomly distributed throughout the sample.

Each evaluation sample was mechanically dry-sieved for 10 min to separate three particle-size fractions:

1. Sand Coarser than Very Fine Sand—Particles in this fraction passed through a 10-mesh sieve, but did not pass through a 140-mesh sieve (particle size less than 2 mm, but greater than or equal to 0.105 mm).
2. Very Fine Sand—Particles in this fraction passed through a 140-mesh sieve, but did not pass through a 270-mesh sieve (particle size less than 0.105 mm, but greater than or equal to 0.053 mm).
3. Silt and Clay—Particles in this fraction passed through a 270-mesh sieve (particle size less than 0.053 mm).

Statistical Evaluation

For all the homogenized evaluation samples, precision for a given size fraction was estimated by the standard deviation of the percent weight of the evaluation samples. The use of percentage data allowed for direct comparisons between samples of different weights.

For the splits of the Bt2 and A evaluation samples, within-split and between sample precisions for a given size fraction were both needed. The standard deviation of the percent weight was used to estimate precision within a split sample. The root mean square, or pooled standard deviation, of the standard deviations of the split samples was used to estimate the precision among the split samples for a given particle-size fraction of the soil horizon. The root mean square (Permutt and Pollack, 1986) of the standard deviations (RMS_s) was calculated as follows:

$$RMS_S = \sqrt{\frac{1}{N} \sum_{i=1}^{N} S_i^2}$$

where

S_i = the standard deviation of the ith of split samples
N = the number of samples that were split (eight for evaluation samples, two for quality control samples)

It was determined *a priori* that standard deviations or RMS_s of less than 5% would indicate data of acceptable precision. When the precision among samples compared was within the acceptable limit, the samples were considered to be homogeneous.

RESULTS

Table 1 presents a summary of the sieve analysis results for the evaluation samples.

The largest standard deviation for any sample is 3.6%. This value lies well within the range of experimental error and the *a priori* condition of 5% imprecision. Based on the sieve analysis results for all samples (100 g to 1 kg), the resolution of each sample appears to be very good despite the initial sample size (150 kg).

The 3.6% standard deviation is associated with the silt and clay fraction of the Bw sample horizon. This sample is approximately 30% silt and clay. Because of its small particle size, the clay material is easily dispersed when shoveled. Thus, it is difficult to contain this material during the cone-and-quartering and sieving procedures. The high rate of loss of these airborne clay particles may account for part of the 3.6% standard deviation.

Table 1. Summary of Sieve Analysis Results of Evaluation Samples, Bw, Bs, C, Bt2 and A Horizons

	Soil Horizons									
	Bw		Bs		C		Bt2		A	
Particle Size	Mean wt.(g) %	Std. Dev.	Mean wt.(g) %	Std. Dev.	Mean wt.(g) %	Std. Dev.	Mean wt.(g) %	Std. Dev.	Mean wt.(g) %	Std. Dev.
<2mm >0.105mm	45.5	1.7	82.3	2.5	87.2	1.6	74.8	1.9	71.5	2.7
<0.105mm >0.053mm	21.8	2.3	10.6	2.5	9.0	1.0	22.5	1.3	25.3	1.4
<0.053 mm	32.7	3.6	7.1	0.8	3.8	0.6	2.7	1.0	4.0	0.6
Total	100.0		100.0		100.0		100.0		100.8	

Tables 2 and 3 show the sample imprecision for the evaluation sample splits of the A and the Bt2 horizons. The standard deviations ranged from 0 to 1.838%, well within the acceptable limit of 5%.

The 1-kg quality control samples were analyzed for particle size distribution in the same manner as the evaluation samples. For the quality control samples that were not split (horizons Bw, Bs, and C), the standard deviations ranged from 0.14 to 1.20% (Table 4). For the split quality control samples (horizons A and Bt2), the RMS_s values, which quantify the imprecision between the split samples, ranged from 0 to 2.9% (Table 5). Therefore, the imprecision within

Table 2. Imprecision Estimates for Homogenized Evaluation Sample Splits, A Horizon. Within Sample, Imprecision Was Calculated by the Standard Deviation(s) of the Percentage Weights

Particle Size Fraction	% Standard Deviation of Subsample Number 1	2	3	4	5	6	7	8
2 mm to >0.105 mm	0.071	0.212	0.707	0.778	0.141	0.566	1.202	1.838
<0.105 mm to >0.053 mm	0.212	0.0	0.354	1.556	0.141	0.212	1.556	1.414
<0.053 mm	0.141	0.212	0.354	0.778	0.0	0.778	0.354	0.453

Table 3. Imprecision Estimates for Homogenized Evaluation Sample Splits, Bt2 Horizon. Within Sample, Imprecision Was Calculated by the Standard Deviation(s) of the Percentage Weights

Particle Size Fraction	% Standard Deviation of Subsample Number 1	2	3	4	5	6	7	8
2 mm to >0.105 mm	0.282	0.071	0.636	0.495	0.566	0.919	0.212	1.131
<0.105 mm to >0.053 mm	0.424	0.354	0.495	0.071	0.636	0.071	1.061	1.202
<0.053 mm	0.141	0.283	1.202	0.495	0.141	0.778	0.919	0.071

Table 4. Standard Deviations (%) Values of Quality Control Samples, Horizons Bw, Bs, and C

Particle Size Fraction	Bw	Bs	C
<2 mm to >0.105 mm	0.92	0.92	0.49
<0.105 mm to >0.053 mm	1.20	0.42	0.14
<0.053 mm	0.28	0.49	0.28

and between splits was below the acceptable maximum level of 5% for all quality control samples. This indicates that the subsample preparation did not affect homogeneity of the bulk samples.

CONFIRMATION OF HOMOGENIZATION

Overview

The final data obtained were derived from pairs of confirmation samples distributed to laboratories as part of a soil survey. Each pair was included in a batch of approximately 40 samples in such a manner as to be indistinguishable from routine samples sent to an analytical laboratory. The confirmation pairs used are from 56 batches of samples analyzed over an approximate 1-year time period. The final standard methods of physical and chemical analysis are outlined in Cappo et al. (1986). For the purpose of this study, only particle-size analysis, pH, and CEC will be discussed.

Particle-Size Analysis

The particle-size analysis focused on three size fractions: sand, silt, and clay. The sand fraction was determined gravimetrically; the silt fraction was determined by difference calculations. The clay fraction was measured by taking an aliquot of suspended soil solution, drying the solution, weighing the clay, and

Table 5. RMS_s (%) Values of Quality Control Sample Splits, Horizons Bt2 and A

	Bt2 Horizon QC Splits RMS_s			A Horizon QC Splits RMS_s		
Particle Size Fraction	1	2	3	1	2	3
<2 mm >0.105 mm	2.0	0.71	1.63	2.9	0.212	2.46
<0.105 mm to >0.053 mm	2.6	0.28	2.18	1.3	0.212	1.08
<0.053	0.64	0.42	0.56	1.7	0.0	1.47

subtracting the weight of a blank. Small measurement errors are multiplied at least 40-fold. Due to the small original sample size of 20 g, samples of 10% clay or less amount to at most 0.05 g of clay in the final weighing. The tare weight of the drying vial is often more than 50 g, so that a 1% variation is on the order of 1/1,000 to 1/10,000 or less of the total weight. Given these restrictions, it is expected that soils of under 10% clay may have very low precision.

Soil pH in Double Deionized Water

For this study, soil pH was determined in double deionized water. Soil pH measured in water is highly variable because the system is not buffered. Hence pH determined in water is very susceptible to the heterogeneity in the soil.

Cation Exchange Capacity

The CEC of a soil represents the ability of the soil to retain cations. Two methods of CEC determination were performed; one used neutral, buffered 1 *N* ammonium acetate and one used unbuffered 1 *N* ammonium chloride. The acetate method produces higher CEC values than the chloride method because the acetate extracts at a higher pH than does the chloride.

Statistical Evaluation of Confirmation Samples

For particle-size analysis and CEC, there are some problems in describing precision among measurements. The magnitude of the error is expected to be roughly proportional to concentration, at least at high concentrations. This proportional relationship, however, breaks down as concentrations approach 0. The percent relative standard deviation (%RSD) was used as the descriptive statistic for the precision of independent measurements for particle-size analysis and CEC. The %RSD is calculated as $(S/\overline{X})100$, where S is the sample pair standard deviation and $\overline{X}$ is the mean sample concentration. The %RSD is, therefore, the ratio of the standard deviation to the mean expressed as a percent. High %RSD values, therefore, indicate low precision relative to the mean. At high concentrations, %RSD is expected to be roughly constant. At low concentrations, however, the %RSD is expected to increase because the standard deviation is larger relative to the mean. For pH in water, the standard deviation is used as the descriptive statistic for precision. Because pH is already on a logarithmic scale, errors are expected to be approximately constant for natural samples.

For each confirmation pair, the %RSD (or standard deviation in the case of pH in water) of the measurements performed by each laboratory for each horizon was calculated. The data from the laboratories were then pooled, and the overall precision statistics for each horizon were analyzed. We made the arbitrary *a priori* decision that %RSD values less than or equal to 20 indicated "acceptable" precision, %RSD less than or equal to 15 indicated "very good" precision and, %RSD less than or equal to 10 indicated "excellent" precision. For pH in water, these limits were standard deviations of 0.20, 0.15, and 0.10 pH units, respectively. Summary statistics for the analyses are given in Table 6.

RESULTS

Particle-Size Analysis

Sand was the major component of each horizon with the exception of the Bw horizon. The Bs and C horizon samples contained roughly 95 and 84% sand, respectively, and there was very high precision within laboratories and between laboratories. For the Bs and C horizon samples, it appears all the samples were statistically identical. Indeed, the overall %RSD for the samples from these two horizons was less than 1.5%. Overall, the Bw horizon samples had "acceptable" precision and had much more variance than samples from any other horizon. This variance was observed primarily between laboratories, not within a laboratory. Laboratories 1, 2, and 3 had excellent precision (%RSD <10) for total sand in all horizon samples. Laboratory 4 had very good precision (%RSD = 13.4) for the Bw horizon samples and excellent precision for the Bs and C horizon samples. For total sand data, the average for Laboratory 4 was nearly identical to the overall mean. The A horizon samples show good precision between laboratories and excellent precision within each laboratory.

On the basis of the precision data for the sand fraction, the majority of the confirmation samples can be interpreted to be homogeneous in nature. There were only a few isolated cases of high %RSDs found for specific horizon samples and laboratories.

The results for total clay strongly support the theory that the lower the concentration, the lower the precision (higher %RSD). Although the standard deviation for the A horizon samples was about the same as for any other horizon samples, the %RSD was much lower. This is because the A horizon samples contained more clay than the other horizon samples. The Bw horizon samples showed a great deal of variation within a laboratory and between laboratories. The imprecision within laboratories ranged from 5 to 27 %RSD for the Bw horizon samples. The C and Bs horizon samples are very low in total clay content (less than 1%), and accordingly they have extremely low precision (35 to 125 %RSD). The amounts of clay in these soils are so low that they are relatively meaningless as a measure of homogeneity. The actual variation is on the order of an absolute 1%, which is about the detection limit of the method.

The value for the silt fraction is found by a difference calculation and is therefore not an independent measurement. Error from the measurements of clay and sand accumulates into the precision estimates for silt. Nonetheless, determinations obtained for silt show that with the exception of the C horizon, precision was excellent overall and within each laboratory. This is an especially powerful piece of evidence for homogeneity. In the C horizon samples, the actual variation on the amount of silt measured is about 2%. Considering that 1 to 2% is the detection limit of the method and considering the number of mathematical manipulations needed to make the silt determination, the precision and variation obtained are very good. The evidence overwhelmingly supports the idea that the confirmation samples were homogeneous at the time they were prepared and shipped.

Soil pH in Double Deionized Water

There was excellent precision overall and within the laboratories for the A, Bs, and Bw horizon samples, much less than the 0.1 to 0.2 pH units of accuracy that the instruments have. For Laboratories 2 and 3, there was also excellent precision for the C horizon samples. The overall "unacceptable" precision in the C horizon samples was due to unacceptable precision for Laboratory 1. We are unsure of the source of the imprecision. In view of this, the A, Bs, and Bw horizon samples appear to be homogeneous within each laboratory and across laboratories. Because imprecision in the C horizon sample was due to the results from only one laboratory, we suspect that the C horizon sample was also homogeneous.

Cation Exchange Capacity

The homogeneity of the soils is evaluated on the basis of precision data obtained from confirmation pairs. If the clay and organic fractions are homogeneously distributed throughout a sample, this will be reflected in a low standard deviation and %RSDs. It should be noted, however, that if the clay and organic fractions are minimal in comparison to the rest of the sample, the CEC values will be too low to be statistically meaningful.

Overall precision was acceptable for the A, Bs, and Bw horizon samples analyzed according to the ammonium acetate method. The overall precision for the C horizon did not meet the *a priori* limit, mainly because of the results from Laboratory 2. Within-laboratory precision was excellent for the Bs and Bw horizon samples in seven cases and was very good in the remaining case (see Table 6). Precision within each laboratory and overall was excellent for the A horizon samples.

For the ammonium chloride method, within-laboratory and overall precision was excellent for the A horizon samples. For the Bs and Bw horizon samples, within-laboratory precision was always very good; however, overall precision for the Bs and Bw horizon samples did not meet the *a priori* limit. For the C horizon sample, neither the intralaboratory precision nor the overall precision met the *a priori* limit. It should be noted that the CEC levels obtained

Table 6. Summary Statistics for Intra-Laboratory Measurements for Confirmation Samples

CEC-NH_4 OAC		Laboratory 1				Laboratory 2				Laboratory 3				Laboratory 4				Overall			
		N	$\overline{X}$	S	%RSD	N	$\overline{X}$	S	%RSD	N	$\overline{X}$	S	%RSD	N	$\overline{X}$	S	%RSD	N	$\overline{X}$	S	%RSD
HORIZON	A	8	18.7	0.40	2	8	15.4	.74	5	—	—	—	—	—	—	—	—	16	17.0	1.77	10
	C	12	1.26	.12	10	8	.76	.34	46	—	—	—	—	10	.85	.15	18	28	1.00	.29	29
	BS	4	27.3	.82	3	10	22	2.22	10	6	26.5	2.01	8	10	25.2	2.16	9	30	24.7	2.8	11
	BW	10	13.9	1.18	8	8	10.4	1.41	14	2	16.4	1.15	7	6	11.9	0.82	7	26	12.5	2.17	17

CEC-NH_4 C1		Laboratory 1				Laboratory 2				Laboratory 3				Laboratory 4				Overall			
		N	$\overline{X}$	S	%RSD	N	$\overline{X}$	S	%RSD	N	$\overline{X}$	S	%RSD	N	$\overline{X}$	S	%RSD	N	$\overline{X}$	S	%RSD
HORIZON	A	8	7.97	.25	3	8	7.33	.46	6	—	—	—	—	—	—	—	—	16	7.65	.49	6
	C	12	.54	.28	51	8	.52	.14	27	—	—	—	—	10	.63	.37	59	28	.57	.29	51
	BS	4	7.54	1.16	15	10	7.53	.38	5	6	12.9	1.40	11	10	8.41	.85	10	30	8.9	2.25	25
	BW	10	5.98	.67	11	8	3.74	.12	3	2	9.62	.59	6	6	4.26	.39	9	26	5.18	1.71	33

pH-H_2O		Laboratory 1			Laboratory 2			Laboratory 3			Laboratory 4			Overall		
		N	$\overline{X}$	S	N	$\overline{X}$	S	N	$\overline{X}$	S	N	$\overline{X}$	S	N	$\overline{X}$	S
HORIZON	A	8	4.53	.04	8	4.52	.03	—	—	—	—	—	—	16	4.52	.04
	C	12	5.19	.27	8	5.56	.08	—	—	—	10	5.34	.09	28	5.32	.24
	BS	4	4.36	.09	10	4.52	.04	6	4.50	.04	10	4.50	.03	30	4.49	.07
	BW	10	5.02	.04	8	5.14	.03	2	5.15	.01	6	5.09	.06	26	5.08	.06

Total Sand

Horizon	Laboratory 1				Laboratory 2				Laboratory 3				Laboratory 4				Overall			
	N	$\overline{X}$	S	%RSD	N	$\overline{X}$	S	%RSD	N	$\overline{X}$	S	%RSD	N	$\overline{X}$	S	%RSD	N	$\overline{X}$	S	%RSD
A	8	54.4	.37	1	8	59.2	.59	1	—	—	—	—	—	—	—	—	16	56.8	2.52	4
C	12	95.1	1.11	1	6	95.9	1.64	2	—	—	—	—	10	95.6	.82	1	28	95.5	1.16	1
BS	4	83.4	.8	1	10	83.4	.93	1	6	84.7	1.13	1	10	84.5	.97	1	30	84.1	1.09	1
BW	10	26	.51	2	8	33.3	2.32	7	2	26.2	.82	1	6	28.7	3.84	13	26	28.1	4.52	16

Total Clay

Horizon	Laboratory 1				Laboratory 2				Laboratory 3				Laboratory 4				Overall			
	N	$\overline{X}$	S	%RSD	N	$\overline{X}$	S	%RSD	N	$\overline{X}$	S	%RSD	N	$\overline{X}$	S	%RSD	N	$\overline{X}$	S	%RSD
A	8	19	.8	4	8	18.2	.71	4	—	—	—	—	—	—	—	—	16	18.6	.82	4
C	12	.17	.13	76	6	1.22	.95	78	—	—	—	—	10	.67	.84	125	28	.57	.76	133
BS	4	.70	.36	51	10	1.69	.86	51	6	.95	.40	.42	10	.71	.25	35	30	1.08	.70	65
BW	10	7.27	.36	5	8	7.70	1.43	19	2	8.85	.92	10	6	5.20	1.38	27	26	7.05	1.51	21

Total Silt

Horizon	Laboratory 1				Laboratory 2				Laboratory 3				Laboratory 4				Overall			
	N	$\overline{X}$	S	%RSD	N	$\overline{X}$	S	%RSD	N	$\overline{X}$	S	%RSD	N	$\overline{X}$	S	%RSD	N	$\overline{X}$	S	%RSD
A	8	26.6	1.04	4	8	22.6	.82	4	—	—	—	—	—	—	—	—	16	24.6	2.3	9
C	12	4.71	1.11	24	6	2.82	1.74	62	—	—	—	—	10	3.8	1.22	32	28	3.98	1.45	36
BS	4	15.9	1.09	7	10	14.8	1.23	8	6	14.3	1.43	10	10	14.8	1	7	30	14.9	1.21	8
BW	10	68.8	.42	1	8	59	1.94	3	2	65	.57	1	6	66.1	3.15	5	26	64.9	4.52	7

from the C horizon sample using the required method are below meaningful levels. Therefore useful precision data cannot be expected at these levels.

CONCLUSIONS

The use of the modified cone-and-quartering method has been determined to be very effective for homogenizing bulk quantities of soil. This conclusion is supported by statistical analysis performed on the physical and chemical results obtained from five homogenized bulk samples. This method can be effectively applied to site comparison samples, natural and synthetic quality assurance samples, and spiked reference materials as a means of obtaining homogeneity.

ACKNOWLEDGMENTS

We are greatly indebted to Brian Schumacher, Mike Homsher, and Roy Cameron for their technical advice. Our thanks also extend to Steve Hern and Ken Hedden for their cooperation and for the use of the U.S. Environmental Protection Agency (EPA) Environmental Monitoring Systems Laboratory, Las Vegas, NV, Greenhouse for the cone-and-quartering, splitting, and sieving procedures. EPA technical support and assistance was offered by Louis Blume, Robert Schonbrod, and James Mullins.

The following people were instrumental in the completion of this project: Peggy Oakes, Jan Engels, Annalisa Haynie, and Marianne Faber of Lockheed Engineering & Sciences Company, Inc., and the Computer Sciences Corporation word processing staff at EMSL-LV.

NOTICE

Although the research described in this article has been supported by the U.S. Environmental Protection Agency under contract number 68–03–3245 to Lockheed Engineering & Sciences Company, it has not been subjected to agency review and therefore does not necessarily reflect the views of the agency and no official endorsement should be inferred. Mention of trade names or commercial products does not constitute endorsement or recommendation for use.

REFERENCES

1. American Society for Testing and Materials. "Standard Methods for Reducing Field Samples of Aggregate to Testing Size," *Annual Book of ASTM Standards,* Vol. 04.02, American Society for Testing and Materials, Philadelphia, PA, 1984, pp. 452–457.
2. Barich, J. J., Jones, G. A., Raab, G. A., and Pasmore, J. R. "The Application of

X-Ray Fluorescence Technology in the Creation of Site Comparison Samples and in the Design of Hazardous Waste Treatability Studies," in Proceedings of the First Internation Symposium, Field Screening Methods for Hazardous Waste Site Investigations, U.S. Environmental Protection Agency, U.S. Army Toxic and Hazardous Materials Agency, Instrument Society of America, 1988.

3. Cappo, K. A., Blume, L. J., Raab, G. A., Engels, J. L., and Bartz, J.K. "Analytical Methods Manual for the Direct/Delayed Response Project," EPA/600/8-87/020, U.S. Environmental Protection Agency, Las Vegas, NV, 1986.
4. Permutt, T. J. and Pollack, A. "Analysis of Quality Assurance Data," National Surface Water Survey Eastern Lake Survey (Phase I—Synoptic Chemistry) Quality Assurance Report, Best, M.D., Ed., EPA 600/4-86/011, U.S. Environmental Protection Agency, Las Vegas, NV, 1986.
5. Soil Conservation Service, "Procedures for Collecting Soil Samples and Methods of Analysis for Soil Survey," Soil Survey Investigations Report No. 1, Soil Conservation Service, U.S. Department of Agriculture, Washington, D.C., 1984.

CHAPTER 4

A Comparison of Soil Sample Homogenization Techniques

B. A. Schumacher, K. C. Shines, J. V. Burton, and M. L. Papp

INTRODUCTION

The need for sample homogeneity prior to laboratory analyses has been long recognized by geologists, pedologists, chemists, and members of other scientific disciplines. Homogeneity is the degree that the material under investigation is mixed resulting in the random distribution of all particles in the sample. Completely homogeneous materials are so rare that they may be considered nonexistent,[10] yet scientists must strive to obtain a homogeneous sample in order to obtain data exhibiting minimal errors attributable to sample heterogeneity.

Care must be taken during the subsampling phase of soil preparation, during sample transport, unpacking, and transfer to other containers in the laboratory to avoid particle sorting via different particle densities, shapes, sizes, and resistance of certain minerals to mixing, such as magnetite.[10,14,18] Numerous methods have been used to obtain homogeneous soil samples. The various methods range from simple grinding and sieving of the sample to a desired particle size to various mixing and splitting devices and machines. Each method will be discussed individually with both advantages and disadvantages being presented.

One of the most common methods used to obtain a homogeneous sample is to grind and sieve the soil to a desired particle size (generally <2 mm), followed by random sampling. This method assumes that the initial material is ground without preference to any given factor, such as color, and that during grinding and sieving, the sample becomes sufficiently homogenized. To further eliminate possible heterogeneity within the sample and to reduce the sample size to the desired quantity for a given analysis, subsamples may be obtained by "spooning"[3] or some other method which involves the random insertion of a spoon or other sampling device into the previously ground and sieved sample. This method is preferably done rapidly and without extensive visual examination of the sample that could lead to a processor preference in certain cases, for example, light catching the shiny surfaces of mica flakes leading to preferential inclusion or exclusion of that part of the sample. The "spooning" type of subsampling markedly reduces the time of sample processing in comparison to multiple, successive splitting operations and has been shown to produce equivalent correlation coefficients between observed settling velocities of sands (median r = 0.994 using other splitting methods and median r = 0.9955 using the spooning technique) when used to obtain 20-g subsamples from initial 1- to 2-kg sand samples.[3]

Two other methods have been devised which are similar to the "spooning" method except they involve an intermediate step between grinding and sieving and subsampling procedures. Gilliam and Richter[5] used an intermediate step of stirring the sample with a spatula before subsampling occurred, presumably until visual homogeneity was obtained. Some analysts content themselves by merely shaking the sample in a bottle prior to subsampling and ignore the risk of constituent segregation.[18]

An additional method of sample subdivision with the goal of obtaining a representative sample was presented by Allman and Lawrence.[1] Their method is similar to the "spooning" method of Carver[3] except that a scoop of ground and sieved materials was divided among four containers. The process was repeated continually, changing the filling order of the containers until the sample had been quartered or an appropriate sample size had been obtained. As with the spooning process, visual bias as to how much sample and into which container the samples were placed is a concern.

Before proceeding to the more elaborate sample splitting schemes, a discussion of methods used simply to mix (homogenize) the ground and sieved sample is warranted. The simplest of the homogenization processes is tumbling the sample on a sheet of paper, cloth, or plastic. This process involves the manual rolling of ground samples such that the sample must tumble upon itself and not just slide along the surface of the sheet.[10,19,21] This method is effective on sample sizes less than 2 kg, yet caution must be taken to ensure that the sheet material does not contain any element which is to be quantitatively analyzed and does not develop static charges that may lead to segregation of the finer particle sizes.

Homogenization may also be achieved through the use of mechanical mixing devices including a spiral mixer, a cement mixer, and a twin-shell V-blender. The spiral mixer involves the rotation of the bottled sample in both horizontal and vertical planes.[21] The cement mixer or similar devices involve the rotation of the sample in a chamber with a series of internal baffles that cause the materials to be thoroughly tumbled and mixed. These two methods are useful for samples ranging from less than one pound to several hundred pounds. The twin-shell V-blender involves the rotation of two hollow cylinders about a horizontal axis such that the apex describes a circle in the vertical plane.[19] Twin-shell blenders are available commercially in sizes ranging from 4- to 16-quarts (approximately 10 to 40 kg of a mineral soil) internal capacity

(Figure 1). However, Ingamells and Pitard[10] expressed serious doubts as to the value of mechanical splitters. They stated that, "In general, machines that use mechanical violence and look and sound as though they were efficient are most likely to cause segregation of heavy and light, large and small, and flat and round particles." Further concerns have been expressed concerning sample breakdown to finer particle sizes due to the violent tumbling in the machines.

The riffle splitter (also called a chute splitter, Jones splitter, or just sample splitter) is perhaps the most common mechanical method for sample homogenization and/or sample size reduction.[9,15] The riffle splitter also provides one of the best general methods of sample mixing to obtain bulk sample homogeneity.[10] A riffle splitter is a device having an equal number of narrow sloping chutes with alternate chutes discharging the sample in opposite directions into two collection bins (Figures 2 and 3). Sample homogenization is achieved by repeated pouring of soil through the splitter and combining the halves between passes. The use of the riffle splitter as a subsampling device is done in a similar manner, with the exception that after the sample is passed through the splitter, one collection pan is replaced with a clean pan. The material in the "replaced" pan, which contains about one-half of the original sample, is then passed through the riffle splitter again, thereby reducing the volume in the clean pan to one-quarter of its original sample volume. This process of sample reduction is repeated until the desired weight or sample size is obtained.

Many variations in the size and construction of materials have been built on the principle of the riffle splitter. For small samples (10 to 50 g), an efficient

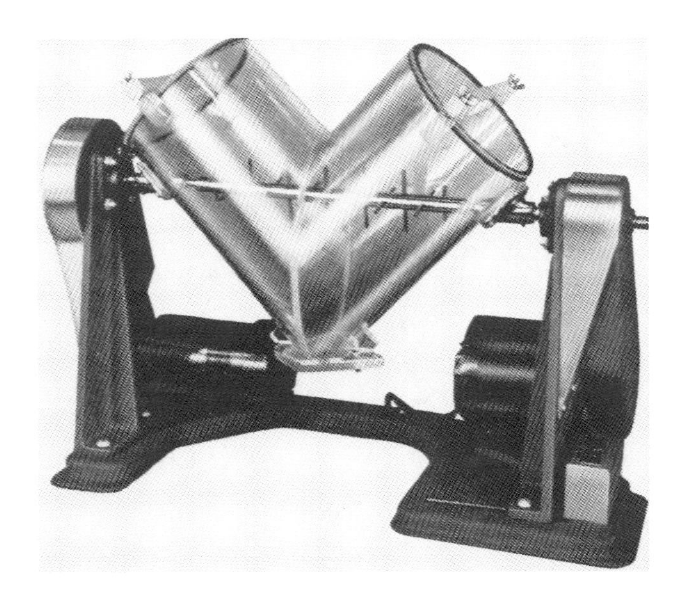

Figure 1. A twin-shell V-blender. (Used with permission of Patterson-Kelley Co., East Stroudsburg, PA.)

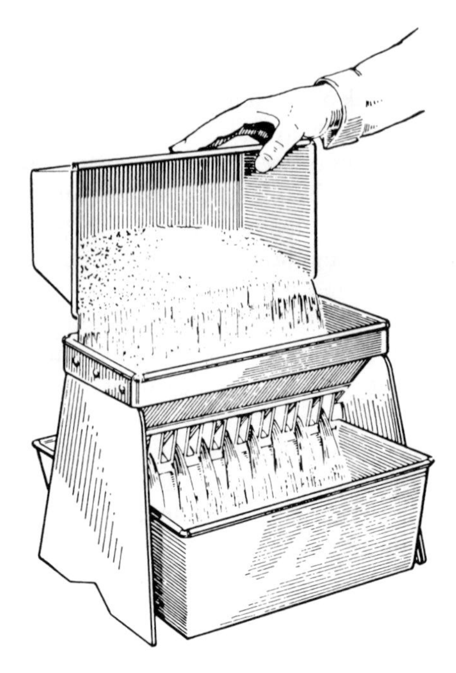

Figure 2. An open-bin riffle splitter. (Reprinted with permission of R. E. Carver.)

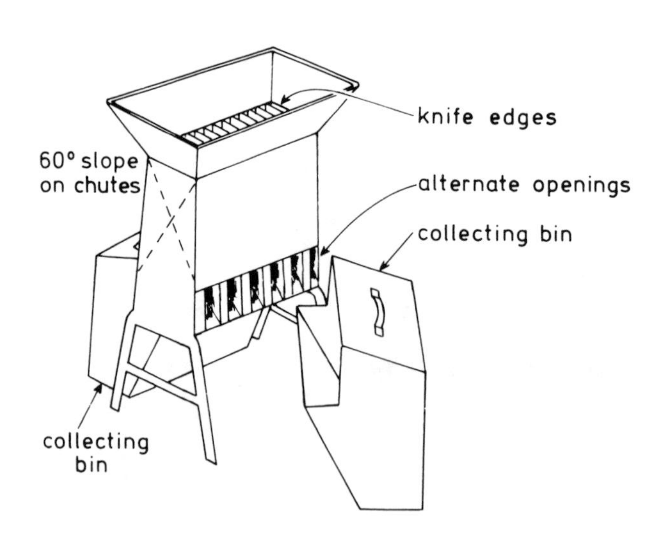

Figure 3. A closed-bin riffle splitter. (From *Geological Laboratory Techniques* by M. Allman and D. F. Lawrence. Reprinted with permission of ARCO Publishing Co. Inc., a division of Cassell CLP, London.)

microsplitter has been designed by Newton and Dutcher[16] using balsa wood and glass microscope slides. These authors conducted experiments using 40-g samples of the fine sand fraction and found only a 1.2% average error when quartering the sample by a two-step halving. The results of Newton and Dutcher[16] are supported by the earlier work of Griffiths,[6] who found that sample homogenization using a riffle splitter produced an overall coefficient of variance (C_v) between median grain sizes of less than 3% in 9 out of 10 samples run on various rock and sand samples. In the one case where the C_v was greater than 3%, the error was attributed to operator differences (16 in the single case vs. five or fewer in the other nine cases). A portable sieving and splitting device for field use has been designed by Ibbeken[8] which uses the riffle splitter to subsample and process coarse-grained sediments. During his research on unconsolidated sediments, no significant differences were found between splits in terms of petrographic mineralogical composition, with the exception of the 125- to 160-mm serpentine fraction (represented by only four pebbles), nor in grain-size distribution. Another variation of the riffle splitter found in the literature was the use of a single piece of tin-plate bent several times to form a riffle splitter.[13] The advantage of this construction method is that free grain flow in the chutes was obtained, without the hindrance of unevenly soldered joints that may be present in other riffle splitters.

The use of riffle splitters and their variations is valuable in splitting samples which range in size from less than 10 g[7] to several kilograms.[21] Ibbeken[8] reported that 0.5 to 1 ton (455 to 909 kg) of sample can be processed daily using a standard riffle splitter to reduce the initial quantity to 5-kg subsamples. Although most authors find the use of the riffle splitter to be an effective, efficient method for sample homogenization and sample splitting, several problems exist.

The major source of error involved in using a riffle splitter is the loss of fine particle sizes via "dusting" when processing air-dried or oven-dried samples which contain fine particle sizes.[21] The process of riffle splitting requires that a uniform stream of material be poured into the mouth or top of the splitter. Dust loss may occur through the chute ends, in the collection bins, and through the mouth openings (Figure 2) in an open-bin system or just through the mouth if a closed-bin system is used (Figure 3).

Several studies have been conducted comparing the effectiveness of the riffle splitter to other homogenization techniques. Wentworth et al.[22] compared the riffle splitter with a rotary splitter (to be discussed) and found that the rotary splitter more accurately split an initial sample of known grain-size distribution into subsamples with similar grain-size distributions than the riffle splitter. Kellagher and Flanagan,[11] in a comparison experiment among a multiple-cone splitter (to be discussed), cone and quartering (to be discussed), and a riffle-based microsplitter found the microsplitter was the worst for both accuracy and precision of grain-frequency percentages for subsampling three different weights (5, 10, and 20 g) of an artificially created very coarse and coarse sand fraction mixture. In a similar type of study, Mullins and Hutchison[15] compared the C_v among sand fraction contents of several soils and ranked, in order of best to worst, the rotary subsampling, riffle splitting, cone and quartering, and spoon sampling in terms of their ability to homogenize a sample. These authors did note, however, that the best and worst methods were significantly different only at the 10% level.

Perhaps the best-known sample-splitting method is the classical cone and quarter technique. This technique involves pouring the sample into a cone, flattening the cone, dividing the flattened cone into four equal divisions (quartering), and then removing two opposite quarters (Figure 4). The remaining two quarters are repiled into a cone and the process is repeated until the desired sample size is obtained. Variations on the process are possible which can enhance the speed of sample size reduction by using just one quarter (chosen at random) to continue the splitting process or which allow this method to be modified to homogenize a large sample. The use of the cone and quarter method to homogenize a sample involves the removal of the first quarter and repiling it into a cone, followed by the subsequent repiling of the opposite quarter and then the remaining two quarters to reform a single cone.[17] This process is repeated several times until sample homogeneity is achieved.

Several sources of error for this method have been identified. Van Johnson and Maxwell[21] reported that during the cone-and-quarter process on large samples (several kilograms), there is the danger of unequal segregation of heavier materials during the flattening and coning of the sample. Similar to the riffle splitting techniques, dusting is also a possible source of error during cone formation. Sample loss from the inability to recollect all the soil from the underlying material, the ability of the sample to "cling" to the underlying material via static charges, and sample embedment are all further sources of

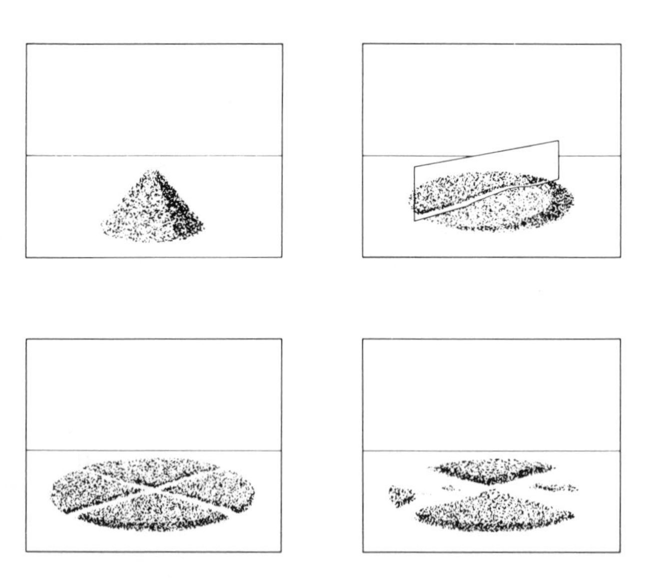

Figure 4. The cone and quarter technique. (Reprinted with permission of R. E. Carver.)

error. Muller[14] placed a limitation on the cone-and-quarter technique to samples greater than 50 g and stated that this method is most successful in the field for sample processing of larger sample sizes. Raab et al.[17] used the cone and quarter method to homogenize large volumes of soil ($\approx$ 150 kg) and then subsampled the repiled cone with a 2-ft long plastic tube (2-in. internal diameter). They then sieved the sample into three particle-size classes (2 to 0.105 mm, 0.105 to 0.053 mm, less than 0.053 mm) and subsampled using a riffle splitter. A resieving to check particle class weights of the three divisions between the six subsamples resulted in the finding of no significant difference (error level = 0.05) among the subsamples, indicating that a homogenous mixture had been obtained.

Several other sample splitting/homogenizing devices found in the literature include a rotary splitter,[1,14,19] a brass disc microsplitter,[2] and a multiple-cone splitter.[11] The rotary splitter involves the pouring of the sample into a feeding hopper located above a rotating disc containing sample bottles or pans (Figure 5). The disc is mechanically rotated and the sample is divided among the collecting bottles or pans. The brass disc microsplitter was designed by Brewer and Barrow[2] for subsampling small particulate samples which range up to 1 g in the particle size range between 10 and 200 μm. This splitter divides the sample by passing it through a fluted brass disc which contains several holes to separate the sample into different collection bottles. The device also has a mechanical vibrator system to ensure complete collection of the fine particles. The multiple-cone splitter of Kellagher and Flanagan[11] consists of three fun-

Figure 5. A rotary splitter. (From *Sedimentary Petrology, Part I, Methods in Sedimentary Petrology* edited by G. Muller. Reprinted with permission of Hafner Press, a division of Macmillan Publishing Company, New York, 1967.)

nels and brass cones vertically mounted which are capable of splitting small samples (10 to 400 g) into 5-mg subsamples, and are then collected in one of four divisions in the base pan (Figure 6). These authors statistically compared the precision and accuracy of grain frequency percentages obtained using their device with a riffle-based microsplitter and the cone and quarter technique and found that their splitter was better in both categories than the riffle splitter while being better than the cone and quartering in terms of accuracy only.

Due to the overwhelming concurrence (although not unanimous) in the literature as to the value of the cone and quarter and riffle splitting techniques in sample homogenization, and the common use of random sampling after

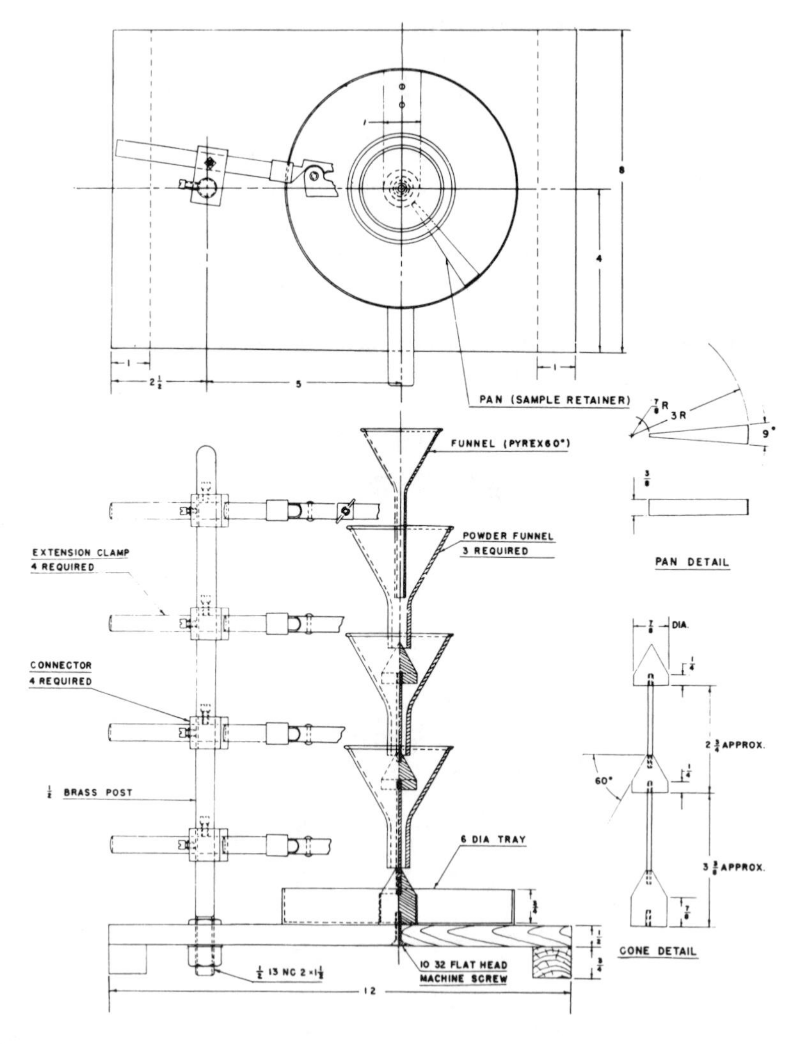

Figure 6. A multiple-cone splitter. (From Kellagher and Flanagan, 1956. Used with permission of SEPM.)

grinding and sieving of the soil, investigations were undertaken to determine the effectiveness (degree of homogenization), efficiency (time consumption), and the extent of the loss of fine particles by these methods during the homogenization of large soil samples.

MATERIALS AND METHODS

Bulk samples (approximately 15 kg) of surface horizons of the Overton clay (fine, montmorillonitic, calcareous, thermic Mollic Haplaquepts), Gila silt loam (coarse-loamy, mixed, calcareous, thermic Typic Torrifluvents), Calico loam (coarse-loamy, mixed calcareous, thermic Typic Torrifluvents), and Jean gravelly loamy fine sand (sandy-skeletal, mixed, thermic Typic Torriorthents) were collected in Clark County, Nevada.[20] Surface horizons with different textures were selected to represent soil systems with varying sensitivities to the loss of fines that may occur during sample homogenization.

Moist bulk samples were split into three subsamples (approximately 5 kg each) and air-dried. Samples were ground and sieved through a 2-mm (10 mesh) sieve with three random samples being collected from each of the subsamples prior to any homogenization procedures. The three original subsamples from each soil series were "homogenized" by either passing the soil seven times through the cone and quarter technique, an open-bin riffle splitter, or a closed-bin riffle splitter. Three samples of approximately 60 g each were collected after the first, third, and fifth passes. Seven samples were collected after the seventh pass through each of the homogenization processes. Samples were collected from the riffle splitters by placement of the receiving bottle under randomly selected chutes prior to the sample being poured through the splitter. Samples from the cone and quarter technique were collected by pouring the sample over the collection bottles during cone rebuilding and subsequently removing the bottles when they were full prior to completion of the cone using all four quarters.

A timed experiment was conducted to determine the efficiency of each splitting technique and involved the passing of the previously "homogenized" soil (from the seventh pass) through the three homogenization procedures seven times without collection of soil samples after any intermediate passes. Seven samples were collected after the timed experiment in which the soil had now been passed 14 times through the homogenization process.

Three parameters were selected to determine the effectiveness of the homogenization method and loss of fines from the system, namely particle-size analyses (for loss of inorganic fines), loss-on-ignition (LOI) organic matter (for organic fine losses), and pH (for bulk chemical changes). Particle-size distribution (<2 mm) was determined using the pipette method described by Gee and Bauder.[4] Soil samples were oven-dried at 110°C overnight and LOI organic matter content was determined gravimetrically after heating at 450°C for a minimum of 6 hr. The pH values were determined in a 1:2 soil to 0.01 M $CaCl_2$

solution ratio.[12] Levels of confidence were determined by analysis of variance (ANOVA).

RESULTS AND DISCUSSION

It should be noted that pass no. 0 represents the random samples collected prior to any sample homogenization process other than preparatory grinding and sieving.

Observed Sources of Soil Loss

Loss of fines via dusting was most apparent during the use of the open-bin riffle splitter. Dust loss occurred from the mouth of the riffle splitter due to the air-dried soil hitting the baffles and sliding down the chutes as well as in the collecting bin from the soil falling upon itself. Apparent fine particle losses from use of the closed-bin riffle splitter were less noticeable than for the open-bin riffle splitter. Dust losses occurred through cracks within the riffle splitter and through the mouth into which the soil was being poured. However, although the overall dust loss appeared to be less, additional soil loss was observed within the closed-bin riffle splitter where soil collected on internal ledges and around the outside of the collecting bins. Little loss of fines, via dusting, was observed using the cone and quarter technique. Dusting occurred only during the piling of the quarters upon each other to form a new cone. Soil losses were observed due to an inability to completely transfer all soil materials during new cone formation.

Influence of Sample Homogenization on pH

No significant differences were found in pH values among replicates regardless of soil texture, splitting method, or the number of the pass from which the subsample was obtained (Table 1). The ranges in pH for the Jean gravelly loamy fine sand and Calico sandy loam were 7.7 to 7.8. The Gila silt loam had pH values of 7.8 in all samples, while the Overton clay had pH values ranging from 7.9 to 8.0.

Influence of Sample Homogenization on Particle-Size Distribution

Standard deviations for total sand, silt, and clay contents ranged from 0.05 to 1.30, 0.05 to 2.59, and 0.00 to 2.00, respectively (Table 2). The lowest standard deviations among replicates were almost always found after the samples had been passed five times through the homogenization process, regardless of homogenization technique used or soil texture (Table 2). When exceptions did occur, the standard deviations were not significantly different from the fifth pass, and occurred in either the random sample or after the first

Table 1. Soil pH Values Determined in a 1:2 Soil to Solution of 0.01 *M* $CaCl_2$ for Homogenization Study[a]

Split No.	Open-Bin Riffle Splitter	Closed-Bin Riffle Splitter	Cone and Quarter Technique
		Jean loamy fine sand	
0	7.7	7.8	7.8
1	7.7	7.7	7.7
3	7.7	7.7	7.7
5	7.7	7.7	7.7
7	7.7	7.7	7.7
T	7.7	7.7	7.7
		Calico sandy loam	
0	7.7	7.8	7.7
1	7.7	7.7	7.7
3	7.7	7.8	7.7
5	7.8	7.7	7.7
7	7.7	7.7	7.7
T	7.7	7.8	7.7
		Gila silt loam	
0	7.8	7.8	7.8
1	7.8	7.8	7.8
3	7.8	7.8	7.8
5	7.8	7.8	7.8
7	7.8	7.8	7.8
T	7.8	7.8	7.8
		Overton clay	
0	7.9	8.0	7.9
1	7.9	8.0	7.9
3	7.9	8.0	7.9
5	7.9	8.0	7.9
7	7.9	8.0	7.9
T	7.9	8.0	7.9

[a]Data presented are the means of three replicates for splits 1 through 5 and means of seven replicates for splits 7 and T.

pass through the homogenization process in which the sample was simply halved via riffle splitting or was effectively a random sample from the first flattened quarter for the cone and quarter technique. At the 90% or greater confidence interval, the random (pass 0), first, third, seventh, and timed experiment passes had 36, 33, 53, 86, and 97% of the samples having significantly greater replicate variances than the fifth pass, respectively. If a 75% or greater confidence limit was used, passes 0, 1, 3, 7, and T had 53, 56, 78, 98, and 100% of the samples, respectively, and had significantly greater variances than the fifth pass. Earlier passes (passes 0, 1, and 3) resulted in fewer samples with significantly different variances among the replicates than in passes after the fifth, yet one-third to more than one-half of the samples had significantly greater variances than found after five homogenization passes. These data indicate that any attempts to further homogenize the soil after the fifth pass through the various homogenization methods markedly increased the variabil-

Table 2. Standard Deviations Between Replicate Analyses for Particle-Size Distribution and Loss-on-Ignition Organic Matter[a]

Split	Open-Bin Riffle Splitter				Closed-Bin Riffle Splitter				Cone and Quarter Technique			
	Sand[b]	Silt	Clay	O.M.	Sand	Silt	Clay	O.M.	Sand	Silt	Clay	O.M.
	(wt. %)											
	Jean loamy fine sand											
0	0.15***	0.08	0.15*	0.05***	0.09	0.05	0.08	0.07*	0.33***	0.38***	0.08***	0.05**
1	0.12**	0.20***	0.08	0.01	0.19*	0.16	0.11	0.03	0.10	0.06	0.04*	0.04*
3	0.22***	0.24***	0.08	0.03***	0.73***	0.63***	0.11	0.07*	0.19*	0.25**	0.07**	0.03
5	0.05	0.06	0.08	0.01	0.09	0.11	0.09	0.04	0.10	0.10	0.02	0.02
7	0.18***	0.13***	0.16***	0.02***	0.25***	0.23***	0.10*	0.14***	0.18***	0.15***	0.07***	0.03**
T	0.17***	0.21***	0.11***	0.03***	0.22***	0.24***	0.13***	0.16***	0.30***	0.35***	0.10***	0.01
	Calico sandy loam											
0	0.10	0.36**	0.41**	0.10***	0.34	0.34	0.00	0.07*	0.27*	0.48***	0.31***	0.08***
1	0.64**	0.63***	0.22	0.03	0.52*	0.43	0.10	0.07*	0.29**	0.13*	0.16*	0.05***
3	0.89***	0.79***	0.20	0.06**	0.40	0.60*	0.21*	0.06*	0.16	0.25***	0.11	0.02*
5	0.23	0.15	0.16	0.02	0.28	0.33	0.09	0.03	0.12	0.07	0.08	0.01
7	0.30**	0.32***	0.30***	0.04***	0.34**	0.36*	0.42***	0.11***	0.47***	0.34***	0.34***	0.05***
T	0.46***	0.74***	0.52***	0.10***	0.33**	0.45**	0.35***	0.06***	0.54***	0.52***	0.24***	0.07***
	Sila silt loam											
0	0.31*	0.41*	0.21	0.02*	0.37	0.30	0.26	0.09***	0.67***	0.38***	0.46***	0.03
1	0.20	0.19	0.19	0.04***	0.48*	0.17	0.32*	0.08***	0.23	0.06	0.28**	0.04
3	0.31*	0.26	0.39**	0.06***	1.30***	0.96***	0.35*	0.06**	0.50**	0.51***	0.14	0.04
5	0.18	0.19	0.17	0.01	0.29	0.25	0.21	0.02	0.19	0.11	0.12	0.03
7	0.34***	0.51***	0.28***	0.07***	0.56***	0.49***	0.22*	0.06***	0.19	0.34***	0.23***	0.07***
T	0.83***	0.88***	0.43***	0.08***	0.37**	0.42***	0.23*	0.07***	0.40***	0.28***	0.39***	0.07***
	Overton clay											
0	1.11***	0.57	0.56*	0.05	0.32	0.24*	0.30	0.24***	0.25**	0.17	0.08	0.10***
1	0.83**	2.59***	1.85***	0.07*	0.07	0.21*	0.26	0.09***	0.59***	0.66**	0.55**	0.06*
3	0.61*	2.07***	1.46***	0.09*	0.51*	0.39**	0.90***	0.12***	0.20**	0.50*	0.44*	0.08**
5	0.28	0.58	0.31	0.04	0.29	0.13	0.24	0.02	0.09	0.23	0.20	0.03
7	0.71***	1.02***	0.57***	0.07***	0.32*	0.37***	0.47***	0.06***	0.51***	0.38***	0.46***	0.07***
T	0.53***	2.33***	2.00***	0.05**	0.53***	0.65***	0.59***	0.08***	0.42***	0.85***	0.66***	0.10***

[a]Data presented are the standard deviations of three replicates for splits 1 through 5 and means of seven replicates for splits 7 and T.
[b]* = 75%; ** = 90%; *** = 95% confidence levels as determined by ANOVA. Significant differences are compared to the fifth pass.

Table 3. Efficiency of Various Homogenization Techniques[a]

Soil	Open-Bin Riffle Splitter	Closed-Bin Riffle Splitter	Cone and Quarter Technique
	(min		)
Jean	4	4	24
Calico	7*	4	44*
Gila	6	5	29
Overton	5	5*	26
Mean	5.50	4.50	30.75
Mean–1[b]	5.00	4.33	26.33

[a] = first sample homogenized by the given method.
[b] = Mean of timed experiment excluding first sample homogenized by the given method.

ity among replicates and thus the heterogeneity of the sample regardless of soil texture or homogenization technique.

Influence of Sample Homogenization on Loss-On-Ignition (LOI) Organic Matter

The lowest standard deviations among replicates for LOI organic matter content were found after the fifth pass, with only two exceptions, similar to the results for standard deviations in particle-size distribution (Table 2). The two exceptions were noted in the Jean gravelly loamy fine sand (after the first pass through the closed-bin riffle splitter and after the timed experiment homogenized by the cone and quarter technique), yet neither standard deviation was significantly less than the standard deviation of the fifth pass. When a confidence interval of 90% or greater was established, 58, 33, 50, 100, and 91% of the replicates from passes 0, 1, 3, 7, and T, respectively, had significantly greater variances than those found in the fifth pass. At the 75% or greater confidence level, two-thirds or more of the replicates (83, 66, 83, 100, and 91% for passes 0, 1, 3, 7, and T, respectively) had significantly greater variances. These results support our earlier findings that the sample appears to have achieved the greatest homogeneity after the fifth pass through the homogenization process.

Homogenization Technique Efficiency

The average time required to homogenize a sample using seven passes was 5.00 (range 4 to 5 min), 5.50 (range 4 to 7 min), and 30.75 min (range 24 to 44 min) for the closed-bin riffle splitter, open-bin riffle splitting, and cone and quartering, respectively (Table 3). The slightly longer time required to perform open-bin riffle splitting compared to closed-bin riffle splitting was attributed to an initial lack of experience in the use of a riffle splitter by the technician. Excluding the first-time use of either riffle splitter, the average time required for open-bin riffle splitting decreased to 5.00 min per sample compared to 4.3 min per sample required for the closed-bin riffle splitter. Excluding the first use of the cone and quarter technique by the technician, the average time required to homogenize the sample decreased from 30.75 to 26.33 min per

Table 4. Loss of Fines Between the Seventh Pass and Timed Experiment

Soil	Open-Bin Riffle Splitter	Closed-Bin Riffle Splitter	Cone and Quarter Technique
	(———	% clay	———)
Jean	−0.03[a]	nc	+0.06
Calico	−0.02	−0.10	−0.56
Gila	+0.09	−0.22	−0.57
Overton	−0.36	−0.19	−0.42

[a] + = clay increase; − = clay loss; nc = no change.

sample. These results indicate that the use of the riffle splitter for homogenization required markedly less time per sample than samples homogenized by the cone and quarter technique and was thus more efficient.

Loss of Fines

Loss of fines was determined by comparison of clay contents between the seventh split and timed experiment. Clay contents generally decreased after the additional seven passes through the homogenization procedures, but the overall clay content loss was very small (less than 0.6%) and could be attributed to expected variance in the homogenization and analytical methods (Table 4). It was interesting, however, to note that the greatest clay content decreases were found when the soils underwent cone and quartering as the homogenization process. This result was due to a twofold effect in which the inability to recover all the soil from the underlying paper led to a greater clay loss from gap filling between the larger sand particles and perhaps due to static charge development on the paper, leading to a retention of the charged clay particles.

CONCLUSIONS

The use of riffle splitting to homogenize a bulk soil sample was more efficient and had less loss of fines than cone and quartering and is therefore the recommended homogenization technique. The use of a closed-bin riffle splitter was preferred to an open-bin riffle splitter due to its greater apparent ability to contain and reduce the loss of fines from dusting. Only five passes, instead of seven, should be used to obtain the most homogeneous sample, in terms of particle-size distribution and LOI organic matter, due to the overwhelming evidence that the least variability among replicates occurred after the fifth pass, using all three splitting techniques and all four soil textures. Random sampling after grinding and sieving was the most efficient homogenization method since no additional sample preparation was involved, yet these samples almost consistently had greater replicate variabilities than the other homogenization techniques after the fifth pass, reducing the value of this technique for soil sample homogenization.

ACKNOWLEDGMENTS

We greatly appreciate the assistance of Gregory A. Raab, William H. Cole III, Conrad A. Kuharic, Gerald L. Byers, Rick D. Van Remortel, and Robert L. Tidwell in the technical review of this document. Our thanks are also extended to Mohammad J. Miah for his aid in producing and interpreting the statistical analyses. U.S. Environmental Protection Agency technical support and assistance was provided by Louis J. Blume.

NOTICE

Although the research described in this article has been funded wholly or in part by the United States Environmental Protection Agency through Contract Number 68-03-3249 to Lockheed Engineering and Sciences Company, it has not been subjected to Agency review and therefore does not necessarily reflect the views of the Agency and no official endorsement should be inferred. Mention of trade names or commercial products does not constitute endorsement or recommendation for use.

REFERENCES

1. Allman, M. and Lawrence, D. F. *Geological Laboratory Techniques.* ARCO Publishing Co., Inc., New York, 1972.
2. Brewer, R. and Barrow, K. J. "A Microsplitter for Subsampling Small Particulate Samples," *J. Sed. Petrol.* 42:485–487 (1972).
3. Carver, R. E. "Reducing Sand Sample Volumes by Spooning," *J. Sed. Petrol.* 51:658 (1981).
4. Gee, G. W. and Bauder, J. W. "Particle Size Analysis," *Methods Soil Anal. Part 1. Agronomy* 9:383–411 (1986).
5. Gilliam, F. S. and Richter, D. D. "Increases in Extractable Ions in Infertile Aquults Caused by Sample Preparation," *Soil Sci. Soc. Am. J.* 49:1576–1578 (1985).
6. Griffiths, J. C. "Estimation of Error in Grain Size Analysis," *J. Sed. Petrol.* 23:75–84 (1953).
7. Humphries, D. W. "A Non-Laminated Miniature Sample Splitter," *J. Sed. Petrol.* 31:471–473 (1961).
8. Ibbeken, H. "A Simple Sieving and Splitting Device for Field Analysis of Coarse Grained Sediments," *J. Sed. Petrol.* 44:939–946 (1974).
9. Ingram, R. L. "Sieve Analyses," in *Procedures in Sedimentary Petrology,* Carver, R. E., Ed., J. Wiley & Sons, Inc., New York, 1971, pp.49–67.
10. Ingamells, C. O. and Pitard, F. F. *Applied Geochemical Analyses,* J. Wiley & Sons, Inc., New York, 1986.
11. Kellagher, R. C. and Flanagan, F. J. "The Multiple-Cone Splitter," *J. Sed. Petrol.* 26:213–221 (1956).
12. McClean, E. O. "Soil pH and Lime Requirement," in *Methods Soil Anal. Part 2. Agronomy* 9:199:223 (1982).

13. McKinney, C. R. and Silver, L. T. "A Joint-Free Sample Splitter," *Am. Mineral.* 41:521–523 (1956).
14. Muller, G. "Methods in Sedimentary Petrology," in *Sedimentary Petrology, Part 1,* Van Engelhart, W., Ed., Hafner Publishing Co., New York, 1967.
15. Mullins, C. E. and Hutchison, B. T. "The Variability Introduced by Various Subsampling Techniques," *J. Soil Sci.* 33:547–561 (1982).
16. Newton, G. B. and Dutcher, R. R. "An Inexpensive Student Sample Splitter," *J. Sed. Petrol.* 40:1051–1052 (1970).
17. Raab, G. A., Bartling, M. H, Stapanian, M. A., Cole, W. H., Tidwell, R. L., and Cappo, K. A. "The Homogenization of Environmental Soil Samples in Bulk," in *Hazardous Waste Measurements,* Simmons, M., Ed., Lewis Publishers, Chelsea, MI, in press.
18. Reeves, R. D. and Brooks, R. R. *Trace Element Analysis of Geological Materials,* J. Wiley & Sons, Inc., New York, 1978.
19. Schuler, V. C. O. "Chemical Analysis and Sample Preparation," in *Modern Methods of Geochemical Analysis,* Wainerdi, R. E. and Uken, E.A., Eds., Plenum Press, New York, 1971.
20. Speck, R. L. "Soil Survey of Las Vegas Valley Area, Nevada," USDA-Soil Conservation Service. U.S. Government Printing Office, Washington, D.C., 1985.
21. Van Johnson, W. M. and Maxwell, J. A. *Rock and Mineral Analysis,* 2nd. ed., J. Wiley & Sons, Inc., New York, 1981.
22. Wentworth, C. K., Wilgus, W. L., and Koch, H. L. "A Rotary Type of Sample Splitter," *J. Sed. Petrol.* 4:127–138 (1934).

SECTION III

FIELD TECHNIQUES AND INSTRUMENTATION

CHAPTER 5

Air Monitoring at Remedial Waste Sites*

Steven P. Levine

The single, most basic problem in industrial hygiene at hazardous waste sites is that the materials being handled are of unknown composition.[1] This is almost always true for drum and tank materials at Superfund sites and frequently true at licensed disposal sites. The consequences of this lack of knowledge, for both the onsite worker and the surrounding community, are that the following cannot be readily chosen:

1. Personal protective equipment (PPE)** for worker protection.
2. Air monitoring methods for worker breathing zones, work areas/unit processes, and the site fenceline and the downwind community.
3. Medical surveillance strategies for site workers and the [potentially] exposed community.

These questions, and related, equally important questions, are dealt with in a number of source materials, which will be extensively quoted in this chapter. These source materials are

1. *Protecting Personnel at Hazardous Waste Sites*, S. P. Levine and W. F. Martin, Eds., Butterworth-Ann Arbor Science, Stoneham, MA, 1985.[1]
2. "Occupational Safety and Health Guidance Manual for Hazardous Waste Activities," NIOSH/OSHA/USCG/EPA Interagency Task Group, U.S. DHHS (PHS/CDC/NIOSH) 85-115, Cincinnati, OH, 1985.[2]
3. "Personal Protective Equipment for Hazardous Materials Incidents: A Selection Guide," U.S. DHHS (PHS/CDC/NIOSH) 84-114, Cincinnati, OH, 1984.[3]
4. A. D. Schwope et al. "Guidelines for the Selection of Chemical Protective Clothing," ACGIH, Cincinnati, OH, 1981 (portions also published by DuPont as individual data sheets on the permeability of Tyvek-based materials).[4]

---

*This chapter is reprinted in part from similar texts written by this author (to be found in References 1, 31, and 32).

**List of abbreviations at end of chapter.

At this point it is important to note that Reference 2, the Interagency Guidance Manual, provides an overview for each of the subject areas discussed in the other reference materials. Equally important is the fact that all of the information contained in this document represents the most recent consensus of scientific opinions.

The structure of this section will be to first discuss the question of the perception of hazard from hazardous waste sites when compared to the "traditional" workplace. Secondly, reasons will be given as to why there is a lack of knowledge of drum and tank material composition. Then the questions of PPE and air monitoring will be discussed and related to information given in the Guidance Document and other reference documents. The question of air monitoring will be given special attention.

PERCEIVED HAZARD

Equally important to the technical questions of personnel and community protection at hazardous waste sites is the question of perspective and perceived danger. There is no fundamental reason why a hazardous waste site cannot be viewed as a collection of unit processes.[5,6] Certainly, if the traditional industrial hygiene triad of recognition, evaluation, and control[7,8] can be applied to such diverse work places as metal foundries, phosphorus furnaces, and pesticide intermediate plants, those principles can also be applied to hazardous waste sites. In my experience as an industrial hygienist, those traditional workplaces have been more hazardous than the hazardous waste sites at which I have worked. Indeed, the recent Bhopal incident resulted in far more serious consequences than has any hazardous waste site incident.[9-11]

However, "several factors distinguish the hazardous waste site environment from other occupational situations involving hazardous substances. One important factor is the uncontrolled condition of the site . . . Another factor is the large variety and number of substances that may be present at the site . . . The combination of . . . these conditions results in a working environment that is characterized by numerous and varied hazards which:

- May pose an immediate danger to life and health
- May not be immediately obvious or identifiable
- May vary according to the location on site and the task being performed
- May change as site activities progress"[2]*

MATERIALS ANALYSIS

One of the basic problems in industrial hygiene at hazardous waste sites is that the materials being handled are of unknown composition[1]. The question is why is this the case when analysis methods for environmental samples are well-

*Quoted by permission of the U.S. EPA, Industrial Hygiene Office, Washington, D.C.

established? In order to answer this question, an understanding of the processes involved in a hazardous waste cleanup should be reviewed. A complete review is outside the scope of this section, but can be found elsewhere (Reference 1, Chapter 6). Briefly, one of the principal unit processes at a hazardous waste site is that of drum bulking (Reference 2, Chapter 11). In that process, drums and tanks that are excavated or otherwise found on a site are brought to a central area called a staging area. The drums are then opened and sampled, and moved to a second staging area until the results of the laboratory analysis are received. At that point, all wastes are bulk recontainerized in compatible chemical groups. Disposal of bulked wastes is far more cost-effective than disposal of individual drums.

When mixed together, hazardous wastes have the potential for producing effects such as fires, explosions, violent reactions, and the release of toxic dusts, mists, fumes, and gases. Therefore, chemical wastes must first be tested for compatibility, and incompatible wastes must be segregated rather than bulked. Chemical compatibility testing involves a group of chemical tests following a flowchart scheme which ultimately classifies the material into general categories such as acid, base, organic, chlorinated organic, oxidizer, sulfide/cyanide, etc. As noted in References 1 and 2, compatibility testing procedures should be carefully chosen and followed. Other reference materials and publications also deal with this question.[13-15]

PERSONAL PROTECTIVE EQUIPMENT (PPE), AIR MONITORING, AND MEDICAL SURVEILLANCE

Personal Protective Equipment

This subject of PPE is discussed in detail in all of the previously cited reference materials.[1-4,15]

The Guidance Document[2] summarizes the subject of PPE:

> "Use of PPE is required by OSHA regulations in 29 CFR 1910 and reinforced by U.S. EPA regulations in 40 CFR 300 which include requirements for all private contractors working on Superfund sites to conform to applicable OSHA provisions and any other . . . requirements deemed necessary by the . . . agency overseeing the activities.
>
> No single combination of protective equipment . . . is capable of protecting against all hazards. Thus, PPE should be used in conjunction with other protective methods. The use of PPE can itself create significant worker hazards, such as heat stress, physical and psychological stress, and impaired vision, mobility, and communication. In general, the greater the level of PPE, the greater are the associated risks . . . Over-protection as well as under-protection can be hazardous and should be avoided . . .
>
> A written PPE program should be established for work at hazardous waste sites . . . a comprehensive PPE program should include hazard identification;

medical monitoring; environmental surveillance; selection, use, maintenance, and decontamination of PPE; and training."[2]

Selection of PPE can best be accomplished by reference to the selection guides from NIOSH[3] and ACGIH[4], which are summarized and extended in Reference 1, Chapter 9, and Reference 2, Chapter 8. The most complete treatments of the subject of heat stress is given in Reference 1, Chapter 10, and Reference 3, Part III.

Schwope[4] developed the permeation testing protocol used by ASTM and NIOSH for PPE garments and gloves. All PPE materials must undergo appropriate testing to the expected challenge chemicals.

Turpin developed the U.S. EPA SOPs[15] for the designation of safe work practices, air monitoring, and PPE at hazardous waste sites; all based on Levels of Protection, with Level A being the highest level and Level D being used when minimal protection is required. These protocols were refined and extended in the Guidance Document.[2] A significant change was the modification of the requirement that airborne THC values be used as criteria for choice of PPE. A complete description of these requirements is given in Reference 2.

This SOP, which was issued in draft form in 1982, became the standard used at all hazardous waste sites simply by default. At the outset, it should be noted that Turpin developed the SOP in "real-time" when it was sorely needed. Had he waited for the normal rule-making process, we might still have no SOP.

In the SOP, the assumption has been made that the use of a full-face respirator (acid gas/organic vapor/HEPA) and a 5-min self-contained air escape pack is inadequate for use by workers at a hazardous waste site, since there is almost never a guarantee that conditions IDLH will not occur.

Air Monitoring

Building on Turpin's SOP[15] and on the NIOSH Guidance Document,[2] OSHA issued a notice of proposed rule making for hazardous waste operations and emergency response.[16] This document essentially codifies into the regulations, the science in the above-cited documents.

The role of air monitoring was addressed in this regulation in Sections "c" (site characterization and analysis), "h" (monitoring), and Appendix B (levels of protection). In Section (c)(1), air monitoring is required to perform preliminary evaluation of a site's characteristics in order to aid in the selection of PPE. Air monitoring is then required to be repeated in a more detailed fashion to help select administrative and engineering controls, and PPE. In Section (c)(2), air monitoring is required to identify all IDLH hazards.

In Section (c)(4), air monitoring is required to ensure that the PPE chosen will protect workers to below the PEL for known or suspected substances. If there is insufficient air monitoring information, Level B protection must be used.

In Sections (h)(1) and (2), air monitoring is called for to specifically identify and quantify airborne substances, especially relating to IDLH conditions. In

Section (h)(3), the requirement for air monitoring is made periodic. That is to say, air monitoring must be repeated since the emissions from many hazardous waste sites are episodic and highly variable in type and quantity, so a single series of measurements is inadequate. In Section (h)(4), personal air sampling is required for highest exposure employees. This sampling must be performed with a frequency sufficient to characterize potential exposures. This requirement is stated, but no guidance is given for the difficult task of deciding what constitutes "sufficient to characterize."

In Appendix B, the requirements for personal air sampling in the setting of PPE is repeated. Essentially, Level C protection (full-face air purifying respirator) cannot be worn unless the types and concentrations of all air contaminants are known. Since this is a rare occurrence on hazardous waste sites, most workers will wear supplied air respirators.

Since real-time air monitoring methods are, at present inadequate for speciation and quantitation, air monitoring data must be developed by nonreal-time methods[1,17,18] which do not provide the information necessary to protect the workers on a daily basis. This forces the use of a self-contained breathing apparatus (SCBA), which cuts down on productivity, and in combination with the protective garment, results in a high probability of the workers suffering from heat stress.[1-3]

Furthermore, the lack of real-time air monitoring data forces the use of very conservative guidelines for community evacuation in case of an airborne release. Unfortunately, the decision to evacuate a community from the downwind area near a hazardous waste site may be made on the basis of the volume and effectiveness of community activists rather than on any sound air monitoring data.

Since, as noted before, one of the main stumbling blocks to any alternative other than the present SOP is the total lack of knowledge of the identity of the contents of the drums and tanks, this might possibly be remedied by fielding a very significantly increased degree of analytical capability at a site. At present, the only significant capability on site is the mobile lab for compatibility testing. It may well be that it is cost effective to field a second mobile lab equipped with GCs, an FTIR, and a GC/MS. The presence of two mobile labs, one for compatibility testing and one for more extensive analysis needs, might yield the critical mass of skilled analytical manpower on a site that should allow, for the first time, some sort of true near real-time air monitoring.

For example, GCs equipped with gas sampling loops and calibrated for a variety of compounds and compound classes, and equipped with FID, ECD, NPD, or FPD could be placed at appropriate unit processes within a site. RS232 lines could be run back to the mobile lab for monitoring, data collection, and/or alarm. This approach will only work when the target analytes are known. GC alone cannot be used in an environment with unknown vapors and gases which may appear as unidentified peaks in the chromatogram. The use of portable GCs equipped with capillary columns and more than one detector, one of which may be semiselective, may help to alleviate this problem.

Alternatively, a high resolution FTIR[19-21] or a MS[22-26] could be used for this purpose. At present, both the FTIR and the MS/MS will probably satisfy most of the air-monitoring requirements at hazardous waste sites. The FTIR may provide the resolution and sensitivity not available to those who presently use portable dispersive IR systems, such as the MIRAN.

The design of portable FTIR systems and the elucidation of the science underlying the use of these systems is underway in our laboratory. It appears, from preliminary results, that the FTIR will provide limits of detection for components of mixtures of organic compound vapors on the order of 0.1 to 1.0 ppm. This will be sufficient to provide adequate monitoring on hazardous waste sites, but it may not meet the requirements of those projects in which downwind ambient air plume monitoring is requested.

The FTIR, equipped with a remote sensing (ROSE) beam attachment,[27,28] has been tested for air monitoring at the Gems Landfill in New Jersey by our group in conjunction with the U.S. EPA Environmental Response Team. The test demonstrated that the FTIR must be close to the hazardous waste site or a significant fraction of the available beam path will be taken just getting to the site. The logistics of aiming the ROSE optics over complex terrain also must be addressed in the planning stages of a hazardous waste site monitoring project.

The use of MS systems for such applications is gaining popularity. A single-stage MS has been designed for use in battle tanks of the NATO forces in Europe. It is sensitive and rugged, and operates on battery power. However, because it is a single-stage instrument, it is prone to interferences, especially by solvent mixtures. It is especially useful in the situation in which a few unique compounds are present, such as in a battlefield where the personnel are under attack by chemical warfare agents. This is not usually encountered on a hazardous waste site, except when that site is an abandoned chemical agent facility.

The problem of interference is obviated by the use of a tandem MS/MS system. This system uses the first MS as a mass filter for the second MS. However, while the instrument is sensitive and specific, it is complex, encompassing the entropy of an entire lab in one system. Stated more simply, it is prone to failure due to mechanical or electrical problems.

The most successful application of MS systems for air monitoring is the ICAMS, which is a double-focusing direct inlet MS that has been used in the space shuttle, nuclear submarines, and in monitoring air in the semiconductor device manufacturing workplace. Again, this system is not ideally suited for situations in which large numbers of known and unknown low-MW gases and vapors may be present simultaneously.

It must be remembered that these instruments are expensive and complex, and the data they produce requires skilled spectroscopists for interpretation. While it is highly likely that this situation will improve in the near future, the use of these systems for real-time hazardous waste site air monitoring applications has not yet been proven in a field trial that could truly be said to represent a validation of these methods.

Certainly, small alarm devices such as the Drager/Ecolyzer Company series 100 compound-specific monitor, or the GCA Corp. Mini-RAM dust monitor could be properly calibrated and placed on workers for continuous use. Similarly, colorimetric indicator tubes may be usable to good advantage under certain circumstances.[30] In our quest for improved methods, these basic devices must not be forgotten.

The question of calibration is central to the question of the use of any air monitoring tool, at hazardous waste sites or elsewhere. Yet this is a subject that is poorly addressed in practice at most hazardous waste sites. An International Guidance Document has been generated that specifically addresses the question of calibration of air monitoring instrumentation for use at hazardous waste sites. Although this document does not have the force of law, since it is meant specifically for guidance and not for regulation, it is anticipated that its guidelines will be adopted by consensus.[29]

ABBREVIATIONS

PPE	=	personal protective equipment
A,B,C,D	=	levels of PPE at hazardous waste sites
IDLH	=	immediately dangerous to life and health
SOP	=	standard operating procedure
SCBA	=	self-contained breathing apparatus
HEPA	=	high-efficiency particulate
FTIR	=	Fourier transform infrared spectroscopy
FID	=	flame ionization detector
ECD	=	electron capture detector
NPD	=	nitrogen-phosphorus detector
FPD	=	flame photometric detector
RS232	=	protocol for transmission of information
MS/MS	=	tandem mass spectrometer/mass spectrometer system
GC/MS	=	gas chromatograph/mass spectrometer
ICP	=	inductively coupled plasma emission spectrometer
THC	=	total hydrocarbon [analyzer]
EPA	=	Environmental Protection Agency
ASTM	=	American Society for Testing and Materials
USCG	=	U.S. Coast Guard
NIOSH	=	National Institute for Occupational Safety and Health

REFERENCES

1. Levine, S. P. and Martin, W. F., Eds. *Protecting Personnel at Hazardous Waste Sites*, Butterworth-Ann Arbor Science, Stoneham, MA, 1985.
2. "Occupational Safety and Health Guidance Manual for Hazardous Waste Activi-

ties," NIOSH/OSHA/USCG/EPA Interagency Task Group, U.S. DHHS (PHS/CDC/NIOSH) 85-115 Cincinnati, OH (1985).

3. "Personal Protective Equipment for Hazardous Materials Incidents: A Selection Guide," U.S. DHHS (PHS/CDC/NIOSH) 84-114 Cincinnati, OH (1984).
4. Schwope, A. D. et al. "Guidelines for the Selection of Chemical Protective Clothing," ACGIH, Cincinnati, OH 1981 (portions also published by DuPont as individual data sheets on the permeability of Tyvek-based materials).
5. Cralley, L. V. and Cralley, L. F., Eds. *Industrial Hygiene Aspects of Plant Operations*, Vol. 1, Macmillan Publishing Co, Inc., New York, 1983.
6. Shreve, R. N. and Brink, J. A. *Chemical Process Industries*, McGraw-Hill Book Company, New York, 1977.
7. "The Industrial Environment—Its Evaluation and Control," U.S. DHEW (NIOSH) (1976), pp. 3-4.
8. Olishifski, P. E. and McElroy, F. E., Eds. "Fundamentals of Industrial Hygiene," National Safety Council, Chicago, IL, (1977), Chap. 1.
9. "Bhopal Report," *Chem. Eng. News*, (Feb. 11); "Bhopal—Special Report," *Chem. Eng. News*, (Dec. 2, 1985); American Chemical Society, Washington, D.C.
10. "Bhopal MIC Incident Investigation Team Report," Union Carbide, Danbury, CT (March 1985).
11. "The Bhopal Disaster: How it Happened," N.Y. Times Series (January 1985).
12. Levine, S. P. et al. "FTIR Applied to Hazardous Waste: I—Preliminary Test of Material Analysis for Improvement of Personal Protection Strategies," *Am. Ind. Hyg. Assoc. J.* 46:181-196 (1985).
13. Puskar, M. A. et al. "Computerized Infrared Spectral Identification of Compounds Frequently Found at Hazardous Waste Sites," *Anal. Chem.* 58:1156-1162 (1986).
14. Puskar, M. A. et al. "Infrared Screening Technique for Automated Identification of Bulk Organic Mixtures," *Anal. Chem.*
15. "Interim Standard Operating Safety Procedures," U.S. EPA, Office of Emergency and Remedial Response, revised (November 1984).
16. "Hazardous Waste Operations and Emergency Response," 29 CFR 1910, Vol. 54, No. 42, 9294-9336 (March 6, 1989).
17. Levine, S. P. et al. "Air Monitoring at the Drum Bulking Process of a Hazardous Waste Remedial Action Site," *Am. Ind. Hyg. Assoc. J.* 46:192-196 (1985).
18. Levine, S. P. et al. "Air Monitoring During a Hazardous Waste Remedial Action at a Drum Bulking Unit Process," *Haz. Waste* 1:573-580 (1985).
19. Levine, S. P. et al. "Advantages and Disadvantages in the Use of FTIR and Filter IR Spectrometers for Monitoring Airborne Gases and Vapors of Industrial Hygiene Concern," *J. Appl. Ind. Hyd.* 4:180-187 (1989).
20. Strang, C. R., Levine, S. P. and Herget, W. F. "Evaluation of the Fourier Transform Infrared (FTIR) Spectrometer as a Quantitative Air Monitor for Semiconductor Manufacturing Process Emissions," *Am. Ind. Hyg. Assoc. J.* 50:70-77 (1989).
21. Strang, C. R. and Levine, S. P. "The Limits of Detection for the Monitoring of Semiconductor Manufacturing Gas and Vapor Emissions by Fourier Transform Infrared (FTIR) Spectroscopy," *Am. Ind. Hyg. Assoc. J.* 50:78-83 (1989).
22. Kieth, C. H., Ed., "Identification and Analysis of Organic Pollutants in Air," Butterworth, Boston, MA, 1984.
23. Burggraaf, P. "Hazardous Gas Safety and the Role of Monitoring," *Semi. Intl.* (November 1987), pp. 56-62.

24. DeCorpo, J. J., Wyatt, J. R. and Saalfeld, F. E. "Central Atmospheric Monitor," *Naval Eng. J.* 42:231 (1980).
25. Helmers, C. T., Jr. "Industrial Central Atmospheric Monitor," *Sensors* (September 1984), pp. 20–25.
26. Krieger, J. H. "System Continuously Monitors Air Quality in Industrial Process Sites," *Chem. Eng. News* (April 12, 1985) pp. 30–32.
27. Herget, W. F. "Analysis of Gaseous Air Pollutants Using a Mobile FTIR System," *Am. Lab.* 72 (1982).
28. Herget, W. F. and Brasher, J. D. "Remote Optical Sensing of Emission," *Appl. Opt.* 18:3404 (1979).
29. Levine, S. D., Chairman, "Portable Instruments for Monitoring Airborne Emissions from Hazardous Waste Sites. Organ. Int. Metrol. Legale., Pilot Secretariat 17, Reporting Secretariat 5 "Measurement of Pollution from Hazardous Waste Sites," published as a draft document by ACGIH, Cincinnati, OH (1988).
30. King, M. V., Eller, P. M., and Costello, R. J. "A Qualitative Sampling Device for Use at Hazardous Waste Sites," *Am. Ind. Hyg. Assoc. J.* 44:615–618 (1983).
31. Gotchfeld, M., Ed. "The Role of Air Monitoring Techniques in Hazardous Waste Site Personnel Protection and Surveillance Strategies," in *Hazardous Waste Site Personnel Medical Surveillance,* Rutgers University Press, New Brunswick, NJ, 1988.
32. Lave, L. B. and Upton, A. C., Eds. "Cleanup of Contaminated Sites," in *Toxic Chemicals, Health and the Environment*, Johns Hopkins University Press, Baltimore, 1987.

CHAPTER 6

Field Measurement of Organic Compounds by Gas Chromatography

Dennis Wesolowski and Al Alwan

Go, measure earth, weigh air, and state the tides, . . .
The starving chemist in his golden view supremely blest, . . .

A. Pope, 1734

Although gathering chemical data outside of the laboratory is part of the history of chemistry, it is only recently that trace analysis of organic compounds under field conditions has become possible.

One of the prime movers in encouraging development of this type of technology has been the U.S. Environmental Protection Agency (EPA). It has been necessary for the EPA to monitor environmental parameters in support of legislation aimed at protecting the environment. Frequently, this involves the need to generate and evaluate analytical data regarding organic compounds in a very short timeframe. Onsite analyses using gas chromatography is one solution to a multidimensional task.

In general, field measurement of analytes implies performing analyses removed from the traditional fixed laboratory setting. Based on an EPA draft document of field monitoring by Fisk et al., the analytical instrument technology associated with this concept has evolved into three categories. They can be descriptively identified as portable, fieldable, and mobile instrumentation. The technological evolution was driven by operational needs in environmental and hazardous waste measurement.

Portable instruments are limited to devices that can be carried and used with little or no ancillary equipment. The range of these instruments spans the need for undifferentiated organic compound detection to fairly sophisticated gas chromatographs (GC).

Fieldable instruments include devices that need to be transported by vehicle and either operated in that vehicle or a suitable building. They contain rechargeable batteries or can be operated from a vehicle's DC power source. They have the capability to operate using AC power from a line source or a portable generator. Also, they may have self-contained gas reservoirs or can be connected to larger gas cylinders for extended use. They need external recorders or computers for data gathering.

The level of versatility of these instruments is greater than the portable instruments in that they can be operated at various temperatures above ambient and, in some cases, temperature programability is possible.

Mobile instruments are those GCs that are sized to readily fit into either a customized recreational vehicle or semitrailer. They have the greatest versatility in that they are fully temperature and method programable. They can be fitted with multiple detectors or be interfaced to a small mass spectrometer (MS) such as an ion-trap or a mass selective detector (MSD). They need external sources of gases and AC power which can be supplied by a generator.

Mobile instruments can be enlarged to include the vehicle itself as in a mobile MS/MS system capable of real-time analyses of selected compounds.

PROGRAM NEEDS

The EPA uses field analyses by gas chromatography to satisfy various program needs. Some early demands for this service were in response to chemical spill episodes. Rapid identification of the extent of contamination as well as determining the progress of amelioration efforts were needed.

An example of this is a situation involving a waste lagoon which was near the point of overflowing into a river upstream of a municipal water supply intake.[1] The upper layer of the liquid waste contained a variety of volatile organic compounds (VOCs) as determined by laboratory analyses. An appropriate action was to remove this top layer which was several feet in depth. It was pumped through a carbon filtration unit and discharged into the nearby river.

A portable GC was used to quickly analyze the effluent from the carbon unit to determine if the carbon was spent. This real–time analysis allowed both qualitative and quantitative data to be obtained. It ensured that new carbon filters were installed when necessary and demonstrated that no significant concentration of VOCs was entering the river.

Other examples include two incidents involving a portable Photovac 10S10 GC system. Both required analysis of soil gas and ground water. Soil gas is a term commonly used to denote the vapor present in the interstitial spaces between soil particles.

The soil gas samples were analyzed by direct injection into the GC column for trichloroethylene and tetrachloroethylene. The great sensitivity of these compounds to the photoionization detector (PID) in this instrument is due to the double-bonded carbons. The headspace above a closed partially filled vial of well water was similarly analyzed.

In one case, the purpose of the analysis was an extended contamination study to define the plume caused by repeated spills. This was needed to determine the possible risk to adjacent water supplies and to trace the source of contamination for possible enforcement action.

In the other case, the site was found to have leaking underground storage tanks. The purpose of the field analyses was to define the area of contamination in relation to the ground water and to pinpoint the source of the leak.

The results for the analytes ranged from less than 100 to 7000 parts per billion (ppb).

Mobile GC equipment was used in a unique international inspection and enforcement event involving the transportation of known and suspected hazardous waste between Canada and the U.S. A Hewlett Packard (HP) 5890 GC with a Hall electrolytic conductivity detector (HECD) was used by the U.S. team. It was temperature programmed from 40 to 290°C. The analytical column was a DB-5 capillary column.

This was a cooperative effort with EPA Region 5 personnel and the National Enforcement and Investigation Center (NEIC) in coordination with the Michigan Department of Natural Resources and the Ontario Ministry of the Environment. The goal of the U.S. team was to analyze waste samples for compliance with the manifested transported waste in 30 min. The target compounds were volatile chlorinated solvents from 1,1,1 trichloroethane to the semi-volatile polychlorinated biphenyls (PCBs). This was achieved with resulting detection limits based on a 0.2-g sample weight of 5 parts per million (ppm) of the solvents to 50 ppm for PCBs. The NEIC performed analyses on 25 samples during the three-day exercise.

The goal of the Canadian team was different in that some inorganic analyses were done, such as a quick scan for metals and cyanide. Also PCBs were not specifically determined, but nonhalogenated solvents such as ketones and acetates were determined. PCBs were indicated, however, by screening with a Fourier Transform Infra-Red Spectrometer (FT-IR), which detected functional groups, such as that signified by aromatic chlorine stretches. This screen was done initially on all tested samples in 5 min.

Additional analysis consisted of using an automated heated headspace sampler with an HP 5890 GC. The analytical column was a DB-5 megabore 30-m capillary column with a splitter coupled to a flame ionization detector (FID) and an electron capture detector (ECD). Thus both halogenated and nonhalogenated target compounds could be detected in a 30-min analysis.

An automated purge and trap system was attached to an HP MSD system to identify nontarget compounds. The column was an HP ultra 2 with a 0.25-mm diameter and a 50-m column length. The system was temperature programmed from 50 to 180°C with a 30-min run time.

The statistically determined detection limit was near 2 ppb for most analytes in each system. However, this was far less than the need to determine the presence or absence of analytes associated with the wastes examined. Large dilutions were made in order to keep the detectors from saturation.

Another demand for field analysis is in the support of the Superfund program. The purpose of this program is to identify, characterize, and remediate abandoned waste sites throughout the country. A site must first be studied to determine if it qualifies as a National Priority List (NPL) site under Superfund. After qualification, remedial investigation (RI) and feasibility studies (FS) are done to characterize the type and extent of environmental or potential environmental contamination and suggest alternate feasible remedies for site cleanup and monitoring.

One of the first things to be done in the initial site investigation stage is to address the health and safety concerns of onsite personnel. Frequently, little may be known about potential health hazards at a site so the level of protection needed by field investigators must be determined.

Ecology and Environment, Inc. (E&E), operating the Field Investigation Team (FIT) contract in Region 5 of the EPA used an HNU 101 total vapor analyzer along with an HNU 301 fieldable GC to determine an area free of contamination on which to place its operations trailer. This provided the safest environment for the crew.

The HNU 101 was used to detect the total amounts of organic vapors. The area with the highest readings were sampled by drawing air into Tedlar bags. These were subsequently analyzed using the HNU 301 fieldable instrument. The results showed the presence of benzene, toluene, xylenes, and several chlorinated hydrocarbons.

Following these initial analyses, a site is sampled to provide information about possible water and soil contamination. The analytes requested are those specified in the Comprehensive Environmental Response, Compensation, and Liability Act of 1980 (CERCLA) as amended by the Superfund Amendments and Reauthorization Act of 1986 (SARA).[2] These include 126 organic compounds and isomeric mixtures and 24 inorganic substances.

Since the results of these analyses are used to "score" a site for inclusion in the NPL, information about all of the analytes is critical.[3] Field analysis of samples is better suited to instances in which specific compounds or a class of compounds are known to be present. Abbreviated extraction and preparation procedures may not be able to provide data with sufficiently low detection limits. Furthermore, information about nontarget analytes which may be of importance in an overall assessment of a site may be limited or nonexistent depending on the field methods used.

Field methods have been used to provide an indication of appropriate sampling stations at a site. For example, the presence of volatile aromatic or halogenated solvents in ground water or soil gas samples would prompt further analyses by more rigorous methods for the broad spectrum of analytes mentioned.

An appropriate example comes from the field work of Cisneros and Clyne et al. of E&E at a site in Indiana which was contaminated with petroleum products. A stainless steel probe was driven 4 to 6 ft into the ground. Soil gas was drawn through an OVA-128A in the survey mode to record single point levels of contaminants.

When a high level was recorded, a 2-mL aliquot was withdrawn from the probe and injected into the 4 ft 1/8 in. SE-30 column of the OVA-128A. The FID detected benzene and xylene, whose retention times (RT) were previously determined in the field.

At this point it was clear that the original readings were due to petroleum products and not to nontarget organic compounds. A selection of soil samples which had the highest concentration of the analytes in the soil gas was sent to the Contract Laboratory Program (CLP) of the EPA for confirmation and further analysis for both volatile and semivolatile components.

In a second study at this site the same system was used to determine which water samples from monitoring wells should be sent to the CLP. In order to do this, headspace near the water surface in the well was sampled by the instrument in the survey mode with subsequent chromatographic injections.

Finally, a different method was used to determine contamination of ambient air at the same site. Calibrated personal air pumps were used to draw air through stainless steel tubes packed with activated charcoal at a rate of 138 to 194 mL/min. The final volume varied from 17 to 34 L of air sampled.

The tubes were desorbed by a Foxboro Programmable Thermal Desorber at 250°C. Although this device allows a portion of the sample to be automatically injected into a GC, a volume was manually removed through a septum port and injected. In this case a Photovac 10S50 with a PID was used. The analytical column was a 30-m, 0.53-mm capillary column coated with an equivalent of SE-30.

The samples were compared with a 1-ppm standard prepared from gas standards diluted with air in a 1-L sampling bulb. Ten microliters were injected to give the chromatogram in Figure 1. Peaks numbered 3, 9, and 23 are benzene (RT 110), toluene (RT 272), and meta-xylene (RT 691).

Figures 2 and 3 are the chromatograms of samples 1 and 2 that had 34 and 32 L of well headspace air drawn through them, respectively. Peaks 4, 9, and 15 in Figure 2 correspond to benzene, toluene, and meta xylene, respectively.

The conditions of operation were the same for the standard as well as each sample except that the injection volume for each sample was 1 mL instead of 10 μL as with the standard.

The first observation regarding these data is that tube 2 has larger peaks with the xylene peak well off scale. Since the sample tube volumes are about the same and the injection volumes are equal, there are higher concentrations of analytes in the sample relative to sample tube 1. There probably is an unresolved peak or peaks associated with the xylene peak.

Another obvious feature is that both samples have more components than the standard. However, these components are not being specifically monitored and are probably hydrocarbons since the site was known to be contaminated with petroleum products.

Since the purpose of these field analyses was to find areas of maximum contamination and remit selected samples for further analytical work, the actual amounts in any particular sample is not important. The standard was used as a retention time and sensitivity marker for subsequent analyses.

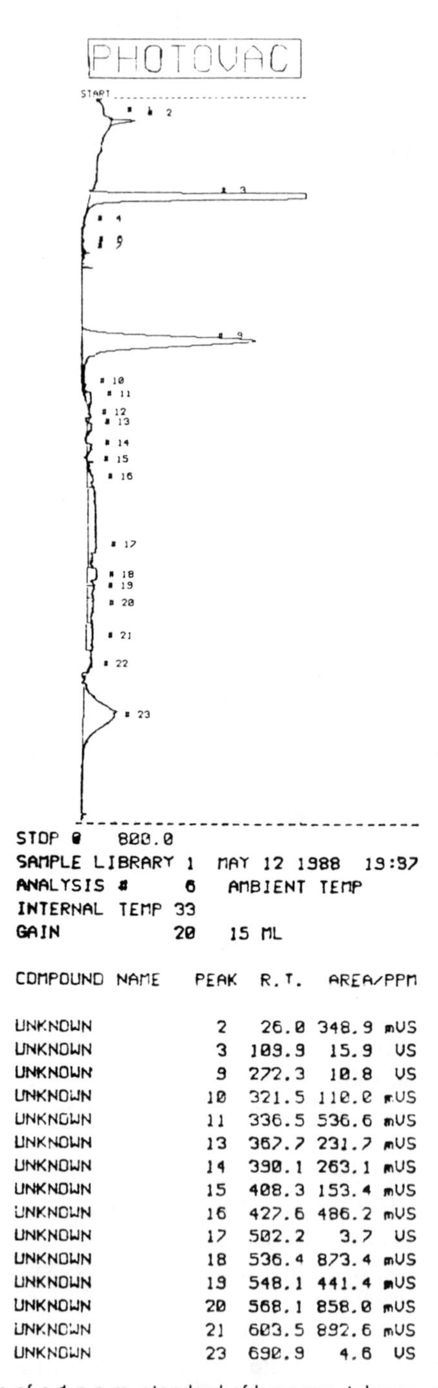

STOP @ 800.0
SAMPLE LIBRARY 1 MAY 12 1988 19:37
ANALYSIS # 6 AMBIENT TEMP
INTERNAL TEMP 33
GAIN 20 15 ML

COMPOUND NAME	PEAK	R.T.	AREA/PPM
UNKNOWN	2	26.0	348.9 mVS
UNKNOWN	3	109.9	15.9 VS
UNKNOWN	9	272.3	10.8 VS
UNKNOWN	10	321.5	110.0 mVS
UNKNOWN	11	336.5	536.6 mVS
UNKNOWN	13	367.7	231.7 mVS
UNKNOWN	14	390.1	263.1 mVS
UNKNOWN	15	408.3	153.4 mVS
UNKNOWN	16	427.6	486.2 mVS
UNKNOWN	17	502.2	3.7 VS
UNKNOWN	18	536.4	873.4 mVS
UNKNOWN	19	548.1	441.4 mVS
UNKNOWN	20	568.1	858.0 mVS
UNKNOWN	21	603.5	892.6 mVS
UNKNOWN	23	690.9	4.6 VS

Figure 1. Chromatogram of a 1 p.p.m. standard of benzene, toluene, and m-xylene in air.

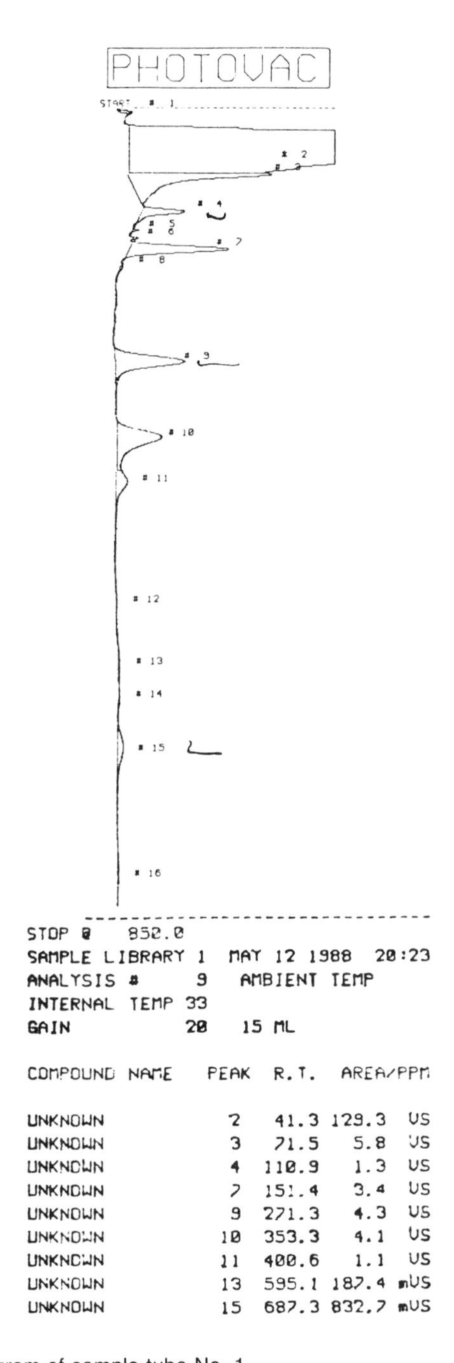

COMPOUND NAME	PEAK	R.T.	AREA/PPM	
UNKNOWN	2	41.3	129.3	VS
UNKNOWN	3	71.5	5.8	VS
UNKNOWN	4	110.9	1.3	VS
UNKNOWN	7	151.4	3.4	VS
UNKNOWN	9	271.3	4.3	VS
UNKNOWN	10	353.3	4.1	VS
UNKNOWN	11	400.6	1.1	VS
UNKNOWN	13	595.1	187.4	mVS
UNKNOWN	15	687.3	832.7	mVS

Figure 2. Chromatogram of sample tube No. 1.

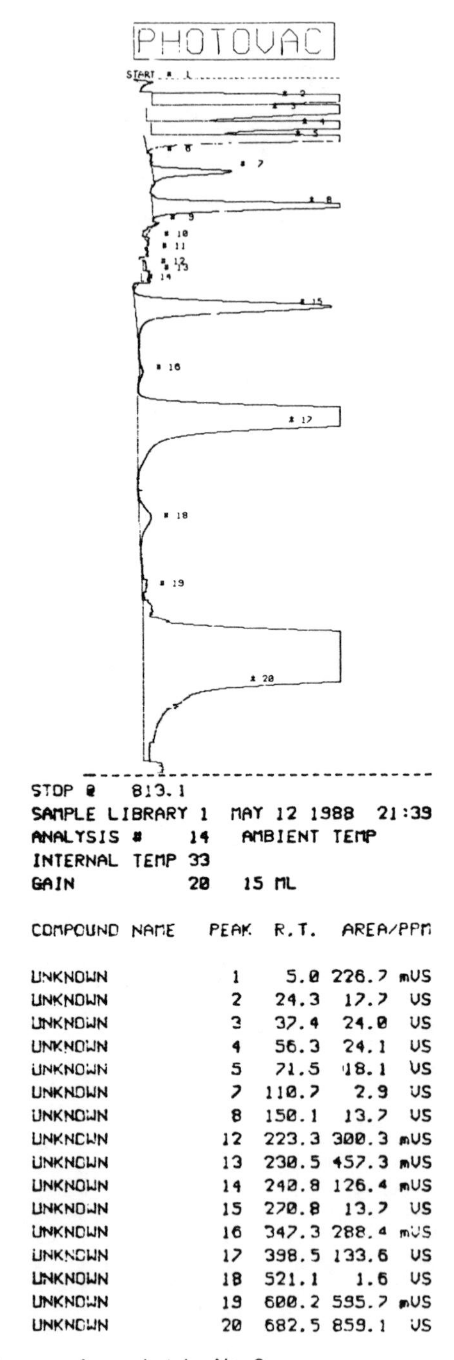

STOP @ 813.1
SAMPLE LIBRARY 1 MAY 12 1988 21:39
ANALYSIS # 14 AMBIENT TEMP
INTERNAL TEMP 33
GAIN 20 15 ML

COMPOUND NAME	PEAK	R.T.	AREA/PPM	
UNKNOWN	1	5.0	226.7	mVS
UNKNOWN	2	24.3	17.7	VS
UNKNOWN	3	37.4	24.0	VS
UNKNOWN	4	56.3	24.1	VS
UNKNOWN	5	71.5	18.1	VS
UNKNOWN	7	110.7	2.9	VS
UNKNOWN	8	150.1	13.7	VS
UNKNOWN	12	223.3	300.3	mVS
UNKNOWN	13	230.5	457.3	mVS
UNKNOWN	14	240.8	126.4	mVS
UNKNOWN	15	270.8	13.7	VS
UNKNOWN	16	347.3	288.4	mVS
UNKNOWN	17	398.5	133.6	VS
UNKNOWN	18	521.1	1.6	VS
UNKNOWN	19	600.2	595.7	mVS
UNKNOWN	20	682.5	859.1	VS

Figure 3. Chromatogram of sample tube No. 2.

The absence of VOCs does not preclude the sampling of certain stations at a site. Compounds of interest could be present without these volatile indicators. However, based on data from thousands of samples analyzed through the CLP, the vast majority showed the presence of VOCs. Consequently, this use of field methods can reduce the number of samples sent to the CLP which give negative results.

This example demonstrates a saving in time and money in the preremedial investigation stage of site characterization. It has become a recurring theme in the Superfund site process.

If a site is found to have satisfied the criteria for inclusion in the NPL, the next step of the process begins. This first phase of a RI deals with site characterization. Initial analytical data gathering determines the degree and extent of contamination. These determinations take the form of assistance in well placement and screening levels, location of "hot spots," plume boundaries, preliminary evaluation of unknown materials in drums, and determination of background levels of onsite contaminants.

Most recently, these types of activities have become part of an extended phase of the initial site investigation and are called Listing Site Investigations (LSIs). In an attempt to provide rapid reliable data for this endeavor, the Field Analytical Support Project (FASP) has been initiated by the EPA.

A recent example of FASP has been the rapid analyses for PCBs of over 2000 soil samples along the streets of a town in Arkansas. The contamination occurred through the oiling of roadways with PCB-tainted waste oil. Samples were collected from near the street and near the adjacent homes. Three different GCs were used in a modified trailer during this ongoing study: a Shimadzu, a Tracor 540, and an HP 5890. The analytical column was a DB-608 wide-bore capillary column 30 m in length. The run time was about 30 min.

The goal of this project was to determine the PCB levels of the soil to 10 ppm. This is the level at which the soil is considered to be clean. However, because of variability in the soil, a 6-ppm level was used to ensure that the maximum amount of PCB did not exceed 10 ppm. The detection limit is about 1 ppm.

The PCB mixture of Aroclor 1260 was the identified analyte. There was interference by chlordane isomers due to termite treatment given to many of the houses. However, with the longer retention times of the Aroclor isomer, much of this was avoided.

As part of a continuing quality assurance program, three samples per day were first analyzed and then homogenized and sent to the EPA Region 6 Central Regional Laboratory in Houston. The field analytical results were to agree to within 25% of the fixed laboratory's results for the analysis to be considered valid.

Another part of the quality assurance (QA) program was to spike four native soil samples per day with Aroclor 1260 and analyze them. More than 90% recovery was normally found.

A further example of this type of field analytical support was given by

Cheatham and Olsen,[4] at a Superfund site in Colorado. A Close Support Laboratory (CSL) was set up and analyses for both volatile and extractable organic compounds were performed. Fieldable instruments were used with a PID, a FID, and an HECD. These detectors were chosen to analyze soils and waters for the target compounds trichloroethylene, toluene, pp′-DDT, and dieldrin.

The methods employed injections of vapor from heated headspace for the volatiles and a hexane extraction for the two pesticides. Method detection limits (MDLs) were in the range of 10 ppb in water to 500 ppb in soil for the pesticides and about 50 ppb for the volatiles in both water and soil.

The results were used to select screening placement in wells and assist in selecting locations for deep soil borings and Phase II wells in order to discover the maximum extent of contaminant migration as well as to determine the level of personnel protection needed.

A second phase of a RI may be initiated in which additional data collection is necessary to refine information already gathered or redefine data needs.[5] Since it may not be possible to identify all of the data needs for a site, a sequential data collection scheme may be necessary. Field GC methods may again be an important analytical tool as in the first phase.

All of the information gathered in support of a RI may eventually be used in the FS done for the site. FS objectives are to develop and evaluate remedial alternatives to cleanup sites.[5] If data gaps are identified during this process, field GC analysis may be called upon to give quick answers.

Following the completion of the RI/FS process which culminates in a Record of Decision (ROD), the remedial design (RD) phase begins. Usually little analytical work is needed for this operation. However, depending on the array of remedies, selected data points may be needed.

For instance, if soil is to be excavated and treated, a well-defined volume of contaminated soil will be needed to project the cost of the procedure. A possibility for field GC analysis presents itself depending on the contaminants involved.

For example, a vacuum stripping process is being used to remove volatile materials from a portion of a Michigan site. The extracted vapors are passed through charcoal columns in series to remove organic constituents.

An HP 5890 GC with a FID was used in a building on site to analyze the vapor before, between, and after passage through the columns. As many as 18 organic compounds were found with trichloroethylene, tetrachloroethylene, and methyl ethyl ketone having the highest concentration.

Direct injection of the extracted vapor from 100 μL to 1 mL was used in the analysis. The column was a SPB-1 wide-bore capillary column.

Presently, the GC analyses are conducted at the beginning and end of each work week to track the progress of the operation. Ongoing GC analysis is no longer needed as it was at the start of the project. The analytical results are correlated with results from a specially adapted HNU detector devise that is continuously sampling the process stream between the primary and secondary charcoal columns.

If the level of organic vapor reaches 10 ppm, the process is automatically shut down and the exhausted primary column is exchanged with the secondary column. This provides for an efficient use of material and manpower in an extended field operation.

The final part of the Superfund process is the implementation of remediation chosen for a site. This remedial action (RA) may need analytical support to assess the progress of the cleanup and when the predetermined acceptable levels of contamination are met.

Since this is a dynamic process that involves field personnel and equipment, turnaround time for results is important. Field GC capability in a CSL setting has been used to minimize costs in this final phase of operation.

DATA QUALITY OBJECTIVES

As seen in the previous example, there must be an awareness in all stages of site investigation from an emergency to the final remedial action of the end use of the data gathered. Each hazardous waste site has its peculiar needs at each stage of the remedial process. Selection of analytical procedures cannot be arbitrary. Data collection must be tailored to each unique facet of that process. Therefore, development of formalized data quality objectives (DQOs) must be incorporated into the project planning process at the start of the RI/FS.

DQOs are defined as qualitative and quantitative statements which specify the quality of the data required to support decisions during remedial response activities.[5] They take into account various appropriate analytical levels for data generation. Broadly speaking, field analyses are considered to be appropriate for site characterization, evaluation of alternative remediation, and monitoring of the implementation phase of remedial actions.

The three stages of the DQO development process is helpful in deciding if field analyses are adequate to fulfill the needs of a project phase. Stage one defines the type of decision to be made and identifies the data user. Stage two specifies the data necessary to meet the need of the data user. Finally, stage three specifies the method and analytical levels needed to satisfy the decision process.

Examining the previous example of the first phase of a remedial investigation in Colorado, the DQO stages can be identified. The data users were the site manager and the health and safety officer, who were making decisions about site characterization, well screening placement, personnel exposure, and decontamination procedures. The data necessary to make these decisions were to be generated for two VOCs and two chlorinated pesticide compounds known to be onsite. Finally, the methods of heated headspace for the VOCs and quick solvent extraction for the pesticides were chosen with heated chromatographic column analysis.

All of these were planned choices before the use of field analysis so that the data obtained could be used to make decisions in the field. Quality controls such as spiked samples, duplicates, and method blanks were used to qualify the data before use. A selection of samples was then made for confirmation in the CLP system.

Obviously, the DQO process is designed to assure that inadequate data or data of limited use will be minimized. This is as important for field analyses as it is for more rigorous methods. The cost effectiveness of field GC work is only as good as the planning that has gone into it.

METHODOLOGY

The methods for field analysis are divided according to volatile and semivolatile compound types as might be expected based on fixed laboratory environmental analyses. However, the analysis of VOCs can be further divided by the ways in which a sample is taken and the manner in which it is introduced into the GC.

As part of a field analytical quality assurance plan, a method of assuring sample chain of custody needs to be addressed. Samples need to be properly marked as to station location, times of sampling, and person taking the sample. This information must be kept along with an analysis log which correlates samples with chromatograms. Besides being just good procedure in avoiding mistakes and delays, this information may be needed at a future date to complete reports on for use in an evidential manner.

Collection methods vary depending on the matrix and the DQOs of the study. For air or soil gas, a sample may be taken directly or dynamically or statically concentrated.

Direct sampling methods include taking aliquots of ambient air or soil gas with gas-tight glass syringes or Tedlar bags. Subaliquots may then be used for analysis.

This may be done by withdrawing the sample directly from pipes or Teflon®* tubes placed in a drill hole. A variation on this is to draw soil gas from a drill hole and allow a portion of the sample to fill a deflated Tedlar bag in an evacuated vacuum desiccator. It is important to note that the bag must be between the point of sampling and the pump in order to avoid any possible contamination from the pump.

Dynamic concentration involves the use of a glass or stainless steel tube filled with either activated carbon or Tenax sorbent material. Again, air or soil gas is drawn through rather than pushed through the tube by a pump.

A combination of direct and dynamic concentration is a procedure in which up to a liter of vapor is captured in a gas-tight syringe and then injected into the trap portion of a purge and trap device. This sample is now concentrated on the trap and ready for subsequent analysis.

*Registered trademark of E. I. du Pont de Nemours and Company, Inc., Wilmington, Delaware.

In the static or passive method, an opened stainless steel can containing an activated carbon organic vapor monitor is inverted and buried at shallow depths. They may be allowed to remain from several hours to several days in many places around a site. The results can give an idea of analyte concentration over a wide area of a site.

Sampling soil itself or water for volatile analysis involves placing a small portion in a septum-sealed vial. In the case of soil, a portion of a core sample may be used. The septum must have a Teflon coating on the side facing the sample.

Sample collection for semivolatile compounds such as PCBs, pesticides, and polynuclear aromatic hydrocarbons (PAHs) is a matter of placing the soil or water samples in uncontaminated glass containers with Teflon-lined caps. Various methods of obtaining the samples exist and will not be mentioned here.

Methods of sample preparation and injection into a GC for volatile analyses depend on their collection method. Direct sampling with a glass syringe or Tedlar bag merely requires injection of a measured aliquot of the vapor. The aliquot size depends on the GC column and conditions used. It can vary from 10 μL to 2 mL.

Sorbent-filled sampling tubes that are thermally desorbed utilize a gas sampling valve which allows a predetermined aliquot to be injected. The remainder is available for subsequent injections. Also activated carbon may be desorbed with a solvent that is compatible with the detector used and the compounds of interest.

In the case of concentrating a sample on a trap, the loaded trap can be flash heated and backflushed onto the analytical column. This may be a good choice when very low levels of volatiles are expected.

Finally, headspace analysis can be relatively simple in that a measured amount of soil or water is allowed to equilibrate at ambient temperature and an aliquot of vapor is removed for injection. To ensure a higher concentration of VOCs in the vapor, an automated heated headspace sampler with a heated syringe can be interfaced to a GC. This would be appropriate for a CSL facility.

Quality control procedures also vary depending on DQOs and the type of method chosen. Usually, blanks are used to ensure there is no contamination problems in the GC system or the injection apparatus including syringes. Many times ultrazero grade air is taken into the field to flush out equipment for air or gas analyses. It can be injected to establish that there is little or no background interference with the analyses. Distilled water is taken into the field to use as an extraction blank for semivolatiles.

Duplicate analyses of samples are often used to show reproducibility. Obviously samples which show no contamination or have concentrations near the detection limits are of no use. Samples having reasonable analyte concentrations should be chosen.

In some cases, samples are analyzed and then fortified with an amount of one or more of the analytes to demonstrate recovery from the matrix. This may be useful when doing headspace injections of soil or water or extractions for semivolatiles.

Calibration of analytes for field analysis is generally not as extensive as would be done in either GC or GC/MS analyses in a fixed laboratory. First of all, most GC runs are made with one analytical column instead of two different columns for confirmatory purposes. This reduces the time for calibrating another column.

Depending on the DQOs, a single calibration point may be done to establish retention times and approximate sensitivity of an instrument. This calibration mixture may be reinjected frequently as a continuing calibration to monitor retention time drift and as a check on potential problems with leaks or loss of sensitivity.

A more comprehensive calibration procedure involves a three-point calibration curve within the linear range of the detector. An average response factor is then calculated. The response factor from continuing calibration runs are checked to see if their percent difference (%D) is less than some predetermined value. The chosen value usually varies between 10 to 20% and is compound dependent within the limitations of the method used. If the %D is exceeded, either a reinjection is made to again check the value or a new calibration curve may be performed for the compound or compounds that are not within specification.

This type of calibration is used only when calculation of analyte concentrations need to be within certain limits of confidence. Of course, reanalysis of the sample must be done if the target compounds are beyond the calibration curve.

Methods for semivolatile compounds have one thing in common in that all samples must be extracted with a solvent. Also, some cleanup of the extract may be necessary.

Soil samples must be either dried or a mixed solvent must be used to allow proper contact between the wet soil particles and the solvent. One mixture is a 1:4:5 ratio of distilled water/methanol/hexane which is used to extract less than 1 g of soil. Another variation is 2 mL of methanol with 10 mL of hexane which extracts less than 2 g of soil. Yet another method uses a 1:1 mix of acetone/hexane with less than 15 g of soil. These techniques have been used for determining both PCBs and pesticides.

Hexane is the solvent of choice for water samples in the analysis of PCBs and pesticides. However, a cyclohexyl-bonded phase sorbent has been used for pesticide extraction using ethylacetate as an elution solvent.

PAHs are extracted by vortex mixing water or soil with methylene chloride. The sample size may vary up to 100 mL of water using 1 mL of solvent and 2 to 3 g of soil with 6 mL of solvent.

Table 1 is based on a catalog of field methods that have been used by the EPA or EPA contractor personnel at various hazardous waste sites across the country.[6] Each of the method numbers refer to the synoptic version of these methods or they appear in the catalog.

Table 1. Summary of Field Instruments and Methods for Organic Compounds

Method Number	Analytes[a]	Matrix	Sampling Procedure[b]	Extraction Procedure[c]	Analysis Time[d]	Instrument Type[e]
FM-5	VOC	Air	I/S	PTD	4–8 hr	P,F,M
FM-6	VOC	Soil/water	D/Vial	AHS	40 min	M
FM-10	VOC	Soil/water	D/Vial	HS	3 min	P,F,M
FM-11	VOC	Soil	I/S	Solvent desorbed	1 hr	F,M
FM-13	VOC	Soil	I/Syr.	Pt	1 hr	M
FM-16	VOC	Soil	I/S	PTD	2 hr	F,M
FM-17	VOC	Soil	D/P/Syr.	Inject	40 min	P,F,M
FM-18	VOC	Soil	D/Syr.	Inject	1 hr	P,F,M
FM-19	PCB	Soil	D	W/M/H; C	10 min	F,M
FM-20	PCB	Soil/water	D	H	25 min	F,M
FM-21	PCB	Soil/water	D	M/H; H; C	10 min	F,M
FM-22	PCB	Soil	D	H/A; C	30 min	F,M
FM-23	Pesticides	Soil/water	D	M/H; H; C	45 min	F,M
FM-24	Pesticides	Soil/water	D	H	25 min	F,M
FM-25	Pesticides	Water	D	BPS	45 min	F,M
FM-27	PAH	Soil/water	D	Mc; C	60 min	F,M
FM-29	PAH/Phenols	Water	D	BPS	45 min	F,M

[a]Analytes types depend on instruments and conditions used. Some VOCs are only low molecular weight compounds.
[b]D = direct; I = indirect; P = pump; Syr. = gas-tight syringe; S = sorbent.
[c]PTD = Programmable Thermal Desorber; AHS = Automated Headspace Sampler; HS = headspace; Pt = purge and trap; W = water; M = methanol; H = hexane; C = cleanup; A = Acetone; BPS = bonded phase sorbent; Mc = Methylene Chloride.
[d]Analysis time per sample may include sampling time which may also include drilling.
[e]P = portable; F = fieldable; M = mobile.

SUMMARY

Many GCs are on the market that are specifically designed for field analyses. Some have been evaluated for specific uses.[7]

Over the last several years field GCs have been used in numerous spill site situations and in various phases of both potential and established Superfund sites.

During this time a number of field analytical methods have been developed and used. Air, water, and soil analytical protocols have evolved. Most have been found to fill a need in the massive amount of data gathering needed in hazardous waste management.

A formalized process of DQO development has emerged from the empirical process of hazardous waste site discovery and remediation. This process has been implicit in field analyses. However, the stages of the DQO process need to be consciously determined and applied to field analyses as they should be for all levels of effort in dealing with hazardous waste sites.

It should be stated here that the personnel involved in doing the analyses and providing the data upon which field decisions are made need to be no less qualified and experienced than those who would be performing similar analyses in a fixed laboratory. Taking, preparing, injecting samples, and operating GCs can be performed by a well-trained technician. However, an analytical chemist experienced in environmental analyses and chromatography should supervise and evaluate data before they are used by decision makers.

Field analytical work will continue to increase in the hazardous waste field in an effort to "work smarter." Getting more information quicker while expending less resources will remain a continuing quest, and field analysis can definitely help. Novel equipment and methods are appearing in various forms to satisfy field analytical needs.[8]

Fixed laboratories will be called upon to provide sophisticated environmental analyses in a much more rapid time frame in the near future. The interest for this comes from the traditionally longer complex analytical protocols which place a time burden on remediation efforts. It also comes from the successes of field analytical work over the last few years across the country. The circle appears to be completed in that rapid analyses designed for a specific need will be coming from the field and back to the laboratory from which they were originally adapted.

DISCLAIMER

Mention of trade name products is not an endorsement of their use by the EPA.

REFERENCES

1. Spittler, T. M. and Clay, P. F. " The Use of Portable Instruments in Hazardous Waste Site Characterizations," *Proceedings of the National Conference on Management of Uncontrolled Hazardous Waste Sites,* HMCRI, Silver Springs, MD, 1982.
2. "Superfund Amendments and Reauthorization Act of 1986," Public Law 99–499.
3. "Uncontrolled Hazardous Waste Ranking System: A User's Guide," National Oil and Hazardous Substance Contingency Plan, Appendix A, 40 CFR 300 (July 16, 1982).
4. Cheatham, R. A. and Olsen, R. L. et al. "Rapid, Cost Effective GC Screening for Chlorinated Pesticides and Volatile Organics at CERCLA Sites," *Proceedings of the National Conference on Management of Uncontrolled Hazardous Waste Sites,* HMCRI, Silver Springs, MD, 1986.
5. "Data Quality Objectives for Remedial Response Activities Development Process," U.S. EPA 540/G-87/003 (March 1987).
6. "Field Screening Method Catalog User's Guide," EPA/540/2–88/005, U.S. EPA (September 1988).
7. Jenkins, R. A., Maskarinec, M. P. et al. *Technology Assessment of Field Portable Instrumentation for Use at Rocky Mountain Arsenal,* ORNL/TM–10542, Natl. Tech. Info. Serv., Springfield, VA (1988).
8. "First International Symposium on Field Screening Methods for Hazardous Waste Site Investigations," U.S. EPA, U.S. Army Toxic and Hazardous Materials Agency, and the Instrument Society of America (October 1988).

CHAPTER 7

Roboticized Analytical Techniques

John G. Cleland

INTRODUCTION

Robots have only recently begun to be seen in the field of chemical analysis. This is particularly true of hazardous waste measurement, where robotics applications are essentially a prospect rather than a reality. However, the analysis of hazardous materials provides a particularly good potential for robot use since safety can be promoted by replacing the human with a machine in dangerous environments. This is borne out by the fact that some of the most important robotic applications have been in such fields as undersea engineering and radioactive materials handling.

As the organization of this volume indicates, hazardous waste measurements can be addressed in two parts: (1) field operations—including sampling, monitoring, and preliminary analysis; and (2) laboratory techniques. There has been interest generated in roboticizing the first area, considering mobile, remote-controlled units to take samples at hazardous waste sites. Such interest took an important step forward with the introduction of a number of developmental units into the Three Mile Island nuclear plant following the accident there in March 1979. Research and development on automated chemical hazardous waste samplers/handlers is progressing in a limited sphere—primarily through university research and the limited developments of major solid waste handling contractors.

However, the following discussion will be concerned with the second area of hazardous waste measurement—laboratory techniques. Much more progress and impact has been made in laboratory robotics than in field applications related to sample preparation and chemical measurement.

About 250 million chemical measurements are taken each day in the United States.[1] As the volume of samples processed in laboratories has increased, so has the complexity of analysis of each sample. Modern instrumentation in some ways contributes to this burden since it has allowed important decreases in error tolerances for given assays.

There are three primary elements in chemical analysis: sample preparation, analytical measurement, and data reduction. Dramatic improvements have been made in the past three decades in analytical measurement and data reduction. However, sample preparation lags behind these developments and still remains the weak link in analytical methods.

One main reason for considering the utilization of a robot system is to handle sample preparation of large batches of samples. A robotic system enables a laboratory to handle large sample loads, not necessarily because the robot works faster, but because it works consistently and constantly. Unlike laboratory staff, the robot does not need breaks, does not tire, and works effectively 24 hr a day, 7 days a week. Successful implementation of this technology has yielded cost and labor savings, improved precision, higher data output, better morale, and improved worker safety.

Laboratory robotics is a relatively young field in chemical analysis. Although a number of review papers[2-5] and a symposium series[6] on laboratory robotics have been published, considerable education related to the use of robots for hazardous waste measurement is needed.

THE PHYSICAL EXTENSION OF COMPUTING POWER

The field of robotics has evolved from science fiction to an important component in science and industry mostly during the last 20 years. There are currently about 125,000 robots operating worldwide, about 30,000 in the United States. Most are dedicated to welding, materials handling, assembly, painting, and/or machining. About 60% of all robots are being applied in automotive and electronics industries.[7]

Robotics is a subset of automation and artificial intelligence with an end goal of smart machines that can redirect themselves to a combination of a variety of tasks without supervision. The Robot Institute of America defines a robot as "a reprogrammable, multifunctional manipulator capable of moving material, parts, or tools through variable program motions for performance of a variety of tasks." The Japanese Industrial Standard (JIS), B0134–1979, defines a robot as a mechanical system which has flexible motion functions analogous to the motion functions of living organisms, or combinations of such motion functions, in which actions respond to the human will. In this context, intelligent functions mean the ability to perform at least one of the following: judgment, recognition, adaptation, or learning. Zenie[8] succinctly defines laboratory robotics as ". . .the extension of programmable computers which allows computers to do physical work as well as process data."

Robots are not yet walking, talking, anthropometrically designed, autonomous units. Today's robots are basically an arm and a hand mounted on a base fixed to a bench or floor. Most common robots are controlled by encoders and servomechanisms which sense the difference between the commanded trajectory of the robot arm and the actual trajectory and act to reduce the difference. Four to seven mechanical degrees of freedom are commonly used to enable the manipulator (or robot arm) and end effector (or robot hand) to achieve the range of positions and orientation desired. Figure 1 illustrates the joint movements and work envelopes for the most common kind of robot configurations. Important components of the robot are given in Table 1. The robot is electronically tied to a microcomputer or other electronic controller

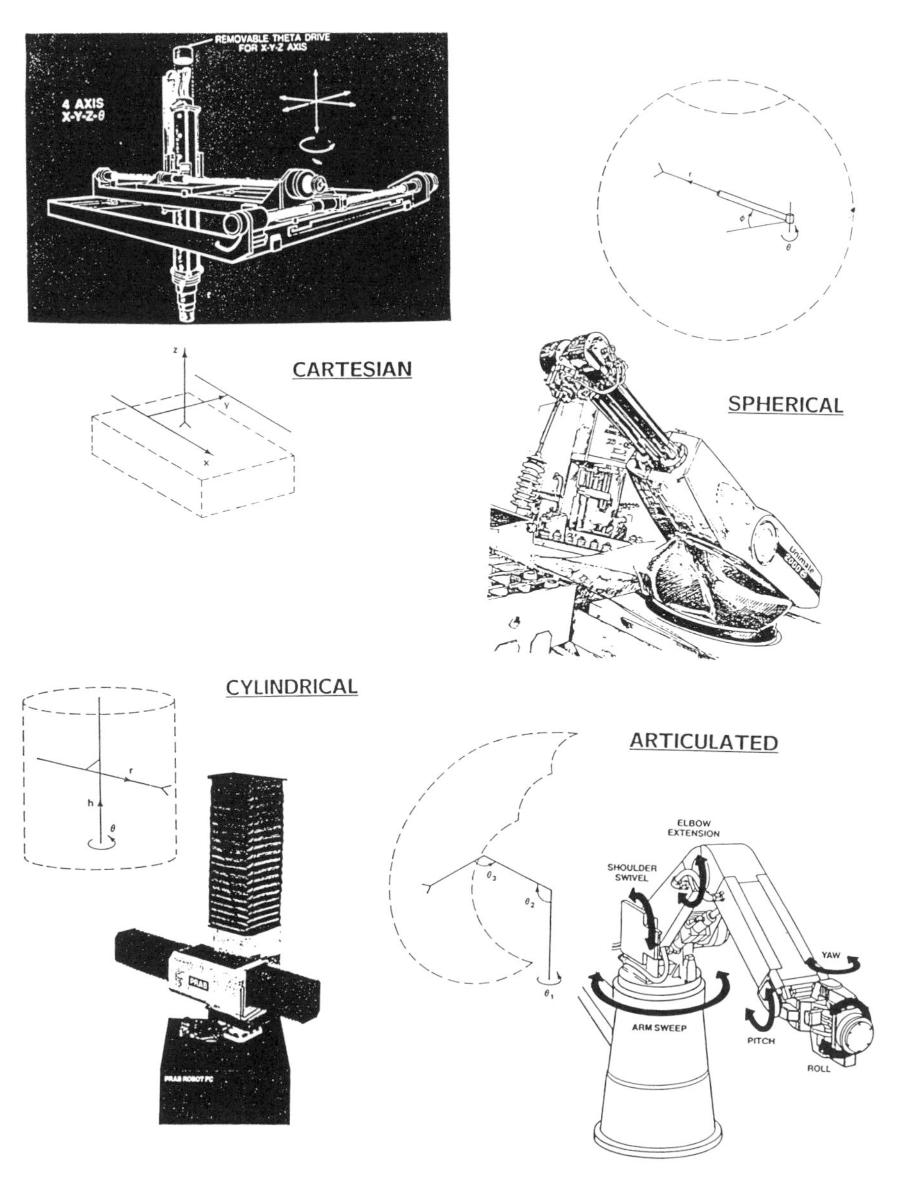

Figure 1. Robots motion geometries and work envelopes. (From Snyder, W. E., *Industrial Robots: Computer Interfacing and Control,* Prentice-Hall, Englewood Cliffs, NJ, 1985. With permission.)

Table 1. Laboratory Robot Characteristics

Attribute	Typical Type or Range
Type	Small industrial type, bench-top or ceiling mounted, joint-articulated or cylindrical, 4 to 7 degrees of freedom, 1 to 2 kg payload
Power	Electric motors (DC servo or stepper) (Gears, chains, timing belts or cables)
Position detection and accuracy	Potentiometer wires, or limit switches and optical encoders, closed-loop feedback, less than ± 1 mm repeatability
Grippers	Interchangeable fingers or hands, programmable gripper position and grip pressure

Source: Adapted from Dong.[9]

which takes signals from the position detection units (encoders) and supplies responding feedback signals to the power units (servos) to control the motion.

Typical robot use begins with arranging the appropriate modules (peripheral equipment and laboratory apparati) within the robot workspace, or envelope. Each communication pathway is defined in a computer file (system configuration file). This process is often termed system configuration.[9]

The operator moves the robot sequentially to all the pertinent positions using a teaching pendant. This may be the computer keyboard, or more often a hand-held unit with push buttons representing each possible robot joint movement. Each position is named by the operator and stored in a directory file. Commands to interact with other peripheral equipment are named and stored. These commands to perform a particular activity may be termed laboratory unit operations. The operations of naming the positions or commands and ordering them in a given fashion, with a sequence of commands stored in the controller/computer, are often referred to as "teaching" the robot.

A subroutine may be created for each unit operation combining several motions such as taking one object from one bin and placing it in another. Finally, a program from the whole operation is built from the subroutines. This is programmed and the operation may be repeated as many times as necessary, but may also be modified to accommodate changes in the environment or objectives of the robot's task.

During individual task evaluations, the system is checked for mechanical reliability. Sensors and verification procedures are added for steps that the robot may find difficult or for steps where a failure to satisfactorily carry out the procedure automatically shuts down the task. Examples of sensing techniques include tactile sense in a robot gripper to assure the presence of an object to be grasped; microswitch closure to verify movement of an object; photocells to verify that an object has been moved in a proper direction, removed or replaced; and weight readings to insure a robot has obtained the right amount of sample.

The problem of avoiding obstacles in the work envelope greatly complicates the task of planning robot movements. A simple solution for this is to have the robot sense the undesirable force of a collision and stop the motion completely. Otherwise, more sophisticated sensors or environment recognition capability is necessary.

What in particular do robots have to offer? Many data handling, processing, and presentational procedures have now been placed on an automatic basis. In consequence, the limits to many automatic procedures center on mechanical manipulation when such steps are needed. A fundamental reason for the interest in robotic systems is that control electronics are now very much cheaper than they were previously, allowing much more versatility from a single mechanical system. This flexibility can justify the cost of sophisticated mechanical devices.[10] The automation of physical manipulation is now possible at reasonable cost.

The generic advantages of robotics include:

- increased productivity,
- improvement in data quality,
- reduction in human exposure to hazardous materials and environments,
- reduction of contamination, and
- elimination of tedious manipulative tasks.

The particulars of how these advantages relate to hazardous material and other laboratory measurements are discussed in the two following sections. Industrial productivity gains and other advantages in manufacturing have been well-documented and are available from the Robotics Institute of America. Typical productivity gains are 20 to 30% on the manufacturing floor.

Of the 563 robot models available in 1987, it was found that the mean price per system was $60,500.[7] A general breakdown of robot production costs shows that the mechanism accounts for about 10% of the cost, electronics for 30%, and software development, 60%. The cost to the user of the robot is often 50% dedicated to software. Other costs include:[11]

- robot (manipulator and base)
- controller
- support equipment
    - feeders
    - sensors
    - end effectors
    - fixtures
- engineering
    - installation
    - training

Laboratory robots are typically on the low end of the cost scale compared to industrial robots. Some estimated system costs are included in the following section.

LABORATORY ROBOTICS

Laboratory robots for chemical analysis were formerly introduced as a commercial option in 1982. The number of robots installed grew to 500 in 1985, to 800 in the United States and western Europe in 1987, and over 1000 worldwide today.[12] Laboratory automation, once limited to computerized data reduction, can now include sample handling and sample preparation along with chemistry procedures and instrumental analysis. Laboratory robots, utilizing programmable computers, can easily be reprogrammed to do a variety of laboratory analyses.

Other, more redundant procedures such as handling a large number of samples in the same way have been addressed by fixed automation or automation of a subsystem of the total process. Such automation is dedicated. Laboratory equipment examples of fixed automation are the Waters Millilab Workstation®, the Hamilton Microlab AT®, the Varian Advanced Automated Sample Processor®, the Teacan RSP 5052 Sample Processor®, and the Beckman Biomec 1000®, most of which perform pipetting, serial dilution, and mixing in small test tubes contained in racks of up to 100 tubes.

The field of laboratory robotics, utilizing a programmable manipulator and end effector, has been pioneered by Zymark Corp., which continues to dominate the market. Zymark has incorporated an approach and product which emphasize close familiarity with chemical analytical requirements, a user friendly system (allowing chemists to become familiar with the robot's operation in a short time), and training and maintenance support which are essential to improved productivity.

However, as sales described here indicate, robotics has only reached a limited number of analytical laboratories. Difficulties in entering this market are indicated by the recent attempts by both Perkin-Elmer and Fisher Scientific to market laboratory robotic systems. While the products were well-engineered, lack of support networks limited sales volumes and both companies have now dropped these product lines. Other robot manufacturers which have indicated interest in laboratory applications (and which have made a few installations) include Precision Robots, Intelledex, TecQuipment, Panasonic, United States Robots, Unimation/Westinghouse, Microbot, Mitsubishi, CRS, Inc., IBM, and Adept. In a survey of laboratory robot candidates in 1988,[13] Research Triangle Institute identified those units in Tables 2 and 3. Those identified in Table 2 had active marketing campaigns at that time for laboratory applications and those in Table 3 incorporated capabilities which would meet many of the requirements for sample preparation for waste analysis. (Costs are estimations and do not represent manufacturers' quotes.)

Certain robot parameters are fundamental to performance and cost of a laboratory robot system. The more important of these are

- *Accuracy and repeatability.* A robot with poor accuracy inevitably results in a system to which the laboratory must be adapted rather than vice versa. In

Table 2. Packaged Automation Systems for Laboratories

Make & Model	Max. Load	Repeatability (in.)	Horiz. Reach (in.)	Vertical Stroke (in.)	No. Axes	Arm Weight (lb.)	Claimed Speed (in/sec.)	I/O Ports	RS232 Ports	Linear Track	Teach Method	Price
Zymark Corp. Z100A	3 lb.	±0.1	24	13	4	40	30	20	Opt.	No	Pendant	$22,800 Arm controller
Fisher Scientific Maxx 5	3 lb.	±0.015	18	24	5	30	51 (w/0.3# load) 6 (w/3# load)	7	8	10 ft.	Joystick	$19,400 Arm Software Joystick Control w/PC & track $30K Track cost $770—$900/ft.
Perkin Elmer Masterlab	2.7 lb.	±0.02	25.7	50	5	60	27 Max. 16 (w/3#)	27	10	Yes	Pendant	$27,800 Arm Software IBM PC
Precision Robots Autobench 3000	15 lb.	±0.002	29.5	21	5	115	20.6	32	1	30 ft.	Pendant	$35,000 Arm Controller 6' track Software Tactile & Optical Sensors Pendant

Table 3. Robot Candidates for Laboratory Automation

Make & Model	Maximum Load	Repeat Ability	No. of Axes	Horizontal Reach (in.)	Vertical Stroke (in.)	Arm Weight (lb)	Claimed Speed	I/O Ports	RS232 Ports	Linear Track	Uses PC	Price
Precision Robots	15	0.002	5	29.5	21	115	20.6 IPS	32	1	Yes	Yes	$35,000
Intelledex 660	10	0.001	6	33	44	240	40 IPS at 5 lb	17	1	No	No	$50,000
Yaskawa V-6	13	0.004	6	36	28	243	70 IPS at 6 lb	95	1	Yes	No	$53,800
TecQuipment MA3000	5	0.08	5	29	18	50	24 IPS	8	1	Yes	Yes	$22,000
Panasonic NM6730	6	0.004	6	30	49	220	71 IPS at 6 lb	34	1	Yes	No	$38,525
U.S. Robots Maker 110	5	0.004	5	45	48	165	38 IPS at 5 lb	34	1	Yes	No	$45,000
Unimation 562	6	0.004	6	36	40	120	20 IPS	64	1	Yes	Optional	$47,000
Typical Low Cost Robots Performance Comparison												
Feedback ESA1071	4	0.01	5(8)	21	21	38.5	35 IPS	24	2	Yes	Yes	$10,000
Autonetics SR450	4	0.001	4	18	24		138 IPS at 4 lb	40	1	Yes	No	$18,000
Universal RTX	4.4	0.2	5	27	36	77	45 IPS at 4.4 lb	12	2	Yes	Yes	$10,000
Zymark Z100A	3	0.1	4	24	13	40	30 IPS	20	Optional	No	Optional	$22,800

other words, existing laboratory equipment must be duplicated and/or arranged to permit a low-grade robot to function.

- *Load capacity.* The maximum payloads recommended by the robot manufacturers are typically not conservative. For example, 3-lb maximum loads covered by the 1988 packaged systems available for laboratory robotics probably overtaxed the robot arm structure in every case. Very low operating speed, plus high maintenance will be the inevitable consequence when the maximum payloads are used.
- *Number of axes.* To perform many laboratory tasks, a minimum of five axes of robot motion are recommended. These include the three axes for movement of the arm in the x, y, and z planes plus two for the robot hand or end effector.
- *Linear motion.* Wherever the laboratory procedures include operations with a required time lapse in excess of the total time for preceding operations, then one robot can serve more than one "work cell," providing the robot has some mobility. For example, a robot may be track-mounted, either overhead or at bench level. Also, if the mobile robot fails, it can be "parked" at one end of the track, permitting laboratory operations to continue on a manual basis until repairs are affected. The laboratory layout can remain virtually unchanged from its format prior to automation.
- *Reliability and maintenance.* To the extent that the robot is overtaxed in any way—payloads, speed, ambient working environment—it is subject to failure and increased routine maintenance costs.

Laboratory robots should be built on an architecture that provides a foundation for continuing technology development and achieves the required system flexibility. Key elements of this architecture are[14]

- *Modular structure.* Modularity is required for flexible adaptation to a wide range of laboratory procedures.
- *Built-in technique.* Quality laboratory results require the consistent use of the best available techniques. Wherever possible, each laboratory robotics module should incorporate valid, tested techniques to eliminate human or procedure variability.
- *Easy programming.* Complex computer applications, which require sophisticated support, reduce control of laboratory applications by the laboratory staff. Since laboratory robotics must be flexible in order to perform a wide range of procedures, programming is best performed by the laboratory staff.
- *Computer and instrument compatibility.* Laboratory robotics must function as an integrated system with general purpose laboratory computers and a wide selection of analytical instruments.

Illustrations of some typical laboratory robots are shown in Figures 2, 3, 4, and 5. Units typically have a set position configured for all peripherals associated with a unit. Through the use of set positioning blocks or screws on the table that holds the robot, it is possible to have some latitude in robot configuration; but, by-and-large, these units are dedicated to a single instrument group or assay. The objects and instruments in the working zone (work enve-

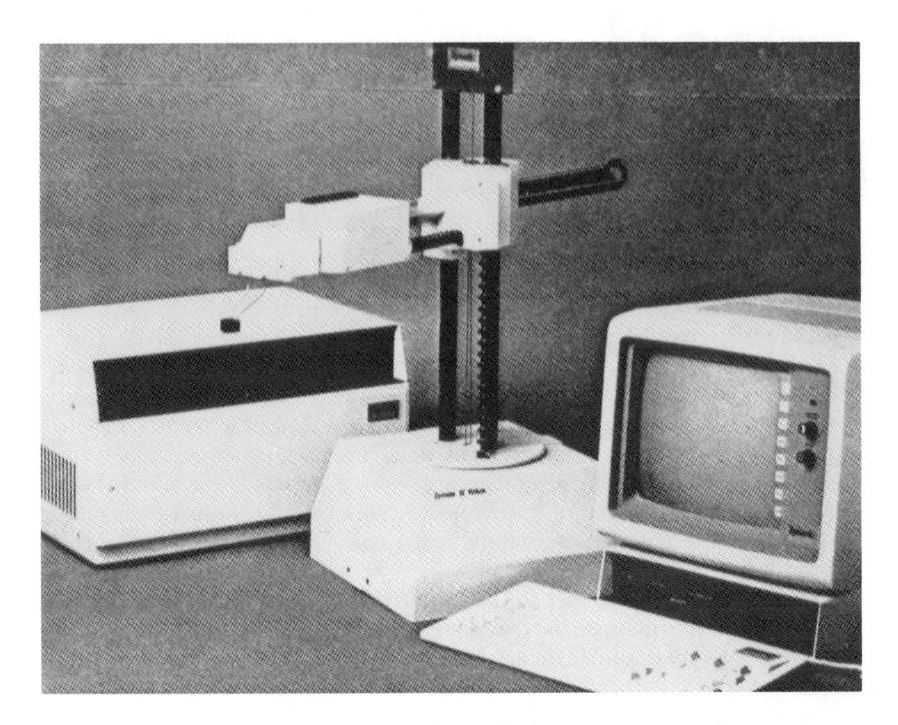

Figure 2. Zymark's Zymate II Robot and Controller. (Courtesy of Zymark Corporation.)

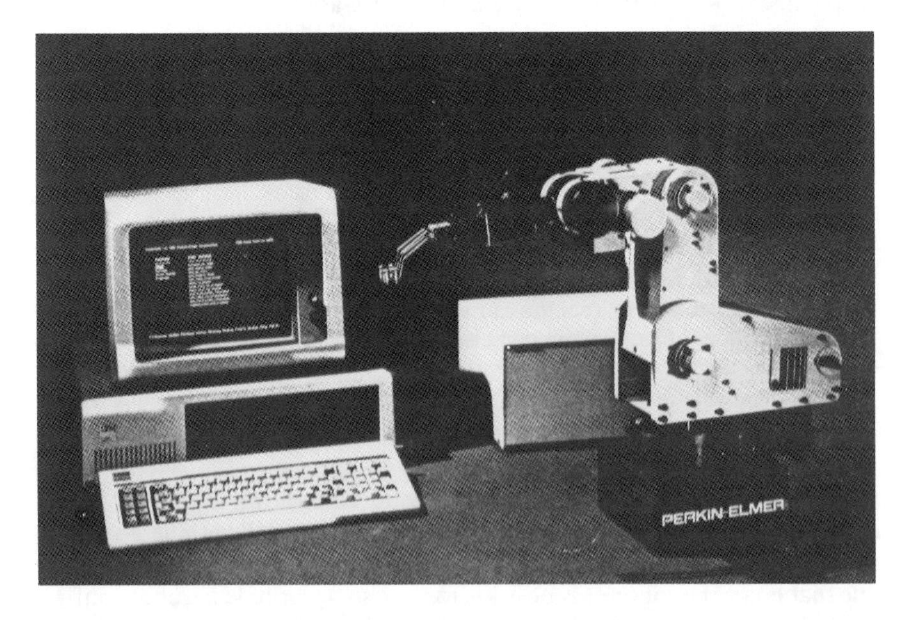

Figure 3. The Perkin Elmer Master Lab System (using Mitsubishi robot arm). (Courtesy of Perkin-Elmer Corporation.)

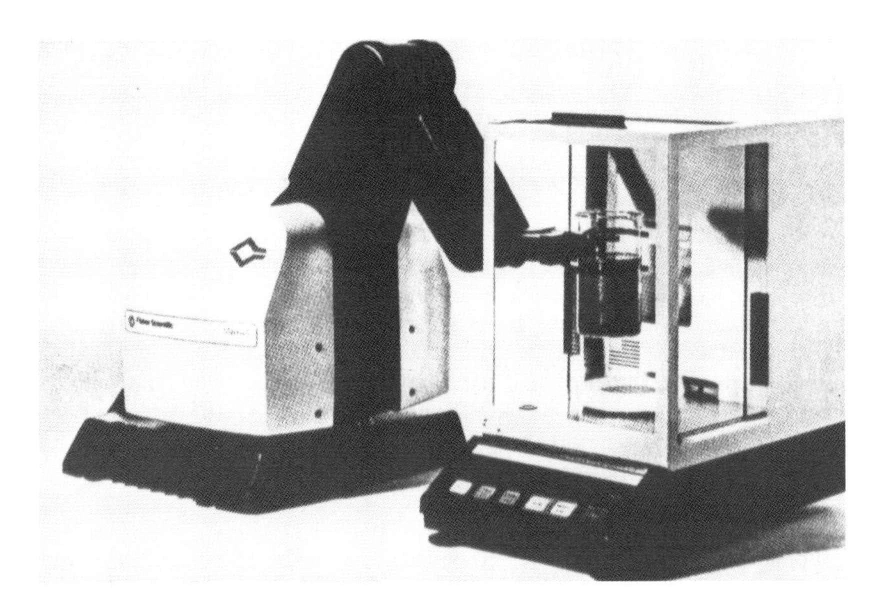

Figure 4. Fisher Scientific Maxx-5 Laboratory Robot. (Courtesy of Fisher Scientific.)

Figure 5. CRS Robot Arm. (Courtesy of CRS, Inc.)

lope) must be rigidly located in space because, otherwise, the robot could not manipulate, since its hand simply does not find the required objects.

Limitations of current machines are overcome, in large part, by adding a number of peripheral devices and instruments to augment the capability of the robotic system. Important components of the robot/peripheral system often include:[15]

- A power and event controller which can bring most standard laboratory apparati under system control. Such a controller provides programmable on/off and variable electric power and switch closures and senses external switch closures and analog voltages.
- Liquid handling, extraction, and partitioning capability station. This automated unit may be programmed to dispense, dilute, and pipette.
- Programmable data acquisition unit to process outputs of laboratory instruments and provide control of other apparatus in the laboratory.
- Capping station for round containers with screw caps (a difficult operation for today's simple robots).
- Automatically changeable end effectors. Different hands can be used for gripping as a syringe for powder dispensing and other laboratory operation. Table 4 gives a typical list of robot peripherals marketed by Zymark and Perkin-Elmer (when their product line was still on the market). Bar code readers have not yet seen their full potential in laboratory robotic applications, i.e., for robot feedback control after identification of items within the work envelope.

Some typical layouts of laboratory robotic systems with peripherals are shown in Figures 6 through 8.

When using laboratory robotics to automate any simple or complex task, following a basic process such as outlined below can reduce the time required and increase the chance of success:[18]

- workstation development
- work cell layout and construction
- robot programming
- system evaluation and modification
- installation (and training) and
- system validation and quality control

The efficient implementation of robotics has been described as occurring in two steps:[12] First, the system is installed, customized, and debugged by experienced automation experts. Second, after successful demonstration of system performance, systems operators take over. The training of these operators and the continued support of the application by the specialists are critical to the success of the project.

Laboratory robots can be easily reprogrammed to do a variety of laboratory procedures and do not require a large quantity of identical, repetitive operations to justify the investment in capital and time. Gas Chromatography (GC) and High Performance Liquid Chromatography (HPLC) applications represent 40% of the current installed base of lab robotic systems.[19] Important

Table 4. Laboratory Robot Peripherals

Perkin-Elmer	Zymark
Balance interface kits	General purpose hand
Barcode readers	Vibrating hand
Cables	Racks
Capping stations	Solvent delivery system
Centrifuge	Pipetting — 1 mL
Crimping stations	Pipetting — 5 mL
Customer implementation support services	Weighing
Device interfaces and AC accessories	Dilution and dissolving
Injector loops for LC injector	Vial crimping
Instrument interface kits	Screw capping
LC injector	Liquid-solid extraction
Linear transports	Liquid-liquid extraction
Liquid handling systems	Linear shaker
Liquid handling system accessories	Centrifuge
Master lab systems	Evaporation station
Master lab system accessories	Membrane filtration
Mixers and mixing accessories	LC injector
Pipette adaptors	GC injector
Pipette apparatus	Spectrophotometer sip
Powder handling accessory	Karl fischer titrator
Pneumatic accessories	Data input and report formatting
Printer and printer supplies	
Racks	
Robot finger accessories	
Solid phase extraction systems	
Syringes	
Test tube feeders	
Test tube support	
Titration accessories	
Verification systems	
Work surfaces	

factors in implementing the robots have been (1) the need for sample preparation of large batches of samples, (2) samples requiring analysis by multiple instruments or techniques, and (3) protection of workers from hazardous conditions or materials. All the steps of an analytical procedure can be automated including sample identification, sample introduction into analytical instruments, data reduction, and subsequent data transfer, report generation, and archiving of the results in an external Laboratory Information Management System (LIMS).

Table 5 lists some uses of laboratory robots, enumerating various laboratory operations that have been automated with robotics. In performing this wide variety of applications, automated laboratory systems including robots build upon a set of what are termed "laboratory unit operations." Table 6 gives one example list of a set of laboratory unit operations (LUOs). A number of these will be combined and repeated for the applications described. Each LUO will require a number of discrete movements of the robot arm and hand, each of which must be programmed. The number of commands for a typical LUO may be 3 to 10, with each command involving 3 to 8 robot moves.

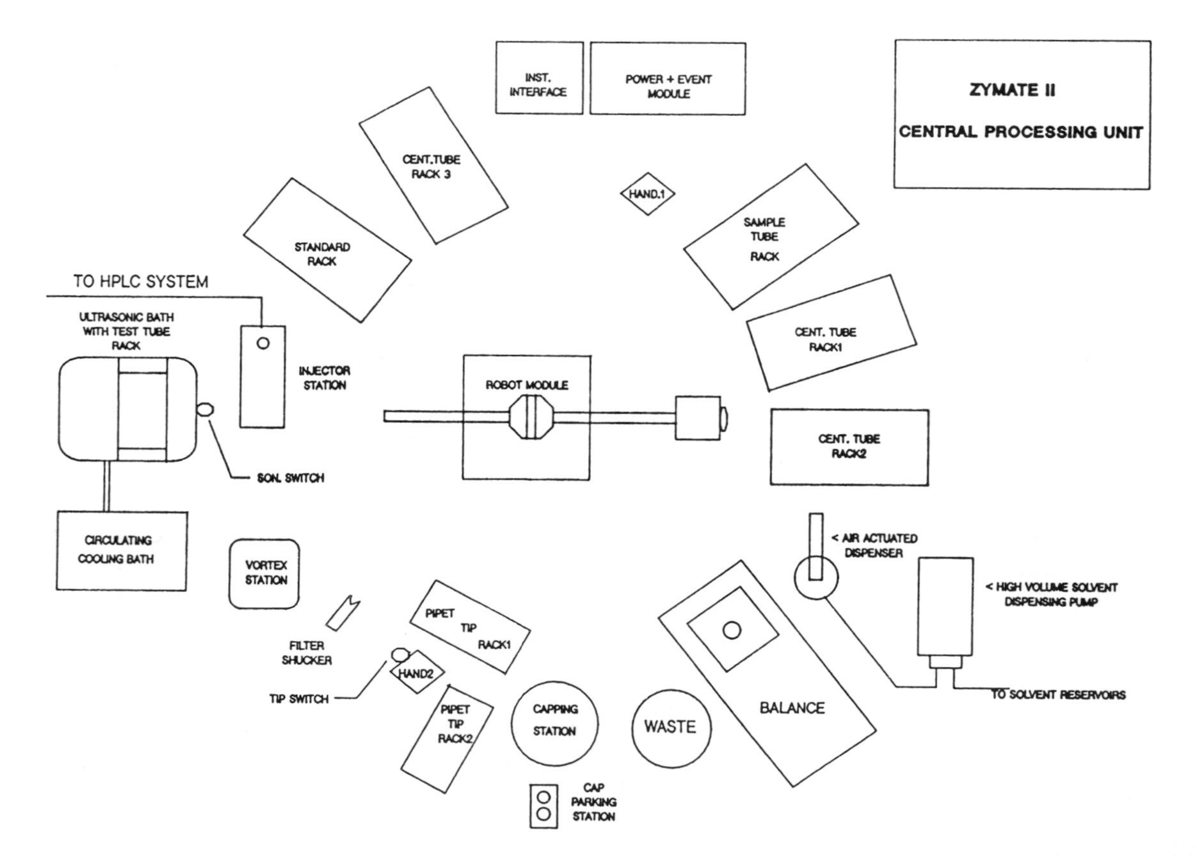

Figure 6. Bench Layout for HPLC analysis of solid dosage forms. (From Zenie, F. H., in *Advances in Laboratory Automation Robotics*, Vol. 4, Zymark Corp., Hopkinton, MA, 1988. With permission.)

Figure 7. Side view of robot table with HPLC underneath. (From Zenie, F. H., in *Advances in Laboratory Automation Robotics,* Vol. 4, Zymark Corp., Hopkinton, MA, 1988. With permission.)

Applications by pharmaceutical companies now appear to dominate well over 50% of the U.S. robot installations. Here samples are small, numerous, lightweight, and can easily be handled within the limitations of current laboratory robot systems. Dissolution testing has been sited by managers as the fastest growing demand on any pharmaceutical testing laboratory.[20] Another example of the successful application of laboratory robots in microassays is

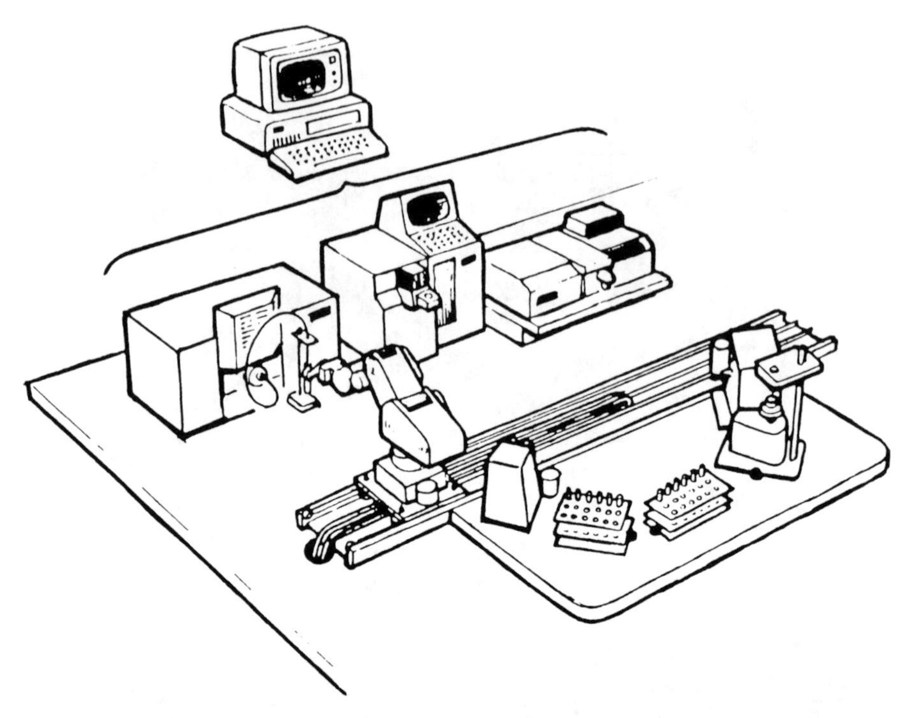

Figure 8. Robot installed on track allowing for movement. (Courtesy of Perkin-Elmer Corporation.)

the configuration called the HP Genenchem Microassay System, produced through the combined efforts of Hewlett Packard, Zymark, and Genenchem. The turnkey product has been targeted for ELISA assays.

Most innovative applications of laboratory robots have been initiated by private companies. The infusion of federal R&D dollars into furthering capabilities and applications of laboratory robots has been very little in evidence. The agencies which sponsor a great deal of analytical chemistry, the Environment Protection Agency and the National Institutes of Health, have demonstrated little ostensible interest in the extension and improvement of laboratory robots applied in their programs. Department of Energy laboratories, such as Oak Ridge National Laboratories and Los Alamos have limited operations with laboratory robots, e.g., in solvent extraction. The Naval Research Lab[21] is utilizing a robot for protein crystal growth and analysis, the U.S. Geological Survey for mineral sample preparation, and there is interest at the National Institutes of Science and Technology in laboratory robotics coupled to expert systems. The U.S. Army Arsenal in Watertown, MA, has instituted an interesting program[22] using robots to automate materials testing and evaluation. Work cells include accelerated environmental testing, physical testing, titration/chemical testing, and mechanical testing. A tabletop robot is now

Table 5. Applications That Have Been Automated Using Robotics

Acid digestion procedure
Allergen formulation testing
Analytical analysis of parasites
Automated feed analysis
Automated methods development
Automated titrations of isocyanates
Autosampler for neutron activation analyzer
(Bio-Rad) protein assay
Blood typing using bar-coded labels
BOD determinations
Bottle washing for AA
Calorimetric procedure and weighing
Carbon residue analysis
Catalyst decomposition
Cellulose hydrolysis
Cholesterol analysis
Clean room solvent wash operation
Cleaning of NMR tubes
Colorimetric endpoint titration
Contact lens solution testing
Content uniformity testing
Determination of drugs in physiological samples
Diethyleneglycol analysis of polymer samples for GC
Dissolution testing (both endpoint and multipoint)
DNA purification
DNA-ninhydrin assay
Dust filter weighing
Elemental analysis of semiconductors
(ELISA) assays
Environmental analysis of priority pollutants by GC
Environmental evaluation of water conditions using metal coupons
Enzyme kinetics
Estimation of Hepatitis B surface antigen concentration
Extraction and sampling for a technicon autoanalyzer
Extraction procedure for tricyclic antidepressants in clinical monitoring
Feed analysis for toxicology
Flavor analysis by GC
Fluoride analysis in toothpaste
Gamma counting
General sample preparation for drug metabolism assays
General sample preparation for food analysis by HPLC
Genetic engineering
Hazardous work environments
Head space analysis
Hybridoma research
Hydrogen and nitrogen analysis
Industrial hygiene preparation using extraction tubes
Introduction of samples into emission spectrophotometer
Ion chromatography preparation and introduction
Karl Fischer titrations
(LASA-P) cancer monitoring test
Limulus amebocyte lysate (LAL) procedure for detection of bacterial endotoxin
Liquid/liquid extractions
Liquid/solid extractions
Lowry protein assay
Letal digestion preparation for ICP
Microbiological inoculation and mixing of cosmetic preservation testing
Microplate manipulation
Microplate testing and reading

Table 5, continued

Multivitamin assay for HPLC
New product formulation
Nitrogen analysis of chemiluminescence
Oil additive formulation
Oil blending
Paint formulation
Paraben sample preparation
Particle-size analysis
Pesticide formulation analysis
Pesticide residue sample preparation for GC
pH measurement and endpoint titration
Pharmaceutical capsule production
Photographic emulsion research
Physical testing of paper and other sheeted material
Physical testing of paper products
Polarography
Polymer moisture analysis
Preparation of fused beads for XRF
Preparation of herbicide samples for HPLC
Preservative testing
Pulsed NMR
Quality control of immunodiagnostic kits
Radioactive waste manipulation
Radioimmunoassay for T4
Radioiodination analysis
Radioligand binding assays
Robotic optimization of organic reactions
Robotics in sterile tissue culture techniques
Sample changer for superconduction NMR
Sample introduction of polymer samples into an Instron
Sample preparation for (IR) analysis
Sample preparation for GPC
Sample preparation for light-scattering spectroscopy
Sample preparation of caffeine analysis
Sample preparation of corn tissue for nitrate analysis
Sample preparation of fermentation broths for HPLC
Serial dilutions
Soil analysis
Specific ion electrode measurements
Spectrophotometric analysis of polymer dyes
Spiral streaking of petri dishes
Stability testing
Sugar analysis by HPLC
Suspended solid determination of environmental samples
Suspended solids in waste water
Tablet dissolution assay
Tensile testing of nonwoven fiber samples
Thermal analysis (DSC)
Thin layer chromatography
Tissue residue analysis
Total nonvolatile extractable analysis
Total solid determination of polymers
Toxicology sample preparation for HPLC
Viscosity measurement of polymers
Vitamin A analysis preparation for HPLC
Wear metals in oil for ICP
Weighing and calorimetry for production of plutonium-238 oxide-fueled mW generators
X-ray fluorescence (powders and oils)

Table 6. Laboratory Operations

LUO Class	Definition
Weighing	Quantitative measurement of sample mass
Grinding	Reducing sample to particle size
Manipulation	Physical handling of laboratory materials
Liquid handling	All physical handling of liquids—reagents and samples
Conditioning	Modifying and controlling the sample environment
Measurement	Direct measurement of physical properties
Separation	Coarse mechanical and precision separations
Control	Use of calculation and logical decisions in laboratory procedure
Data reduction	Conversion of raw analytical data to usable information
Documentation	Creating records and files for retrieval

employed at the U.S.D.A. Agriculture Research Service Nutrient Composition Laboratories in Beltsville, MD.[23]

General advantages utilizing robots in the laboratory have been described earlier. An important contribution to productivity is the fact that robot systems can perform analyses in a serialized mode versus a batch mode. Serialization allows each sample to have the same time history, important in derivitization experiments where possible side reactions can occur causing error. Serialization also increases the throughput of samples versus a batch mode since the robot is always busy and never waiting for one operation to finish before going on to the next. Laboratory robotic systems typically improve precision by a factor of 2 or 3 and maintain that precision over extended periods of time. An additional consideration is that robots can reduce sample contamination relative to human operations, with special design features allowing high-quality sterile testing and clean-room operation over extended periods of time. Automation also permits the economic use of frequent replicates, standards, and controls to verify precision.[24]

Table 7 shows a generic example in which an analytical instrument is automated for extended use, as opposed to purchasing a second instrument. In this case, payback occurs in less than 6 months. Robotics will allow a laboratory to utilize equipment on a more efficient basis and will allow one to utilize equipment at times when it would usually be idle. As a corollary, robotics allow the researcher to gain full usage of the current capacity rather than purchase additional units. Robots, in some methods such as gravametric ones, often provide higher accuracy than available from experienced analytical assistants.

A few examples of related performance of laboratory robotics systems follow. Robots have:

- Provided an avenue to 10- to 20-fold precision enhancements during some extraction/derivitization procedures.[12]
- Saved more than 30 hr of laboratory technicians' time for each hour of setup time spent preparing the robot for automated dissolution testing. Over a 6-month period, approximately 1100 dissolution tests were run on a single robotics system (45 tests per week). Continued at this rate, the labor saving amounted to about 6500 technician hours in 1 year.[20]

Table 7. A Comparison and Justification for a Laboratory Robot

	Formula	Present Method	Robot System	
A.	Number of samples per sample group	Input	30	30
B.	Total time per sample group (hr)	Input	7.0	7.0
C.	Number of sample groups per day	Input	1.0	2.0
D.	Opertaing hours per day	B×C	7.0	14.0
E.	Technician cost per hour	Input	$15.00	$15.00
F.	Technician hours per operating day	Input	7.0	3.0
G.	Technician hours per sample group	F/C	7.0	1.5
H.	Technician cost per sample group	E + G	$105.00	$22.50
I.	Instrumentation cost	Input	$100,000	$25,000
J.	Estimated user setup and programming cost	Input	$0	$10,000
K.	Total investment	I + J	$100,000	$35,000

- Improved relative standard deviation from 5% using a manual sample preparation injection to 2% using a robot for an analysis of OPA derivitized (orthophthaldehyde) amino acids by liquid chromatography.[25]
- Performed 2 years of operation seeding 70,000 petrie dishes and freeing a microbiologist from repetitive tasks. A complete system payoff resulted over 3 years.[26]
- Performed determinations on rocket propellants and reduced personnel by 60% on each shift. Data reporting was considerably speeded to meet the time constraints of the manufacturing process.[27]
- Conducted the same amount of work required of three persons in performing a UV spectrophotometric determination. The robot performs about 60 to 80 determinations (two per sample) every 24 hr under the control of one technician. Relative standard deviation was reduced by 20 to 40% below that for manual methods.[28]
- Reduced labor requirements and variability (by 50%) for determination of glycols in polyester polymers. The automated system treated pellets, powders, films, fibers, and liquids identically with less than 5% downtime required for maintenance.[29]
- Reduced relative standard deviation by 30% for analysis of antipyrine in human serum. Sample preparation by robot was essentially identical to the manual procedure. Robot throughput is about 18% higher than the manual method and requires only 1 hour of work daily, saving about 7 person hr a day.[30]
- Supported pharmaceutical preparations including sampling, extraction, HPLC, and calculation with three different types of samples: tablet, granule

and capsule. Sample preparation was reduced from 5 hr to 1 hr and total assay time was reduced from 10 hr to 6.5 hr. The mean values and the coefficient of variations for the robotic procedure agreed well with those for the manual procedures.[31]

- Performed fluoride analyses involving over 20,000 pH and soluble fluoride determinations. Measurements were made with the same procedures and accuracy as manual determinations. Throughput of the laboratory was increased by 30% while reducing the amount of analyst time by 40%.[12]

One of the major justifications for using a laboratory robot can be found in relative operational costs of human labor ($20/hr) to a robot's operating costs ($5/hr).[15]

Robots also have a number of limitations, some of which can be overcome by further research and development. Current robot systems lack sensors and the intelligence to allow the true flexibility of human experimenters. Most general purpose robots require large working areas. One malfunction in a complex setup can disable the entire system, and the hardware requires specialist backup. Systems operate sequentially, yielding a long cycle time in some cases.[32] Through malfunction or erroneous programming, a robot may do unforeseen harm to apparatus in its working zone and also to itself. Programming a robot requires considerable time, care, and effort, more so than for a computer. Simplicity of interfacing the robot system to the wide range of available microcomputers has not yet been reached. A robot is inefficient for single analyses and when the number of analyses is very large, an especially designed analyzer would be more effective.

As mentioned earlier, placement of a robot on a series of overhead tracks also increases flexibility and allows robot access to a larger number of instruments. Means to move additional work to the robot through innovative racks, drawers, or a motorized cart could be effective. In some cases, lab benches have been redesigned, e.g., a fume hood with a circular bench is currently available for use with laboratory robots having a cylindrical envelope.

"Smart robots" go beyond the simple problem detection/systems-shutdown mode of operation. Strategically embedded throughout smart robot programming would be decision-making, problem-handling subroutines which would often lie dormant during normal operation. Programming will be extended to allow understanding of a given robot's assigned duties and a knowledge of the many ways it might conceivably fail to execute them. Collision avoidance or shutdown features result from the feedback obtained from sensors. The key to crafting robot intelligence is (1) to program the system to ask questions, (2) to provide it with a means to answer those questions, and (3) to encode appropriate responses for each answer.[34]

A number of software routines have been developed for specific application in robotic operations, including:[9,12]

- a data-logging package written in Better BASIC (Summit Software Technology, Norwood, MA) can be acquired from Bio-Tek

- spreadsheet portion of SYMPHONY (LOTUS Development Corp., Cambridge, MA) providing offline data analysis capabilities applied to a laboratory robot system
- Zymark EASYLAB software routines programmed in a FORTH-like language
- ARL (A Robot Language from IBM), PERL, a Perkin-Elmer robot language, and the capability to perform robot programming with a number of common languages such as BASIC, FORTRAN, or PASCAL

Expert systems have been considered for application in laboratory robotics. Application software such as 1st-Class can be utilized to make the robotic system more intelligent. Expert systems provide an approach to experimental design using the robot, i.e., applying SIMPLEX-type routines to analyze data in near-real-time to do experimental planning while their operations are being conducted, and to optimize robot manipulations based upon data analysis.

The possibility of forming robot-integrated local area networks (LANs) is quite feasible. The LAN will allow independent devices to communicate with each other. For example, in the 1985 Pittsburgh Conference, there was an example of a LAN in which an HPLC, FT-IR, and robot were interconnected through the use of a DEC MicroVAX. Considerable work continues in the interfacing of robots with micro-and minicomputers. The true benefit of a lab robotic system cannot be fully realized until it becomes an integral part of a work cell architecture capable of interfacing to a wide variety of lab instruments and laboratory information management systems.[35] Archiving of all results can be done on an external laboratory information management system (LIMS). In an integrated system, the efficiency of the instrument can be maximized, with a sample always ready to be introduced for analysis (provided that analysis time is greater than preparation time).

ROBOTICS AND HAZARDOUS WASTE MEASUREMENTS

Hazardous waste measurements have been affected little, if at all, by the field of robotics. This is primarily due to the fact that hazardous waste sample preparation requires larger sample volumes and more complex preparation than do most of the current areas of lab robot application, e.g., pharmaceutical preparations. Also, as stated earlier, there have been no institutional directives by such agencies as the EPA to promote the use of robots for hazardous waste sample preparation.

However, the benefits of robotic systems for protecting workers from hazardous samples, noxious fumes, radioactivity, and biohazards cannot be understated. Robots can be placed in fume hoods or radiation environments. At the same time, just as humans can be harmed by dangerous samples, many samples are subject to contamination from handling by humans, which may cause erroneous results. In the future, as safety requirements become more stringent, it is conceivable that robots will be required by code for certain applications.[25]

One industry which has made significant contributions to the development and use of robotics in handling of hazardous materials is the nuclear industry. This is usually manifested in the application of telerobots or simple master-slave mechanisms. These units always have a human in control, remotely operating a gripper which handles the "hot" materials. An illustration of a teleoperator is shown in Figure 9. The nuclear industry has also made recent advancements in mobile robotic uses, as mentioned earlier in relation to the Three Mile Island accident. The Electric Power Research Institute and the Department of Energy nuclear laboratories have also been conducting research and development on units which are mobile and can perform observation or maintenance functions in nuclear power plants.[36] Robotic applications at chemical hazardous waste sites would be remotely operated, mobile sampling units, sniffers, and heavy equipment for onsite disposal.

A robotic (rather than teleoperated) system for handling radioactive materials has been developed at Los Alamos National Laboratory. Because of safety considerations, however, the robot is under the supervision of a human operator at all times by speaking commands into a microphone. This is more convenient than entering commands at the keyboard because the operator's hands remain free for other tasks.[30]

A literature survey conducted for the purpose of this chapter indicated no publications in the National Technical Information Service (NTIS), Chemical Abstracts, or Compendex (Engineering Index) databases directly relating robots to chemical hazardous waste measurements. This is not to say, however, that no consideration is being given to the problem.

The EPA Office of Solid Waste recently commissioned the Research Triangle Institute to conduct a preliminary study[13] into the applications of robotics to the SW-846 methods designed for assessing waste materials. These SW-846 methods for characterization of environmental or process waste samples are contained in "Test Methods for Evaluating Solid Waste: Physical/Chemical Methods," EPA, SW-846, 3rd Edition. SW-846 methods address process water and ground water, oil, organic liquids, sludges, solids, multiphase samples, and Extraction Procedure (EP), and Toxicity Characteristic Leaching Procedure (TCLP) samples. It is estimated by the EPA that at least 2000 laboratories are conducting SW-846 or directly related methods for determinations of waste, with volatile, semivolatile and nonvolatile, and organic and inorganic contaminants in the waste being analyzed.

Automation has had some isolated successes in hazardous waste measurement. A Soxtec™ apparatus is on the market for performing six simultaneous Kuderna-Danish-type concentrations which are an important part of sample concentration for many hazardous waste measurements. ABC Laboratories has developed an Autovap unit which offers potential for replacing such techniques as rotary evaporation and Kuderna-Danish in some instances. More development and validation is required for application to EPA methods for

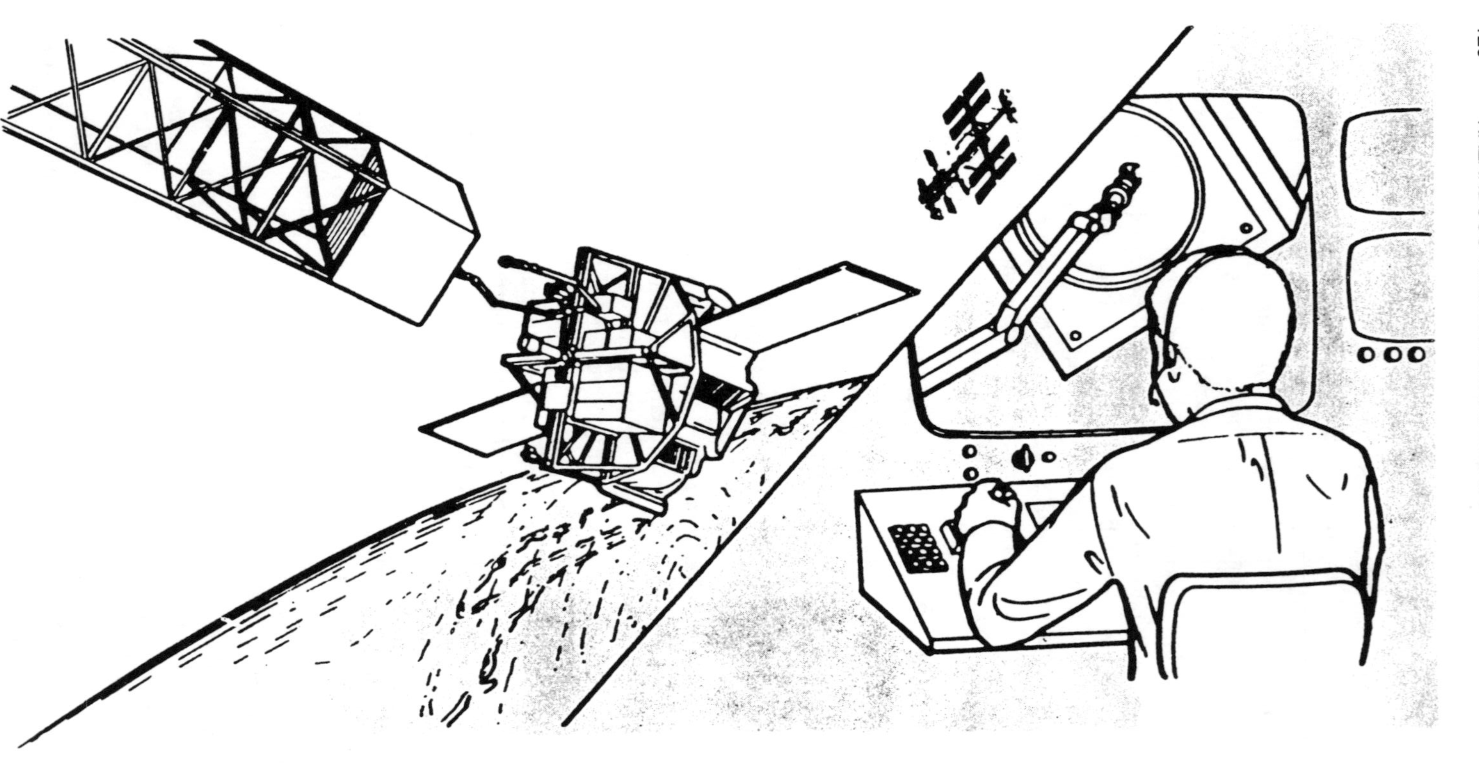

Figure 9. Remote operations (teleoperations) can be conducted under human control with the human operator feeling present at the worksite (telepresence). (Courtesy of the National Aeronautics and Space Administration.)

hazardous waste sample preparation. A cooperative effort between Metatrace Laboratories and Fisher Scientific was planned to provide a roboticized dioxin measurement method. This method has not been completed since the retraction of Fisher from the laboratory robotics market.

Robotic performance of both purge and trap and microwave digestions has been performed, e.g., by BP America (SOHIO). Kidd Creek Mines Ltd., Ontario, also have a robotic microwave procedure in use for geological and ore samples. Haile and Brown[38] reported a "robotic preparation of water samples for trace herbicide analysis." An automated solvent evaporator, an automated plug filtration apparatus and an automated separatory funnel utilizing Zymark robots are described by Kramer and Fuchs.[39]

A robotic system was developed to perform solvent extractions[40] for an actinide separation laboratory. The robot proved to be at least as precise and as accurate as the manual method. The ability of the robot to prepare and equilibrate solutions proved to be "invaluable." Eli Lilly has also used a robot in automatic preparation of protein crystals. The primary impetus for developing this system arose from a requirement to conduct safe crystallization experiments on toxic materials.[41]

Under the SW-846 protocols for organics and inorganics, there are 21 techniques defined specifically for sample preparation. About one third of the 90 analytical methods also incorporate sample preparation methods. Typical sample preparation under SW-846 may include 10 to 12 separate laboratory operations, some of which may require two or more repetitions.

Methods such as SW-846 provide new challenges to laboratory robotics which have not yet been met. An example is the consideration of sample size. Rather than samples ranging from microliters to a few milliliters, the environmental laboratory technician may have to handle liter containers of sample and solvent and similarly large extraction and concentration apparatus. Other challenges to robotics by waste measurement methods include:

- variability of sample matrices
- dirty and hazardous materials
- very long sample preparation times
- many laboratories of different sizes
- standardization of lab approaches to meet EPA regulations and guidelines, i.e., limitations must be placed on customizing robot approaches.

The study conducted by RTI was intended to provide information leading to a possible demonstration of a roboticized SW-846 method to illustrate feasibility to the user community and to robot developers. The methods evaluated are essentially divided into three categories:

- extraction and sample preparation procedures
- cleanup procedures
- determinitive procedures

Table 8. Summary of Laboratory Responses to Survey

Near-Term Method	No Problem	Problem	Robot Solution
3510	0	10	Yes
3520	1	7	Yes
3540	2	2	No
3550	1	5	Yes
Purge & trap	1	3	Yes
VOST	0	1	Yes
Column cleanups	1	8	Yes
K-D	0	6	Yes
EP	0	7	Partially
TCLP	0	6	No
Inorganics	1	4	Yes

Method Descriptions
3510: separatory funnel liquid-liquid extraction
3520: continuous liquid-liquid extraction
3540: soxhlet extraction
3550: sonication extraction
Purge & trap: for capture of volatile organics to be analyzed by GC/MS
VOST: volatile organic sampling train
Column cleanups: to eliminate interferences for chromatography
K-D: Kuderna-Danish evaporative concentration
EP: extraction procedure toxicity characteristic test method
TCLP: toxicity characteristic leaching procedure
Inorganics: analysis including microwave digestion and other sample preparation for metals analysis

A survey of 25 laboratories conducting these methods was completed with emphasis on obtaining opinions about automating some of the protocols. Where well-informed opinions on incorporating laboratory robots could be obtained, the laboratories' assessments were tabulated for comparison. Tables 8 and 9 briefly summarize results from the review and survey. In Table 8, an attempt is made to quantify responses to inquiries about specific methods. About 70% of the laboratories had familiarity with laboratory robotics, at least through direct contact with a vendor. About one third of the 25 laboratories were using robots somewhere within their organization, while only three of the labs applied robots or planned to apply robots in protocols related to EPA methods.

Most emphasis was placed by the laboratories on improving sample preparation/extraction methods. Column cleanup was next emphasized. The estimates of operating parameters (Table 9) associated with integrating a laboratory robot were provided by two of the larger commercial labs applying SW-846 methods. Factors of cost, time, space, and procedure are critical in assessing robot applications. The bench space requirements are given for a typical batch of samples. Analyst time required and cost per sample are estimated on a single sample basis. Analyst time can be reduced for particular methods on a per sample basis when analyzing samples in batches. Cost per sample will decrease somewhat for large batches of samples. It is estimated that bench

Table 9. SW-846 Methods Operating Parameters

Method	Cost per Sample ($)	Analyst Time (hr)	Total Time (hr)	Space Required (ft of bench)	Major Problems
3510	150	2–2½	3	20 for 20 samples	Volumes, shaking, venting, K-D, decanting
3520 (nonreflux)	150	2–2½ samples	48 extraction	16 for 6	Volumes, time, K-D
3540	160	1½–2	26	12 for 8 samples, including 8-ft hood	Time, space, titration, soxhlet K-D
3550	160	1½–2	2½	12 for 12 samples (minimum); acoustic hood	Decanting, centrifuging K-D
Column cleanups	90	1–2	2½	12 for 10 columns	Column size, Hg, CN, dioxin
GPC	90	2	2½	$4\frac{1}{2} \times 5$ ft^2 for auto–apparatus	None
TCLP	300 (semivolatiles) 400 (volatiles)	2	20–48	15 for 5 samples	ZHE, sample preparation
EP	125–200	4½	24–48	15 for 6 samples	Many

space required for each sample is often an order of magnitude greater than for most sample preparations currently being done by robots.

A laboratory unit operation is one of the step functions a chemist must perform in a laboratory before going to the next step or a function which provides particular information. For example, weighing a sample on a balance, conducting a sequence of surrogate additions, or performing sonication with a microprobe for 3 min can be considered single laboratory unit operations.

RTI found that waste sample preparation techniques usually require considerably more unit operations than those techniques currently being performed by robots. For example, for Method 3510, there are about 51 to 57 LUOs which must be performed. Each LUO may involve a few, typically 3 to 10 robot commands and each command will often involve 3 to 8 robot moves. Method 3510 was estimated to require a total of about 210 robot commands with about 6 to 10 seconds required to perform most commands with an existing robot system. It was therefore indicated that careful and clever integration and serialization of sample handling would have to be planned for the robot functions in order to considerably reduce the time per sample. Transport peripherals such as rotating carousels, conveyors, or multiple sampling handling fixtures should also be employed to reduce operational time. For other test methods cited as important by analytical laboratories, RTI found a requirement of about 51 LUO for Method 3550 (including the 18 Kuderna-Danish steps) and that about 31 LUOs would be required for Method 3520.

Some consensus recommendations from the waste analysis laboratories related to robotics included:

1. Sensors are needed for detection of liquid levels or interfaces in such equipment as Kuderna-Danish concentrators, sonicators, chromatographic columns, and separatory funnels. Supervision of such operations by lab technicians is one of the least cost-effective parts of protocols.
2. Mobility for automated devices, such as a track system, would be very useful in the context of the laboratory layout for such methods as SW-846, especially considering the universal problem of lab space. Two supervisors of major labs indicated that they would have wasted space if a robot worked in one place. Extraction sample preparation is a particularly congested situation.
3. Large sample volumes handled under some SW-846 protocols might be too large for existing laboratory robot systems.

Some laboratories also indicated that the EPA CERCLA, or Superfund, program requirements would be better suited to automation. It was also pointed out, however, that such an approach would *not* solve the *most* time-consuming and labor-intensive problems.

Research laboratories seemed less supportive of laboratory robotics than the commercial processing laboratories analyzing large numbers of samples. There is a widespread perception that improved quality control and assurance from utilizing a robot was as important or more important than labor savings. Again, processing labs felt this was more important since research labs usually have the benefit of more intensively trained personnel and less turnover.

CONCLUSION

Hazardous waste measurement is a new area for robot application. It has been estimated, however, that a worldwide market for up to 50,000 laboratory robots exists. Justification for such robots may be based upon:

- high-quality results—better precision than manual results, leading to higher quality products, more effective research, and faster assessments
- multiple application flexibility—ability to do multiple procedures, transfer proven procedures between laboratories, and the flexibility to reconfigure when needs change
- faster turnaround of results—more rapid availability of information leading to more timely decisions and research, quality control and regulatory approaches

- safety—capability to isolate people from hazardous environments and protect critical experiments from human contamination.

Zenie[14] discusses laboratory robot trends and justifications. He states that "It is unreasonable to ask that a lab guarantee a reduction in personnel upon introduction of a robotic system. Such a reduction should not be expected, at least during the first 6 to 12 months after the system has begun functioning. What I do guarantee is the fact that, over a period of time, the use of lab robotics will result in significant technician time savings and increased efficiency. These time savings will lead to a reduction of personnel involved in routine procedures that will allow the lab to take on additional projects."

Zymark[15] and others[42-44] provide systematic approaches to define the cost benefits of robot systems and economic analysis of their installation.

For the scientist wishing to automate a method, there are two basic routes forward. A system can be purchased completely from a manufacturer or it can be developed "inhouse."[32] In the instance where a system is purchased from a manufacturer, two further classes can be identified—systems using a flexible manipulator and devices incorporating a dedicated motion system serving a fixed operating area. "Inhouse" developments are often characterized by being over budget, overdue, and incomplete. However, inhouse development may be appropriate for facilities which operate in such confined areas as a glovebox or involve a harsh environment.

The significant demands of hazardous waste measurement may also require considerable augmentation of existing systems and even complete redesign to meet the requirements. Implementing laboratory robots is a project requiring goals, planning, resources, milestones, and leadership. Experience shows that all successful implementation projects include the following steps:

1. manual validation of automated techniques
2. system configuration and functional tests
3. method programs for single sample
4. laboratory validation of the automated method
5. method programs for multiple samples (serialized)[14]

Some laboratories surveyed by RTI were quite candid in their skepticism about the benefits of robotics. It is perceived that no good fallback is provided when a robot fails and that typically a robot would replace the low end of the personnel scale, thus reducing cost benefits of robotics. Those who were skeptical about the savings were less familiar with the documentation and testimonials on labor-and quality-associated savings from robot utilization. On the other hand, those interviewed who were utilizing robots were almost uniformly enthusiastic.

Needs recommended to be met as a result of the RTI survey on robots for waste analysis laboratories were

- higher payload capabilities than with most existing laboratory robot systems
- robot mobility for accessing the extensive bench space distribution of methods equipment such as that employed in SW-846 methods
- supervisory capability and feedback (e.g. related to position location, liquid levels, balance readings, emulsion formation, and analysis results comparisons)
- higher speed and precision
- easier software development
- better teaching techniques
- lower costs
- more support and better maintenance

Utilization of robots for hazardous waste measurement could offer much improved standardization for measurement techniques, with more uniform results from one lab to another. This requirement, the labor intensive nature of hazardous measurements, and the requirements for personnel safety mean that robotics should be gaining considerably more attention. However, this will require institutional support from those agencies responsible for hazardous waste control.

REFERENCES

1. Heylin, M. *Chem. Eng. News* 64(12):22 (1986).
2. Schmidt, G. J. and Dong, M. W. "Robots: Applications in Automated Sample Analysis," *Am. Lab.* (February 1987), pp. 62–72.
3. Franzen, K. H. "The Next Step in Laboratory Automation—Robotics," Fresenius' Z. *Anal. Chem.* 323:556–559 (1986).
4. Lochmuller, C. H., Cushman, M. R., and Lung, K. R. "Laboratory Robotics: The Advent of Soft Instrumentation," *J. Chromatog. Sci.*, 23:429–436 (1985); *Nachr. Chem. Tech. Lab.*, 33:482–492 (1985).
5. McGratten, B. J. and Macero, D. J. "Laboratory Robotics—Past Performances, Present Considerations, and Future Trends," *Am. Lab.* (September 1984), pp. 16–24.
6. Smith, H. A. and Burn, J. *J. Am. Chem. Soc.* 66:1494–1497 (1944).
7. Cleland, J. "Robotics in the Advanced Manufacturing Equation," Proceedings of the Advanced Manufacturing Technology Conference '88 (Jackson, MS: Institute for Technology Development, 1988).
8. Zenie, F. H. "Trends in Laboratory Automation-Technology and Economics," in *Advances in Laboratory Automation, Robotics* (Hopkinton, MA: Zymark Corp., 1984), pp. 1–16.
9. Dong, M. W. "The Use of the Laboratory Robot for Sample Preparation," *J. Liquid Chromatog.* 9:3063–3092 (1986).
10. Pierce, T. B. "Application of Robotic Principles to Laboratory Automation," *30th IUPAC Congress. Anal. Proc. (London)* 23:318–319 (1986).
11. Economist Intelligence Report 135, Chips in Industry, 1982.
12. Owens, G. D., Eckstein, R. J., and Franz, T. P. "Laboratory Robotics—Past, Present and Future," *Mikrochim. Acta.* 2:15–30 (1986).

13. Cleland, J. G. "Development of Roboticized Analytical Methods," 1st Interim Report for EPA Office of Solid Waste, Contract No. 68-01-7075 (October 1987).
14. Zenie, F. H., "Trends in Laboratory Automation-Technology and Economics," in *Advances in Laboratory Automation Robotics*, Vol. 4 (Hopkinton, MA: Zymark Corp., 1988).
15. Hurst, S. J. and Mortimer, J. W. *Laboratory Robotics. A Guide to Planning, Programming, and Applications.* (New York: V.C.H. Publishers 1987).
16. Davidson, W. A. and Lam, K. "Robotic HPLC Analysis of Solid Dosage Forms," in *Advances in Laboratory Automation Robotics,* Vol. 4, J. R. Strimaitis and G. L. Hawk, Eds., (Hopkinton, MA: Zymark Corp., 1988), pp. 74, 76, 85.
17. Hamilton, S. D. "An Integrated Robotic Sample Preparation and HPLC Analysis of Biosynthetic Human Insulin," in *Advances in Laboratory Automation Robotics,* Vol. 4, J. R. Strimaitis and G. L. Hawk, Eds. (Hopkinton, MA: Zymark Corp., 1988), pp. 199-203.
18. Cross, J. "Laboratory Robotics: Making Them Work," *Advances in Laboratory Automation Robotics,* Vol. 4., J. R. Strimaitis and G. L. Hawk, Eds. (Hopkinton, MA: Zymark Corp., 1988), 323, 325, 333.
19. Little, J. N. "The Role of Robotics in the Laboratory of the 80s," *J. Res. Natl. Bur. Stand.* 93(3):191 (1988).
20. Paul, A. M., Good, K. S., Tucker, K. A., and Strimaitis, J. R. "Automated Dissolution Testing Using a Laboratory Robotic System," *Am. Lab.* 21:80, 82-84, 86, 88, 90-91 (1989).
21. Zuk, W. M., et al. "Automated Preparation of Protein Crystals: Integration of Automated Visual Inspection Station," in *Advances in Laboratory Automation Robotics*, Vol. 4., J. R. Strimaitis and G. L. Hawk, Eds. (Hopkinton, MA: Zymark Corp., 1988), pp. 74, 76, 85.
22. Harrison, R. J. and Roth, L. D., Eds. "Artificial Intelligence Applications in Materials Science," in *Proceedings of Symposium* (Metallurgical Society, Inc., 1987).
23. U.S. Department of Agriculture. "Tabletop Robot Assesses Nutrients in Food," NTIS Tech Notes (May, 1987).
24. Brown, R. K., Zenie, F. H., and Johnson, G. "Robotics in Chemical Laboratories," *Chem. Br.,* 22, (1986).
25. Salit, M. L., and Cirillo, M. G. "The Decision to Buy a Laboratory Robot," *Am. Lab.* 18 (1986).
26. Lebec, A. "Automatic Microbiological Testing of Cosmetics Using Robotics," in *Advances in Laboratory Automation Robotics,* Vol. 4., J. R. Strimaitis and G. L. Hawk, Eds. (Hopkinton, MA: Zymark Corp., 1988), p. 194.
27. Bergman, R. K., Lofgreen, L. R., Hendrickson, K. A., and Koecher, D. R. "An Automated Analytical Laboratory," in *Advances in Laboratory Automation Robotics,* Vol. 4, J. R. Strimaitis and G. L. Hawk, Eds. (Hopkinton, MA: Zymark Corp., 1988), p. 276.
28. Millier, A. "Automation of the Quality Control of Vitamin A," in *Advances in Laboratory Automation Robotics*, Vol. 4, R. Strimaitis and G. L. Hawk, Eds. (Hopkinton, MA: Zymark Corp., 1988), p. 158.
29. Oestreich, G. J. "Automation of a Gas Chromatographic Procedure for the Determination of Glycols in Polyester Polymers with a Laboratory Robot,"*J. Chromatog. Sci.* 25:214-218 (1987).

30. Fouda, H. G., and Schneider, R. P. "Robotics for the Bioanalytical Laboratory: Analysis for Antipyrine in Human Serum," *Am. Lab.* 20:116–121 (1988).
31. Yoshida, T., Ito, Y., Handa, M., Kasai, O., and Yamaguchi, H. "An Automated System for the Simultaneous Determination of Several Ingredients in Pharmaceutical Preparations Using HPLC and a Laboratory Robot," in *Advances in Laboratory Automation Robotics,* Vol. 4., J. R. Strimaitis and G. L. Hawk, Eds. (Hopkinton, MA: Zymark Corp., 1988), pp. 127, 133.
32. Huddleston, J. "The Use of Robots for Automation in the Radiochemical Laboratory," *Anal. Proc. (London)* 25:178–181 (1988).
33. Schmidt, W. A., Rollheiser, J. J., Campbell, D. P., and Stelting, K. M. "Capping and Other Opportunities for Custom Automation Robotics," in *Advances in Laboratory Automation Robotics,* J. R. Strimaitis and G. L. Hawk, Eds. (Hopkinton, MA: Zymark Corp., 1988), pp. 439–440.
34. Jones, R. D., Pinnick, H. R., Cross, J. B. "Making Laboratory Robots Smart," in *Advances in Laboratory Automation Robotics,* Vol. 4., J. R. Strimaitis and G. L. Hawk, Eds. (Hopkinton, MA: Zymark Corp., 1988), pp. 402–415.
35. Martin, A. and Tung, J. "Using Lotus 123 to Control and Collect Data From Your Lab Robotic System," in *Advances in Laboratory Automation Robotics,* Vol. 4., J. R. Strimaitis and G. L. Hawk, Eds. (Hopkinton, MA: Zymark Corporation, 1988), p. 535.
36. Bartilson, B. M. et al. "Automated Maintenance in Nuclear Power Plants," EPRI Final Report NP-3779 (November 1984).
37. Beugelsdijk, T. and Phelan, P. "Controlling Robots with Spoken Commands," in *Advances in Laboratory Automation Robotics,* Vol. 5 (Hopkinton, MA: Zymark Corp., 1988), p. 439–440.
38. Haile, D. M. and Brown, R. D. "Robotic Preparation of Water Samples for Trace Herbicide Analysis," in *Advances in Laboratory Automation Robotics 1985,"* G. L. Hawk, and J. R. Strimaitis, Eds. (Hopkinton, MA: Zymark Corp., 1985) p. 123.
39. Kramer, G. W. and Fuchs, P. L. "Robotic Automation of Some Common Organic Laboratory Techniques," in *Advances in Laboratory Automation Robotics,* Vol. 4., J. R. Strimaitis and G. L. Hawk, Eds. (Hopkinton, MA: Zymark Corp., 1988) pp. 347, 350, 355.
40. Nekimken, H. L., Smithm, B. F., Peterson, E. J., Hollen, R. M., Erkkila, T. H., and Beugelsdijk, T. J. "The Development of Robotics for the Determination of Actnide and Lanthanide Distribution Ratios," in *Advances in Laboratory Automation Robotics,* Vol. 4., J. R. Strimaitis and G. L. Hawk, Eds. (Hopkinton, MA: Zymark Corporation, 1988).
41. Ward, K. B., Perozzo, M. A., and Zuk, W. M. "Automatic Preparation of Protein Crystals using Laboratory Robotics and Automated Visual Inspection," *J. Crystal Growth* 90:325–339 (1988).
42. Jenkins, J., Smith, P., and A. Raedels. "The Financial Evaluation of Robotics Installations," in *Safety and Business. Robotics 3,* (New York: Elsevier North-Holland Inc., 1987), pp. 213–219.
43. Gainer, F. E. "Robotics in the Analytical Laboratory—A Management Perspective," *Advances in Laboratory Automation Robotics 1985,* Vol. 2, J. M. Stimaitis and G. L. Hawk, Eds. (Hopkinton, MA: Zymark Corp. 1985), p. 1.
44. Abundus, M. "Making Robots Count," *Robotics Eng.* (December 1986), pp. 17–20.

BIBLIOGRAPHY

Barette, R. C., Labrecque, J. M. "Microwave Acid Pressure Dissolution of Mineral Ores Using Robotics," in *Advances in Laboratory Automation Robotics,* Vol. 4, J. R. Strimaitis and G. L. Hawk, Eds. Zymark Corporation, Hopkinton, MA, 1988, p. 525.

Beals, P. C., Wildman, W., and Timmoney, P. "Trace Level Analysis of Herbicides Performed on a New Robotic Sample Preparation System," in *Advances in Laboratory Automation Robotics,* Vol. 4, J. R. Strimaitis and G. L. Hawk, Eds., Zymark Corporation, Hopkinton, MA, 1988, p. 44.

Benedicenti, C. and Beretta, M. A. "Computers and Robots at the Service of Chemical Researchers: Data Processing and Automation in the Laboratory," *Chim. Oggi.*: 27-9 (1986).

Beugelsdijk, T. and Phelan, P. "Controlling Robots with Spoken Commands," in *Advances in Laboratory Automation Robotics,* Vol. 4, J. R. Strimaitis and G. L. Hawk, Eds., Zymark Corporation, Hopkinton, MA, 1988, p. 446.

Beugelsdijk, T. J. and Knobeloch, D. W. "Laboratory Robotics in Radiation Environments," *J. Res. Natl. Bur. Stand.* 93:268-269 (1988).

Brodack, J. W., Kilbourn, M. R., and Welch, M. J. "Automated Production of Several Positron-Emitting Radopharmaceuticals Using a Single Laboratory Robot," *Appl. Radiat. Isot.* 30:689-698 (1988).

Brown, R. K., Zenie, F. H., and Johnston, G. "Robotics in Chemical Laboratories," *Chem. Br.* 22:640-642 (1986).

Brunet, J. M. "First National Symposium of Laboratory Robotics. Some Concrete Cases of Robotization," *Spectra 2000 [Deux Mille]* 127:51-53 (1988).

Denton, M. B. "Concepts for Improved Automated Laboratory Productivity." Interim Report, Arizona University, Tucson (May 1987).

Fisher, J. J., Ward, C. R., and Schuler, T. F. "Development of a Telerobotic System for Handling Contaminated Process Equipment," International Topical Meeting on Remote Systems and Robotics in Hostile Environments, Pasco, WA, March 29, 1987.

Godfrey, O., Raas, A., and Landis, P. "The Application of a Robotic Work Station to the Handling of Microbial Colonies," in *Advances in Laboratory Automation Robotics,* Vol. 4, J. R. Strimaitis and G. L. Hawk, Eds., Zymark Corporation, Hopkinton, MA, 1988, p. 165.

Herndon, J. N., Babcock, S. M., Butler, P. L., Costello, H. M., and Glassell, R.L. "Laboratory Telerobotic Manipulator Program," NASA Conference on Space Telerobotics, Pasadena, CA, January 31, 1989.

Herndon, J. N., and Hamel, W. R. "Telerobotic Technology for Nuclear and Space Applications," American Institute of Astronautics and Aeronautics, NASA, and U.S. Air Force Symposium on Automation, Robotics and Advanced Computing, Arlington, VA, March 9, 1987.

Hurst, W. J., Kuhn, K. C., and Martin, R., Jr. "Comparison of First Generation and Pytechnology Robotic Systems for Laboratory Operations," in *Advances in Laboratory Automation Robotics,* Vol. 4, Zymark Corporation, Hopkinton, MA, 1988, pp. 361-369.

Hurst, W. J., McKim, J. M., and Martin, R. A. "Integration of Laboratory Robotics with an Automated HPLC Sample-Injection System," *Liquid Chromatog.* 5:49- 50 (1987).

Jones, R. D., Cross, J. B., and Pinnick, H. R., Jr. "Making Laboratory Robots More Productive by Expanding Their Effective Work Envelope," in *Advances in Laboratory Automation Robotics,* Vol. 4, J. R. Strimaitis and G. L. Hawk, Eds., Zymark Corporation, Hopkinton, MA, 1988, p. 545.

Katori, N., Aoyagi, N., and Takeda, Y. "Use of Robot for Automated Content Uniformity Tests of Ajmaline Tablets," *Bunseki Kagaku.* 37:633–635 (1988).

Koel, M. and Kaljurand, M. "Robots in Analytical Laboratory," *Zh. Anal. Khim.* 42:947–951 (1987).

Kropscott, B. E., Coyne, L. B., Dunlap, R. R., and Langvardt, P. W. "Alternative Task Performance in Robotics," *Am. Lab.* 19:70, 72–75 (1987).

Kuban, D. P. and Hamel, W. R. "Application of a Traction-Drive Seven-Degrees- of-Freedom Telerobot to Space Manipulation," American Astronautical Society Annual Guidance and Control Conference, Keystone, CO, January 31, 1987.

Li, C., Potucek, J., and Edelstein, H. "Automation of Drugs of Abuse Testing with Fisher's MAXX 5™ Robot & GC/MS Analysis," in *Advances in Laboratory Automation Robotics,* Vol. 4, J. R. Strimaitis and G. L. Hawk, Eds., Zymark Corporation, Hopkinton, MA, 1988, p. 31.

Little, J. N. "Precision in Laboratory Robotic-HPLC Systems," *J. Liq. Chromatog.* 9:3033–3062 (1986).

Lochmueller, C. H. "The Role of the Robot in the Chemical Laboratory," *J. Res. Natl. Bur. Stand.* 93:267–268 (1988).

Lochmuller, C. H., Lloyd, T. L., Lung, K. R., and Kaljurand, M. "Evaluation of the Use of a Laboratory Robot in a Study of Esterification Kinetics," *Anal. Letters,* Marcel Dekker, Inc., New York, (1987).

Lochmuller, C. H., and Lung, K. R. "Applications of Laboratory Robotics in Spectrophotometric Sample Preparation and Experimental Optimization," *Anal. Chim. Acta* 183:257–262 (1986).

Lorenz, L. L., Sochon, H. R., Zeiss, J. C., and Buell, S. L. "An Automated Station for the Collection and Analysis of Metered Dose Aerosol Products," in *Advances in Laboratory Automation Robotics,* Vol. 4, J. R. Strimaitis and G. L. Hawk, Eds., Zymark Corporation, Hopkinton, MA, 1988, p. 146.

Matsuda, R., Ishibashi, M., and Uchiyama, M. "Simplex Optimization of Reaction Condition Using Laboratory Robotic System," *Yakugaku Zasshi* 107:683–689 (1987).

McCarthy, J. P. "Automation of USP Dissolution Apparatus I, the Basket Method," in *Advances in Laboratory Automation Robotics,* Vol. 4, J. R. Strimaitis and G. L. Hawk, Eds. Zymark Corporation, Hopkinton, MA, 1988, p. 61.

McKinney, W. J. "Automation and Data Handling in the Residue Analysis Laboratory," *Pestic. Sci. Biotechnol. Proc. Int. Congr. Pestic. Chem.* 6:317–324 (1987).

Meacham, S. A. "Oak Ridge National Laboratory's Robotics and Intelligent Systems Program," Roane-Anderson Economic Council Meeting, Oak Ridge, TN, January 23, 1987.

Nekimken, H. L. Smith, B. F., Peterson, E. J., Hollen, R. M., and Erkkila, T. H. "Development of Robotics for the Determination of Actinide and Lanthanide Distribution Ratios," International Symposium on Laboratory Robotics, Hopkinton, MA, October 18, 1987. Los Alamos National Labl., NM.

NTIS Tech Note, "Robot to Transfer Wafers in New Integrated Circuit Facility," NTIS, (October 1987).

NTIS Tech Note, "Tabletop Robot Assesses Nutrients in Food," Dept. of Agriculture, Washington, D.C. (May 1987).

Pesheck, C. V., O'Neill, K. J., and Osborne, D. W. "Automation of Surfactant Phase Behavior Determination by Laboratory Robotics," *J. Colloid Interface Sci.* 119:289–290 (1987).

Roberts, K. M. "Robots in the Analytical Laboratory for Sample Preparation," Chemsa:149–151, 153 (1987).

Sample, R. M. "Tablet Dissolution Testing with a Laboratory Robot," *Anal. Proc. (London)* 23:266–267 (1986).

Schlieper, W. A., Isenhour, T. L., and Marshall, J. C. "A Flexible Laboratory Instrument Control Language," *J. Chem. Inf. Comput. Sci.* 27:137–143 (1987).

Sommer, D., Koch, K. H., and Grunnenberg, D. "Laboratory Robot System in Oil Analysis," *GIT Fachz. Lab.* 32:1075–1076, 1079–1080 (1988).

Stewart, J. A., Schuler, T. F., and Ward, C. R. "Demonstration of a Remotely Operated TRU Waste Size-Reduction and Material Handling Process," ERD/A/052001, Department of Energy (November 1986).

Thompson, C. M., Sebesta, A., and Ehmann, W. D. "A Robotic Sample Changer for a Radiochemistry Laboratory," *J. Radioanal. Nucl. Chem.* 124:449–455 (1988).

Wass, J. C., and Packham, N. J. C. "Laboratory Robotics for Gaseous Sample Collection and Injection," *Liquid Chromatg.* 6:420, 422–423 (1988).

Weisbin, C. R., Barhen, J., and Hamel, W. R. "Research and Development Program Plan for the Center for Engineering Systems Advanced Research (CESAR)," ORNL-TM-10232, Department of Energy Report (January 1987).

Zeigler, B. P., Cellier, F. E., and Rozenblit, J. S. "Design of a Simulation Environment for Laboratory Management by Robot Organizations," Fourth Conference on Artificial Intelligence for Space Applications, (October 1988) pp. 313–321.

Zenie, F. H. "Trends in Laboratory Automation-1987," in *Advances in Laboratory Automation Robotics,* Vol. 4, J. R. Strimaitis and G. L. Hawk, Eds., Zymark Corporation, Hopkinton, MA, 1988, pp. 388–389, 391, 394–395.

CHAPTER 8

The Application of Infrared Spectroscopy to Hazardous Wastes

Douglas S. Kendall

INTRODUCTION

The analysis of hazardous wastes is important for many reasons, including enforcement efforts, hazard assessment, and remedial actions. While the EPA and other agencies have a larger number of standard methods for the analysis of hazardous waste, these methods are far from being the complete answer for every situation. Measures which can reduce the time and cost of the analysis of complex and perhaps completely unknown samples are worthwhile. There is a need for methods which can determine analytes missed by the usual techniques. Present methods usually focus on target compounds. Techniques which can help characterize an entire sample and identify all major components would be an improvement. A technique which can contribute significantly to the analysis of hazardous wastes is infrared spectroscopy, which is one of the most widely applicable techniques for chemical analysis and compound identification (Kendall, 1966). Among the many advantages of infrared analysis are the often minimal sample preparations and the ability to qualitatively identify a wide variety of both organic and inorganic species. The application in this laboratory of infrared spectroscopy, in addition to the more standard techniques, to the analysis of hundreds of hazardous waste samples has greatly increased the quality of the analytical results and the number of components identified. The purpose of this chapter is to show how IR can be applied to the analysis of hazardous wastes.

Despite its widespread use in industry, infrared spectroscopy is little used for environmental analyses because it lacks sensitivity compared to mass spectrometry. However, in many ways the analysis of hazardous wastes is more like industrial chemical analysis than environmental analysis. When analyzing drum samples, the major components or those above about 1% are almost always of most importance. Many samples of contaminated groundwater and waste ponds have significant concentrations of contaminants. For these types of samples infrared spectroscopy has more than enough sensitivity. Furthermore, continual progress is being made in increasing the sensitivity of IR spectrometers, particularly Fourier transform infrared spectrometers (FTIR), and the associated sampling devices, such as GC interfaces.

Infrared spectroscopy can identify many compounds which will not gas chromatograph and so are missed by GC/MS. This is important because most of the organic content of hazardous wastes will not chromatograph. For numerous reasons, including toxicological assessment and disposal considerations, it is often necessary to identify the major components of a waste. Even if compound identification is not possible, infrared has a role to play in classifying materials and in functional group analysis. Such classification can serve to screen samples and to more efficiently direct subsequent analyses. A field laboratory at a remediation site or at a treatment and disposal site may need to quickly screen a material before further action is taken. An enforcement laboratory may screen samples in order to select only those techniques for further work which will yield useful information. Because of its simple sample preparation and broad applicability, IR has much to offer in this regard.

Infrared spectrometry has the ability to identify many materials which are abundant in hazardous wastes, but which are not by themselves highly toxic. The identification of these materials is, however, very important in many cases. In assessing the health hazards of a particular site or situation, it is desirable to know everything that is present in significant amounts. If the identified components only add up to 1 or 10%, doubt remains as to the effects of the remaining material. An enforcement laboratory can use an unusual compound to trace a waste to its source or to tie a particular generator to a contamination incident. The toxic compounds may be relatively widespread and hard to trace, but an accompanying marker compound may be much easier to trace even if by itself it is harmless.

The infrared spectrum of polystyrene is shown in Figure 1. This spectrum illustrates the richness of data available in an infrared spectrum. Much information is contained in the absorption band positions, intensities, and shapes. An infrared spectrum can be considered as a collection of a large number of physical constants for a compound. By matching the spectrum of an unknown material to the spectrum of a reference material, a very positive identification can be made.

This chapter is divided into a number of sections. A very brief exposition of infrared fundamentals is provided. A section on sample preparation follows which outlines basic techniques and provides some suggestions for hazardous waste samples. Sections on instrumentation and GC/IR are followed by a section on interpretation. This is followed by a number of examples of the use of infrared spectroscopy for the analysis of hazardous wastes. This section concentrates on applications other than GC/IR and attempt to illustrate the wide range of applicability of IR spectroscopy.

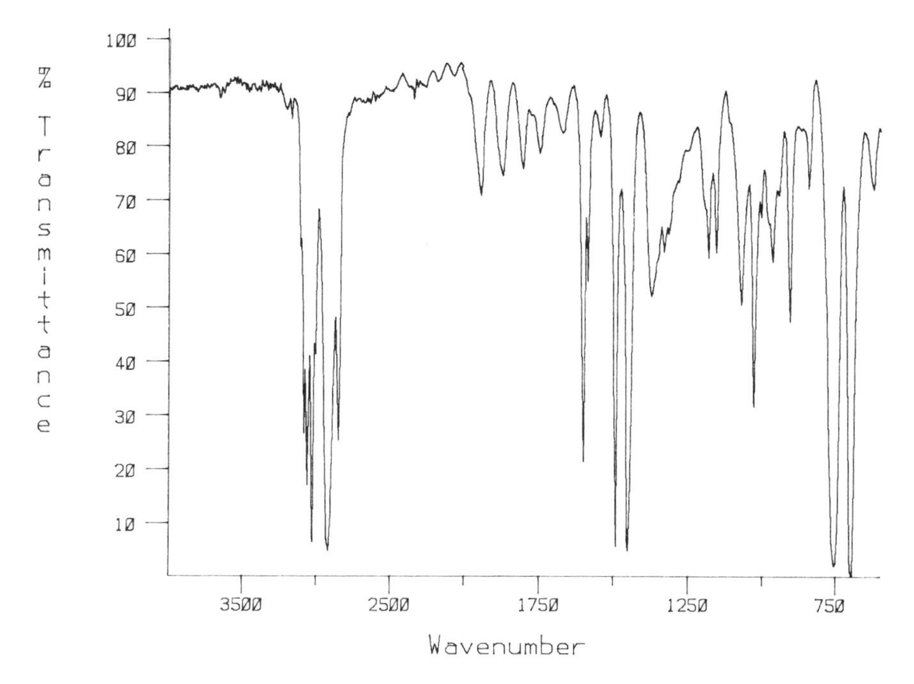

Figure 1. The infrared spectrum of polystyrene, illustrating the richness of data in an infrared spectrum. Much information is contained in the band positions, intensities, and shapes.

INFRARED FUNDAMENTALS

A basic understanding of the interactions between infrared radiation and matter provides a basis for understanding spectrometer operation, selecting and optimizing sample preparation techniques, efficiently and properly using different sampling accessories, and interpreting spectra in the most informative manner. While this is not the place for a thorough treatment of the theory of infrared spectroscopy, such treatments are available and are well worth study (for example, Harris and Bertolucci, 1978). A brief treatment of infrared fundamentals will be provided as a review and a stimulus to further study.

The midinfrared region is the most useful portion of the infrared spectrum for compound identification and functional group classification. This region includes the wavelengths from about 2 to 25 μm or from 4000 to 400 cm^{-1} (wavenumbers, which are proportional to frequency). Wavenumbers are now used more commonly, and the conversion is easily made by dividing a number in one type of units into 10,000 to obtain the value in the other type of units.

The primary and most important interaction is between IR radiation and the vibrations of covalent bonds. Absorption of infrared radiation occurs when a molecular vibration causes a change in the dipole moment of the molecule. Thus, the stretching vibrations of diatomic oxygen or nitrogen do not result in infrared absorptions. Similarly, the symmetric stretching of the polar carbon-chlorine bonds in a symmetric molecule such as carbon tetrachloride do not lead to infrared absorption, since there is no overall change in the molecular dipole moment (of course, asymmetric vibrations of carbon tetrachloride do lead to absorptions). The stretching vibrations of highly polar bonds, such as carbon oxygen bonds, lead to strong absorptions. Most molecular vibrations change the dipole moment and are IR active.

Most molecular vibrations can be divided into two types: stretching vibrations and bending vibrations. In the former, bond lengths change and in the latter, bond angles change. A useful relationship for stretching vibrations states that vibration frequency ($\bar{\nu}$ in cm^{-1}) is proportional to the square root of the bond force constant (f) and inversely proportional to the square root of the reduced mass [(μ = ml · m2/(ml + m2)] of the vibrating atoms:

$$\bar{v} \sim \sqrt{\frac{f}{\mu}}$$

This equation explains why the stretching vibrations of OH, NH, and CH, which have a small reduced mass, absorb at high wavenumbers, and why carbon oxygen double bonds, which are stronger, absorb at higher frequencies than carbon oxygen single bonds.

While the preceeding considerations may seem relatively straightforward for simple molecules, the situation for large molecules might appear to be very complex. An N atom molecule has 3N-6 (3N-5 if linear) vibrations. Fortunately, this large number of vibrations falls into manageable patterns. Although many molecular vibrations are characteristic of the entire molecule, a significant fraction are characteristic of a particular substructure. Vibrations which are characteristic of a small atomic grouping, and which remain reasonably constant from molecule to molecule, give IR spectroscopy its great utility as a functional group identifier. Some absorption bands give functional group information while others are characteristic of an entire molecule and serve to fingerprint the molecule.

The method of identifying functional groups from their characteristic patterns has been thoroughly developed and is the empirical approach to spectral interpretation. The infrared spectra of thousands of molecules have been examined and compared. Many correlations between structure and spectra have been made. The results of these studies have been tabulated in group frequency tables. Further information on such tables and their use is included in the section on interpretation.

SAMPLING

Proper sample preparation is the most important aspect in obtaining high-quality infrared spectra, since current spectrometers are highly automated and easy to use. Good sample preparation starts with selecting the proper method and the proper sample cell or optical materials. A representative portion of the sample must be obtained and put in the proper form for the chosen technique. Good sample preparation is both an art and a science, and patience and skill are required. This section presents some of the basics of sampling. The references should be consulted for more information.

The fewer the number of components, the easier the interpretation. Thus, it may be worthwhile to at least partially separate a sample by solubility, filtration, etc. Computer-aided interpretation is much more successful for pure compounds than for mixtures.

In order to acquire good spectra, one must be able to recognize them and know how to deal with problems. A good spectrum has a baseline above about 85% transmission and the baseline is flat. The strongest absorption band should be close to 10% transmission. Noise should be less than about 1% transmission and should not obscure small absorption bands. In particular, the absorptions of water vapor and carbon dioxide should be unobstrusive. The absence of solvents which were used in sample preparation and which were then evaporated should be confirmed by looking for their strongest absorption bands. A sloping baseline, particularly with KBr pellets, often indicates insufficient particle size reduction and additional grinding is recommended. Usually simple adjustments of concentration or path length suffice to produce satisfactory spectra. It must be emphasized, however, that extra effort in sample preparation is well rewarded in terms of spectral quality and ease of interpretation.

The most straightforward way to obtain spectra from liquids is as a drop between salt plates. Two flat crystals of sodium chloride, potassium bromide, or other infrared transparent materials are used to form a capillary film of the sample. A device which holds the salt plates, permits pressure to be applied in varying amounts, and helps achieve a uniform distribution of sample is used to mount the plates in the spectrometer. By varying the amount of sample and the pressure applied to the plates, the proper sample thickness can be achieved. If sufficient pressure is applied, materials such as viscous liquids and greases can be sampled.

For solids the classic techniques are the KBr pellet and the mineral oil mull. Both require that the sample have dimensions less than 2 μm, the shortest wavelength used in midinfrared spectroscopy, in order to minimize light scattering. A sloping baseline indicates insufficient grinding. For preparing mulls, a sample is mixed with mineral oil and ground with a mortar and pestle (a flat pestle and a flat piece of glass are best) to form a paste which is spread between salt plates. The hydrocarbon absorptions of the mineral oil are, of course, present in the spectrum. A second mull prepared with Fluorolube or other fluorinated oil allows scanning of those spectral regions obscured by the mineral oil. Mulls require a measure of experience and skill, but have been widely used. Spectra in the Aldrich collection (Pouchert, 1985) have been prepared this way.

KBr pellets offer a way to easily prepare solids for infrared examination. For routine use, pellets can be made with the small dies to which pressure can be applied with wrenches. About 1 mg of sample per 100 mg of KBr is used for a pellet 8 mm in diameter. The pellet dies usually can be easily cleaned with water and quickly dried in an oven. A ball mill or similar device is used to grind and mix the sample KBr mixture.

A most useful technique, especially for difficult samples, is based on total internal reflection and is commonly called ATR (attenuated total reflectance). If a sample can't be ground to make a KBr pellet or can't be pressed into a capillary film between salt plates, then ATR is the technique to try. It is especially valuable for very viscous liquids and soft solids. Although, strictly speaking, ATR spectra are not transmission spectra, for qualitative purposes transmission reference spectra are fine. The basis of the technique is total internal reflection and the concept of the critical angle. If an infrared beam or any light beam is moving from a medium of higher refractive index to one of lower refractive index (such as air), the beam will not enter the second medium if the angle of incidence is greater than the critical angle. Instead, it will be totally reflected and will remain in the initial medium. As shown in Figure 2, after entering the ATR crystal, the infrared beam is reflected a number of times before exiting the crystal. These internal reflections are total because the ATR crystal has a much higher index of refraction than air. As long as the angle of incidence is greater than the critical angle, the internal reflection is total and the beam intensity is not reduced by reflection. The usefulness of this

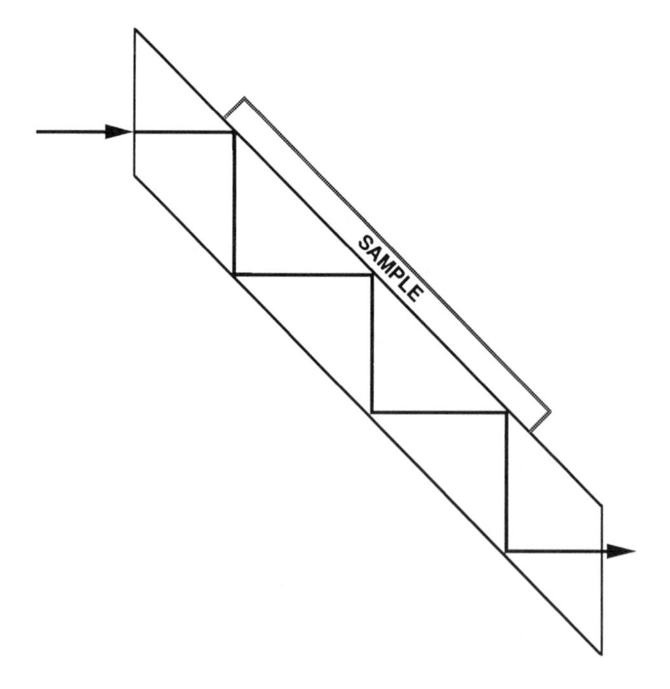

Figure 2. The path of the infrared beam through a typical ATR crystal. The sample is placed on the outside of the crystal.

arrangement for obtaining spectra follows from the fact that internal reflection takes place at the interface of the crystal and air. The sample is placed on the outside of the crystal, while the infrared beam is inside the ATR crystal. The important point is that at wavelengths where the sample has an absorption band, the reflection is attenuated or, in other words, the reflection is less than total. This attenuation is recorded as a function of wavelength or frequency by the spectrometer. The resulting spectrum is for most purposes the same as a transmission spectrum. It differs slightly with respect to relative intensities.

As mentioned before, the ATR crystal must have a rather high index of refraction. Typical materials are germanium, zinc selenide, and KRS-5 (thallous bromide iodide). Although these materials are somewhat expensive and require careful handling and cleaning, the ATR technique is not difficult to use. ATR accessories are available for most spectrometers. Samples such as a piece of rubber or other soft solid are clamped to the outside of the crystal. The intensity of the spectrum can be varied by changing the coverage on the ATR crystal and the applied pressure. Greases and similar materials will often stick on their own accord to the ATR crystal. Optical arrangements in which the ATR crystal is horizontal are available and these can be used for liquids and powders. ATR equipment in which the crystal is placed so that liquid samples can surround much of it are also available. These cells are particularly useful for aqueous solutions. The absorptions of water are at the correct intensity to allow subtraction of the water spectrum. The spectrum of analytes with a concentration of less than 1% can be observed with these cells. Water is such a strong infrared absorber that transmission measurements of aqueous solutions are difficult and require a very short path length.

Fourier transform spectrometers are particularly advantageous for ATR because of their high energy throughout. The usual ATR accessory allows less energy to reach the detector than when the spectrometer is used for transmission spectra, but with the FTIR instruments, this is not a problem. Another FTIR advantage of the usual ATR arrangement follows from the single beam nature of most FTIR instruments. The background spectrum is scanned with the clean ATR crystal, the sample applied, and then the sample is scanned. When the transmission or absorption spectrum is calculated from the ratio of the sample to background spectra, a flat, smooth baseline results.

A procedure useful for scanning a complex sample, such as a paint waste, involves the following four spectra. A hazardous waste treatment and disposal facility, or a field laboratory at a Superfund or cleanup site could use such a procedure to quickly characterize a waste. The liquid sample is first scanned and received as a capillary film between salt plates. The plates are then separated and the volatile components evaporated, perhaps in an oven. A spectrum of the dried sample is then obtained. The difference between the as-received spectrum and the spectrum of the nonvolatiles provides information on the volatile components. The difference spectrum can be computer-generated or quite effectively estimated by visual comparison. In order to study the nonvolatile components, a portion of the sample is dried. For a paint sample or similar material, the soluble resins or polymers are extracted with toluene, MEK, or a similar solvent (best if done with boiling solvent). After filtering and perhaps concentrating by boiling, a film is cast on a salt plate by evaporating the solvent from the extract (the solvent is evaporated in an oven), and a spectrum of the nonvolatile, soluble components obtained. The insoluble material from the filter is dried and then scanned as a KBr pellet or by ATR techniques. This last spectrum contains information on pigments, fillers, and similar materials. A useful reference contains spectra of solvents, resins, pigments, and other common industrial compounds (Anderson et al., 1980). Modifications of this procedure for other sample types are easily made.

The objectives of the study and the available equipment and time usually determine the effort involved with sampling. The well-equipped enforcement laboratory will attempt to separate mixtures and try different sampling techniques. A laboratory at a waste facility or a Superfund site will adopt a sampling method of wide applicability and ease of use. The techniques discussed here are a good starting point. Other techniques, such as diffuse reflectance and photoacoustic spectroscopy, can be profitably explored by the experienced spectroscopist. An extensive literature is available for assistance (for example, Miller and Stace, 1972), and is reviewed annually in *Analytical Chemistry*.

INSTRUMENTATION

A well-maintained and knowledgeably operated infrared spectrometer is important if quality spectra are to be obtained. While this is not the place to describe the components and operation of infrared spectrometers, an understanding of IR instrumentation is desirable. A minimum requirement is that the analyst be able to verify that the spectrometer is operating properly and is correctly configured for the type of sample being scanned.

Modern infrared spectrometers are wonderful instruments which are easy to use and require only minimal maintenance. It is important that correct operation be periodically verified. Many types of laboratories, including enforcement, private contract, and in-house testing laboratories, must have records indicating that instruments were operating correctly when important measurements were made. Successful participation in legal hearings, government inspections, etc. often require such records. Often it is sufficient to scan a film of polystyrene, widely available from instrument vendors, and verify that the observed band maxima are close (usually 1 to 3 cm^{-1}) to the standard positions. In any case, standard operating procedures, which should be in written form, should include checks of proper operation.

Two distinctly different types of spectrometers are now in use. Traditional dispersive instruments use a grating to disperse the infrared radiation from the source. The dispersed radiation is then sequentially directed toward the sample and then on to the detector. In contrast, Fourier transform infrared spectrometers direct all wavelengths of radiation from the source at the sample. An interferometer imposes an interference pattern on the IR radiation. The interferogram produced by the detector can be transformed by a computer into a conventional spectrum (Griffiths, 1986).

For many of the IR applications described in this chapter, either type of spectrometer will do fine. A laboratory purchasing a spectrometer should look at Fourier transform instruments most closely. However, there are many good dispersive instruments still in use, and if one is available, it can serve quite well. The two advantages of Fourier transform instruments are a faster scan speed and a higher energy throughput. Fast scans are of most use in GC/IR work. Complete scans can be done in less than 1 sec, fast enough to do several scans while a peak elutes from a capillary column. Some sampling accessories, such as ATR equipment, do not transmit a large fraction of the incident radiation, and the high energy throughput of FTIR instruments results in a better signal-to-noise ratio.

GC/IR

One of the more intensely studied applications of infrared spectrometry to hazardous waste analysis is gas chromatography/Fourier transform infrared spectrometry (GC/FTIR). Most of the studies have been done with lightpipe systems. The GC effluent is passed through a gold-coated capillary glass cell and the IR spectrum is obtained on the fly. Lightpipes are available for capillary GC columns. Spectra of good quality can be obtained, and the EPA has a collection of reference vapor phase spectra which many vendors have formatted for their instruments. Vapor phase spectra are slightly different from condensed phase spectra and it is best to use reference spectra obtained from the vapor phase. This is not absolutely necessary if one is cognizant of the possible differences. For instance, the absence of hydrogen bonding in the gas phase results in sharper and weaker OH stretching bands. Lightpipe GC/FTIR is much less sensitive then GC/MS. While 10 to 20 ng of a compound may give spectra with several peaks, good quality spectra require 100 ng or more.

The author believes that an important application of GC/FTIR to hazardous wastes will be to the identification of major and minor solvents in wastes. Solvents are common in hazardous wastes and should be identified. For instance, an enforcement laboratory must often identify the solvents in ignitable wastes. Sensitivity is not a problem for GC/FTIR when looking at concentrations over 1%. Lightpipe GC/FTIR equipment should be much more trouble free than GC/MS equipment. With wide-bore capillary columns, many wastes can be injected with little or no dilution. The GC effluent can be directed to an FID after the lightpipe for quantification or the FTIR signal can be used for quantification. These considerations would seem to warrant serious consideration of GC/FTIR for solvent identification in hazardous wastes.

Hewlett Packard now sells an FTIR spectrometer configured as a GC detector which seems well suited to this application.

The published GC/FTIR hazardous waste studies have concentrated on the analyses now done by GC/MS. In many ways, these applications of GC/IR compete with GC/MS. Given the greater sensitivity of GC/MS and the large investment of the time and money in GC/MS equipment, it is hard to envision a change in the predominance of GC/MS for environmental analyses. Infrared's forte in environmental analysis may be rapid screening and the analysis of nonchromatographable materials. Nevertheless, GC/FTIR has much to offer as a complement and an extension to GC/MS. For compounds such as aromatic isomers, IR is more selective than MS. Gurka has offered a protocol for semivolatile organic compound determination by GC/FTIR (Gurka, 1985). The sensitivity and applicability to environmental samples of GC/FTIR continues to improve (Gurka and Pyle, 1988).

By combining gas chromatography, infrared spectroscopy, and mass spectrometry in either a serial or parallel fashion, advantage can be taken of the complementary nature of MS and IR (Wilkins, 1987; Shafer, 1984). The information from both spectroscopies can be combined to yield both additional identifications and more positive identifications. This will be important as organic environmental analyses move beyond the present emphasis on priority pollutants or similar lists of target compounds. As the emphasis shifts to the complete characterization of complex samples such as contaminated groundwaters, the benefits of complementing MS with IR should become apparent.

A fundamentally different method of combining GC with IR is matrix isolation GC/FTIR (Reedy et al., 1979). Analytes in the GC effluent are trapped in a frozen matrix of argon. The sampling surface is a rotating cylinder so that the trapped analytes are spread out in order of retention time. The spectrum is scanned after the GC run by reflecting the IR beam off the gold-coated sampling surface. The absorption bands are especially sharp since molecular rotation is inhibited and the sample molecules are isolated by the solid argon matrix. A matrix isolation system is about one hundred times more sensitive than a lightpipe system. A commercial model is marketed by Mattson Instruments and can be configured with a mass spectrometer. The specificity of IR for aromatic isomers and the sensitivity of matrix solution GC/FTIR have been used to determine dioxin isomers (Wurrey et al., 1986). Since MS has trouble determining dioxin isomers, this use of matrix isolation GC/FTIR is a very promising approach to an important environmental analysis. Studies of complex hazardous waste samples should benefit from the application of this powerful GC/FTIR technique.

INTERPRETATION

Once acceptable spectra are in hand, there is still much work to do in order to assure that the maximum amount of information is obtained from the available data. Although progress has been made in computer-assisted interpretation, the expertise of a skilled analyst is required if the spectra are to yield all their information.

An often overlooked, but most essential, factor in spectral interpretation is the gathering and use of all available information about the sample. If possible, the source of the sample and as much information about its origin, uses, possible components, and physical and chemical characteristics should be available. This knowledge can be used to limit the possible interpretations and the reference spectra that are searched. In most cases hazardous wastes are composed of widely used industrial chemicals. This greatly reduces the number of possible compounds and suggests using references which emphasize industrial compounds. When interpreting a spectrum, the physical state of the sample, the method of sample preparation, and any solvents used during preparation should be kept in mind. There is little point in using reference spectra for solvents when interpreting the spectrum of a solid film.

Once the sample history is in hand, the spectrum should be examined for the presence of functional groups. This should be done using one of the group frequency correlation tables. These are widely available in texts and other books; only a small selection will be cited here (Bellamy, 1975; Colthup et al., 1975; Nakanishi and Solomon, 1977; Lambert et al., 1976; Silverstein et al., 1981). The strongest absorption bands should be examined first. It may not be possible to uniquely assign an absorption, so the likely possibilities should be listed. It is convenient to divide the midinfrared region into two parts. The functional group region is at higher frequencies (wavenumbers) and the fingerprint region is at lower frequencies or higher wavelengths. The dividing line is at about 1350 cm^{-1}. In the functional group region, most of the bands will be characteristic group frequencies and all of the strong bands in this region should be examined. A possible procedure is to start at the highest wavenumber absorption band and proceed downward in an orderly fashion. The OH and NH stretching region is first examined for the presence of saturated or unsaturated CH. The region between 2700 and 1800 cm^{-1} is usually free of absorption bands. Any bands in this region are unusual and their origin should be sought. Double bands occupy the region between about 1430 and 1800 cm^{-1}. Carbonyl groups and aromatic rings have especially characteristic absorptions and the spectrum should always be checked for these groups. Strong bands in the fingerprint region may be characteristic group frequencies and should be checked.

It is not necessary or desirable to interpret every band in a spectrum or to unduly limit the possibilities. It is much better to retain a number of options and not over interpret the spectrum. The absence of a band can provide useful information and the use of negative information should not be neglected. Many groups have several characteristic absorptions and all of these should be checked before turning to reference spectra. For instance, if an OH stretching absorption leads one to suspect that an alcohol is present, then the spectrum should be checked for single-bonded carbon oxygen stretching vibrations.

When the examination of the spectrum for functional groups is complete, a list of possible classes of compounds should be prepared. The analyst is then ready to compare the unknown with reference spectra. The analyst should proceed to look through the spectra of compounds which best fit the interpretations. Usually as the comparison continues, additional ideas are generated and checked. Matching a spectrum with an authentic reference spectrum is the only way identification can positively be made. A useful exercise is to scan some pure compounds and then compare the spectra with published reference spectra. This will show how closely the spectra of a given compound will match when scanned in different laboratories. In favorable circumstances, a reference spectrum will be found that matches the unknown spectrum. If this does not happen, then there are several possibilities to consider. The unknown may not be in the reference library, and the analyst must be content with identifying functional groups. Since almost all hazardous wastes are large volume industrial chemicals, it is likely that the constituents of the waste will be in a library. For a waste sample, it is more likely that the failure to find a match is due to the presence of a mixture. While mixtures can be complex and hard to interpret, it is surprising how often only two or three components predominate in hazardous waste mixtures. With experience, knowledge of the sample history and some diligence, it is possible to recognize the spectrum of a major component in a mixture. The spectrum of the first compound identified can be subtracted from the mixture spectrum, either by computer or by mental subtraction. The remaining bands can then be identified. If necessary, separation procedures such as those outlined in the sample preparation section can be tried. For some types of samples, such as polymers, it will only be possible to identify the type of compound present with infrared. Likewise, infrared provides little molecular weight information for such samples as hydrocarbons and polymers.

There are programs available which will search a spectral library. Aldrich has an IBM PC program available which will search their FTIR library (Pouchert, 1985). Sadtler also has an extensive library of computerized spectra with a search program. Many instrument vendors have search programs. These search programs work well for pure compounds which are in the database, but often have trouble with mixtures. They can be useful but should not be considered the final word on interpretation.

An important exercise for anyone wishing to become proficient at interpreting infrared spectra is to look at a wide variety of spectra, perhaps in one of the many available reference collections, and to observe how group frequency correlations are shown by real spectra. In this spirit, a number of spectra which show commonly observed features are presented with some commentary on important absorptions. The wavenumbers mentioned are from Bellamy (1975) and Pouchert (1985).

The spectrum of mineral oil in Figure 3 shows the common features of alkane spectra. CH stretching frequencies of the methyl group are at 2962 and 2872 cm^{-1}, while those of the methylene group are at 2926 and 2853 cm^{-1}. Both

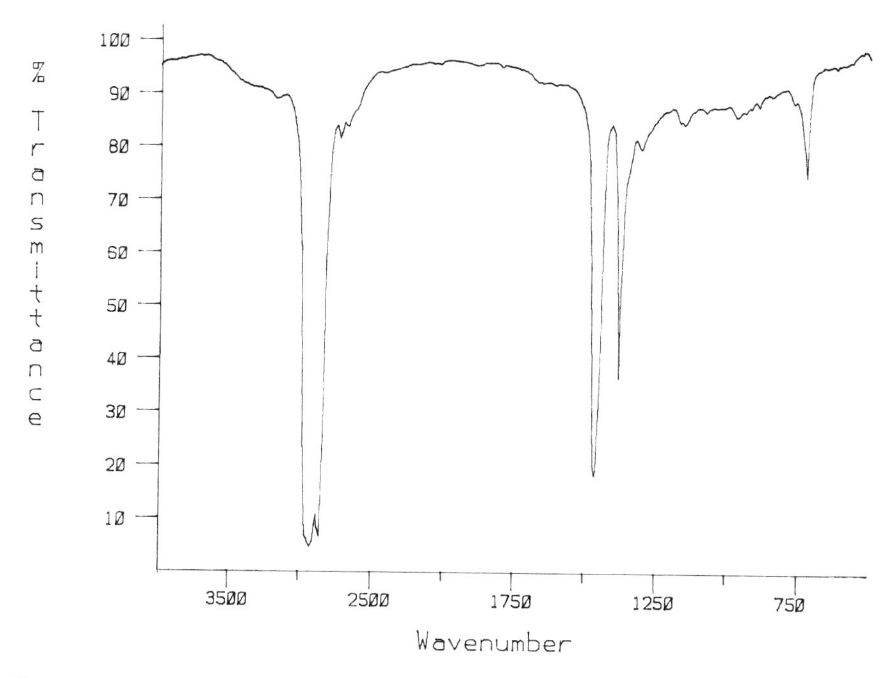

Figure 3. The infrared spectrum of mineral oil.

groups have bending vibrations near 1460 cm^{-1} and the methyl group has another bending vibration near 1375 cm^{-1}. Particular CH configurations have characteristic group frequencies. One that is useful is the absorption near 720 cm^{-1} arising from four or more methylene groups in a straight chain, and this is seen in Figure 3. A spectrum very similar to this is produced by all sorts of alkanes and these features are seen in the spectra of compounds which contain saturated hydrocarbon groups. All of the stretching bands may not be resolved and the bending vibrations may be overlapped with absorptions from other sources. Waste oil is often seen at hazardous waste sites. An infrared spectrum will quickly identify the oil. No information is available as to molecular weight ranges or particular compounds, but seldom is this necessary. In contract, a GC/MS run might identify many individual compounds and perhaps lead to the need to clean the instrument. Usually, it is sufficient to know that oil or saturated hydrocarbons are present, and the infrared analysis is a great savings in time and effort. Percent levels of most of the common chlorinated solvents can be observed in the spectrum of waste oil, so that infrared examination can serve to check waste oils for the presence of waste solvents.

Figure 4 is a spectrum of meta-xylene and has features common in the spectra of aromatic compounds. The stretching vibrations of the aromatic CHs are at frequencies higher than 3000 cm^{-1}, while the aliphatic CHs absorb at less than 3000 cm^{-1}. Stretching vibrations of the aromatic carbon carbon bonds, the ring breathing modes, are observed at 1614 and 1492 cm^{-1}. Sharp

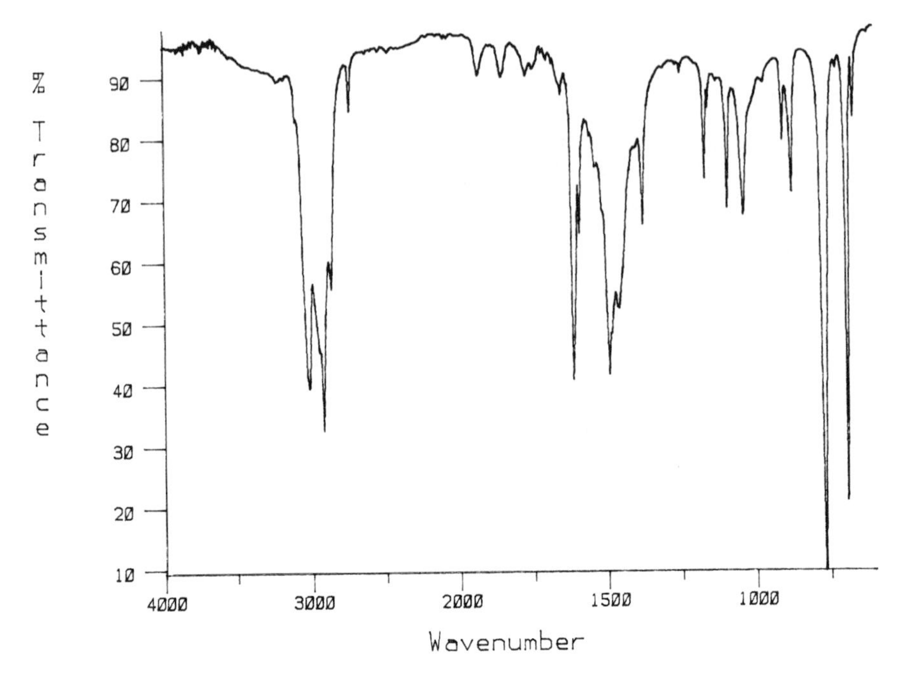

Figure 4. The spectrum of meta-xylene.

bands such as these near 1500 and 1600 wavenumbers are highly characteristic of aromatic compounds. The intensities can vary widely and one of the bands may be absent, but it is almost always possible to recognize the presence of aromatic compounds from bands in this region. The two strongest absorption bands in m-xylene are at 691 and 768 cm^{-1}. These are out-of-plane bending vibrations of the hydrogen atoms on the aromatic ring. Bands like these are also characteristic of aromatic compounds and the type of ring substitution can be ascertained from their position. The patterns for the different types of substitution can be found in group frequency charts. The patterns are usually independent of the type of substituent on the ring and depend only on the number and arrangement of ring hydrogens.

The spectrum of isoamyl acetate is shown in Figure 5. The 1743 cm^{-1} absorption is due to the stretching vibration of the carbonyl group. The carbonyl group is perhaps the group most easily recognized by IR. The carbonyl absorption is intense and occurs in a part of the spectrum with few interferences. The exact frequency of the carbonyl absorption offers valuable information about the type of carbonyl compound present. Saturated esters absorb near 1740 cm^{-1}, aromatic esters near 1730 cm^{-1}, and ketones absorb near 1710 cm^{-1}. These are probably the most common carbonyls in hazardous wastes. However, careful studies, summarized in correlation charts, have shown that each type of carbonyl group has its own characteristic pattern. The physical state, method of sample preparation, and the presence of other compounds will

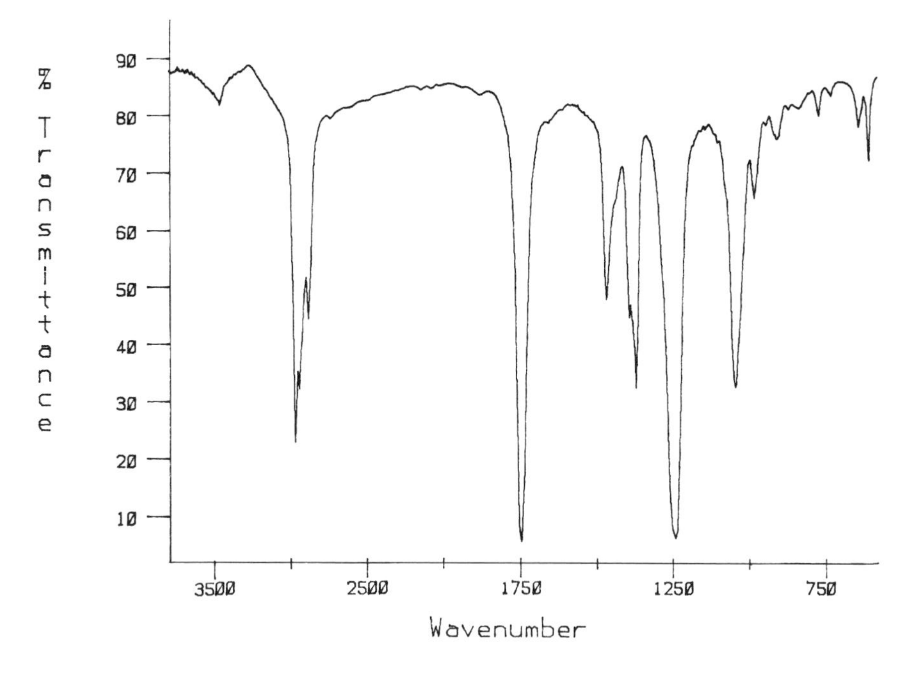

Figure 5. The spectrum of amyl acetate (mixture of isomers).

effect the carbonyl band position, as well as those of other bands. Nevertheless, these effects are usually small and the type of carbonyl can be deduced. In addition to the position of the carbonyl stretching band, the presence of other absorptions allow conclusions about the type of carbonyl. Esters, of course, will also have a carbon oxygen single bond stretching absorption. The position of this band conveys information about the type of ester that is present. Acetate esters, common in industry and in wastes, have such a band and show a strong absorption near 1240 cm^{-1}, which is very characteristic.

The environmental impact of chlorinated solvents is an important concern, and chlorinated solvents have distinct IR spectra. The IR spectrum of tetrachloroethylene is shown in Figure 6. The high symmetry of the molecule results in only a few strong infrared absorptions. The intense absorption bands involve carbon chlorine stretching vibrations. Figure 7 shows the spectrum of trichloroethylene. The spectrum of trichloroethylene is more complex than that of tetrachloroethylene because the former is less symmetric and contains a hydrogen atom. The CH stretching vibration is at 3083 cm^{-1}; such bands suggest unsaturation. At 1586 cm^{-1}, an absorption due to the stretching of the carbon carbon double bond is present. The intense bands involve carbon chlorine stretching vibrations or an out-of-plane hydrogen bending vibration.

Figure 8 shows the spectrum of ethylphenol as a neat liquid. The broad band near 3400 cm^{-1} is the stretching vibration of a hydrogen-bonded OH group. Both aromatic and aliphatic CH groups are present, above and below 3000

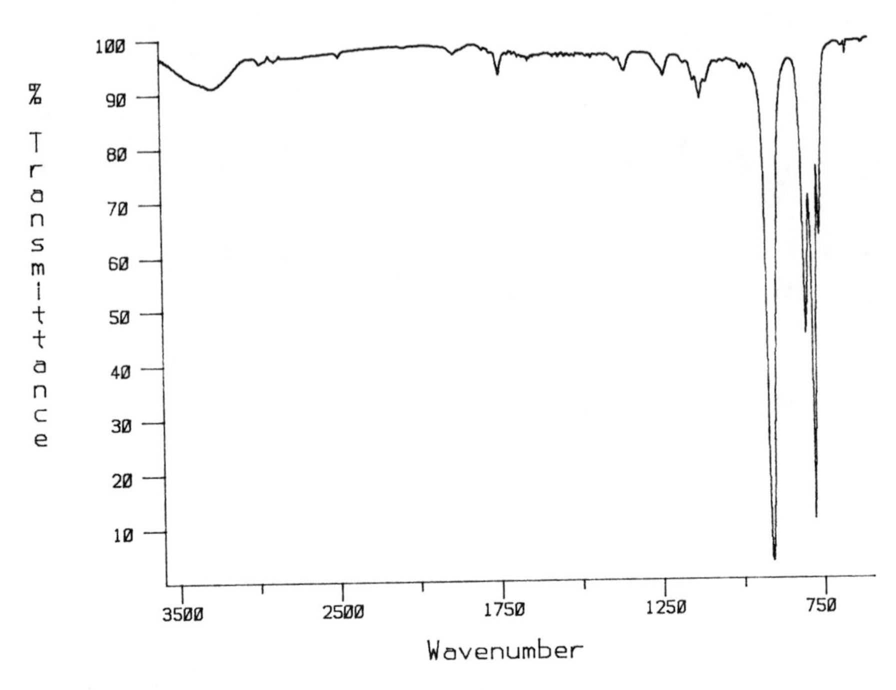

Figure 6. The spectrum of tetrachloroethylene.

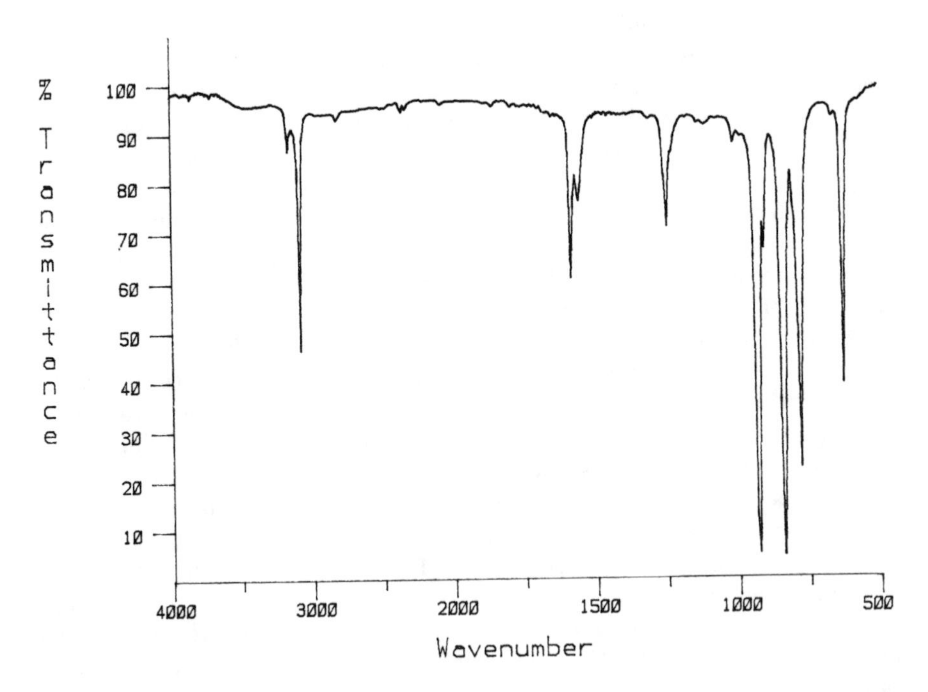

Figure 7. The spectrum of trichloroethylene.

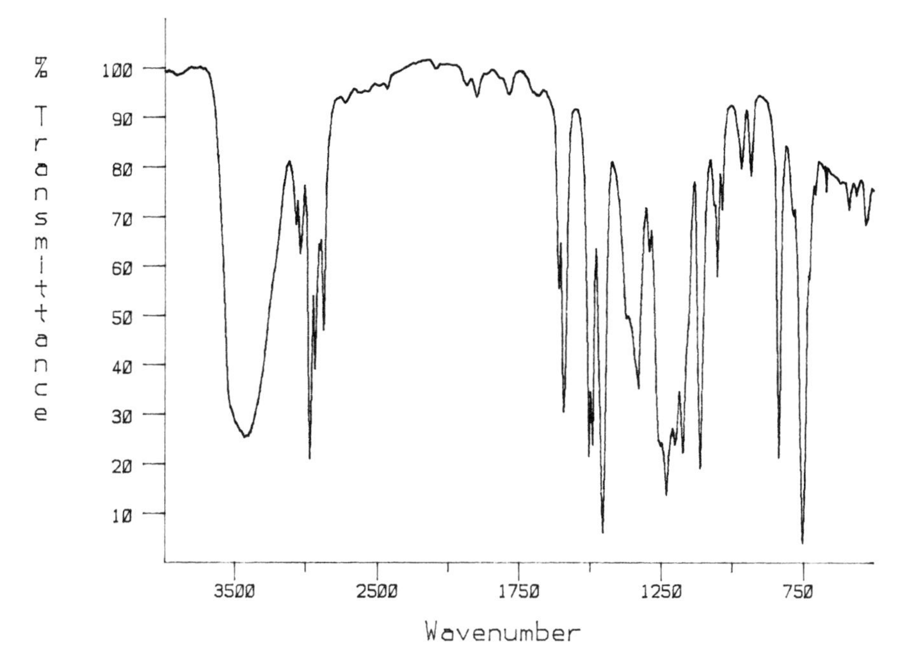

Figure 8. The spectrum of ethylphenol.

cm^{-1}, respectively. The sharp bands near 1500 and 1600 cm^{-1} indicate the presence of an aromatic ring. The 752 cm^{-1} band is due to the out-of-plane bending vibrations of the hydrogens on the aromatic ring. The carbon oxygen stretching band is a strong absorption near 1200 cm^{-1} in phenols and the spectrum of ethylphenol has such an absorption.

Inorganic compounds which contain covalent bonds have characteristic infrared spectra. Many anions, such as sulfate and carbonate, can be identified from their infrared spectra. In favorable cases it is possible to identify the cation from its influence on the anion spectrum. Study of some reference spectra will indicate the utility of the IR spectra of inorganic compounds (Miller and Wilkins, 1952; Nyquist and Kagel, 1971). Our enforcement laboratory has found IR to be useful for identifying paint pigments, such as chromates, and fillers for polymers, such as mica and diatomaceous earth. Also, significant amounts of cyanides can be observed by IR, including cyanide complexes which are insoluble or do not distill. Figure 9 is a spectrum of a KBr pellet of potassium ferrocyanide. The CN stretching band at 2043 cm^{-1} occurs in a region of the spectrum with relatively few absorption bands. Such bands in waste samples are almost always cyanides. The position of the CN stretching band and the band at 587 cm^{-1} can serve to identify this particular cyanide. The absorption at 1621 cm^{-1} is from water, including waters of hydration and water absorbed by the KBr pellet.

Infrared spectroscopy is widely used to identify polymers. Although most

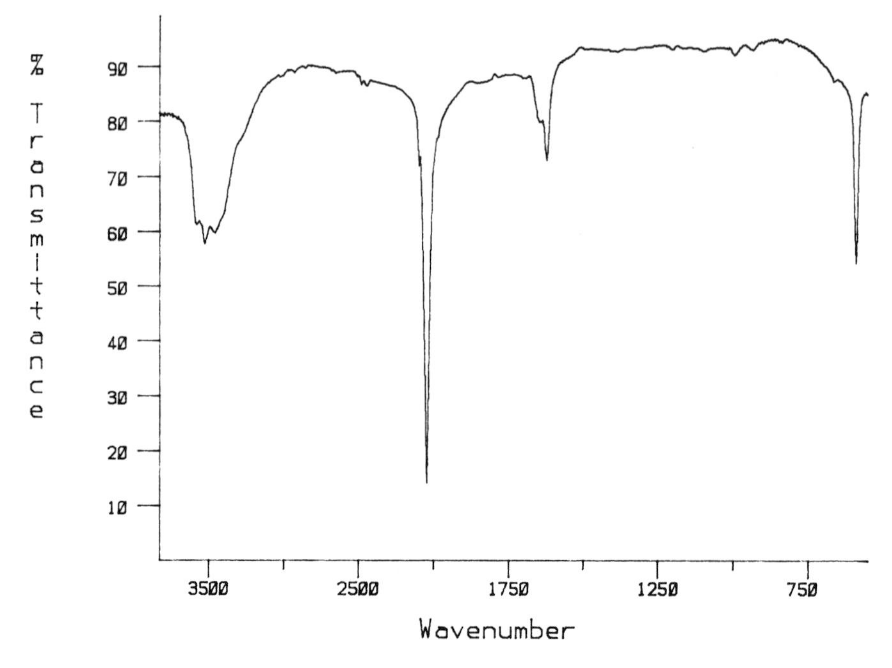

Figure 9. The infrared spectrum of potassium ferrocyanide, $KFe(CN)_6 \cdot 3H_2O$.

polymers are not hazardous, they are often found in the presence of hazardous compounds. By identifying the polymer, the analysts can answer questions about other components in a sample and perhaps identify the source of a waste more definitively. Figure 10 shows the spectrum of an o-phthalate-based alkyd resin. In addition to the o-phthalate ester moieties, these polymers usually contain a glycol and a long-chain oil. The bands just above and just below 3000 cm^{-1} show that both aromatic and aliphatic CH are present. The small, broad band near 3500 cm^{-1} is from residual OH groups. The strong carbonyl band near 1735 cm^{-1} shows the presence of an aromatic ester. The sharp bands near 1600 and 1500 cm^{-1} are from the aromatic ring. The doublet at 1580 and 1600 cm^{-1} is particularly characteristic of o-phthalate esters. The carbon oxygen single bond stretching vibration of the aromatic ester has an intense absorption near 1270 cm^{-1}. The band at 1125 cm^{-1} is from the stretching of the carbon oxygen single bonds in the glycols. The bands at 1710 and 1740 cm^{-1} are due to bending vibrations of the ring hydrogens and are characteristic of ortho ring substitution. This spectrum is very similar to that of almost all o-phthalate alkyd resins, even though the glycol or oil that is used differs from one resin to another. With experience an analyst learns to recognize many such common patterns. A spectrum such as this contains much information in addition to the position and intensity of band maxima. Most computer interpretation and search programs use only these two types of information and thus lose much of the information contained in a spectrum. Most current computer programs

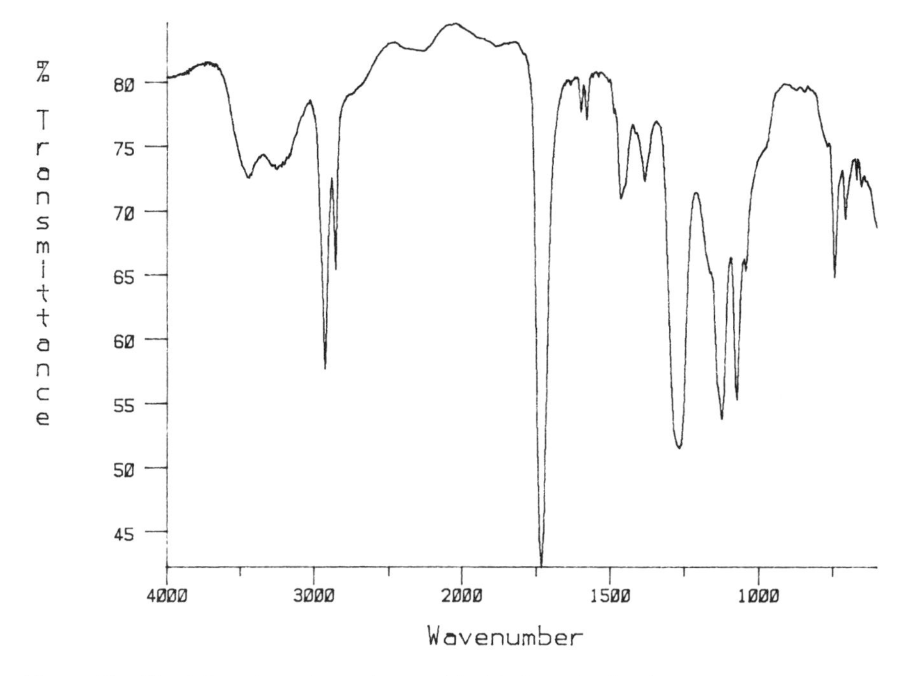

Figure 10. The infrared spectrum of an o-phthalate-based polyester.

have trouble with mixtures, while an experienced spectroscopist can recognize patterns when interpreting the spectra of mixtures.

Considerable effort has been put into designing computer programs for the interpretation of spectra. Most of these involve searching a data-base of reference spectra. Many work well for pure compounds, such as found in GC/IR work, and not so well for mixtures. These programs are available from vendors for those instruments which come with computers. Further information on such programs can be sought in the reviews published in *Analytical Chemistry*. A different approach is that of an expert system in which a set of interpretation rules are applied by a computer to spectral data. A program called PAIRS has been developed as an expert system for the interpretation of infrared spectra (Woodruff and Smith, 1980). The program includes interpretation rules, which reportedly can be easily modified and extended by the user, and a way of systematically applying the rules to the spectrum of an unknown. This is the most promising approach at present to computer interpretation of the spectra of mixtures and of compounds not present in a database.

Levine and co-workers have applied the PAIRS program and its logic to hazardous waste samples (Ying et al., 1987). They are addressing an important problem, which is the rapid classification and identification of wastes at remedial action and similar sites. The work takes advantage of the fact that only a relatively small number of industrial compounds are abundant at hazardous waste sites. A computer program was developed which can look at a training set of reference spectra and automatically develop a set of rules. The PAIRS program of Woodruff's then applies the rules to the interpretation of the spectrum of an unknown. Promising results have been obtained for mixtures.

APPLICATIONS AND EXAMPLES

The remainder of this chapter presents examples of the application of infrared spectroscopy to the analysis of hazardous wastes. A variety of applications are described in order to show the many possible uses of IR spectroscopy. The applications of IR to hazardous wastes will be divided into two groups. First is the screening of samples; second is the identification of components best determined by infrared.

SCREENING

Screening samples by IR is an integral and key step in the composition analysis of hazardous wastes by our laboratory. Rather than look only for a selected set of compounds, the approach is to determine the components of a waste, generally those present at the percent level. In almost all of the samples we have examined, the nature of the hazard was determined by the components present in amounts greater than 1%. When determining the major components, it is most efficient to rapidly screen the sample and then use the results of the screening to guide further analyses. Screening in our laboratory includes not only infrared analysis, but also energy-dispersive X-ray fluorescence analysis and a few specific spot tests. These results are then used to direct further analysis, if necessary. The screening eliminates unneeded work by eliminating some samples and some tests from further consideration. Screening by IR will first be outlined and then the examples which follow will illustrate the approach.

Usually, the screening and subsequent analyses are done on the separated phases of the sample, after which the results are combined to give a view of the entire sample. After phase separation, a portion of each solid phase is dried. The percent weight loss upon drying is measured and is valuable for making semiquantitative concentration estimates. If grindable, the dried samples are scanned as KBr pellets; if not, ATR techniques can often be used. All liquid phases have their water content measured by an automated Karl Fischer titration. Phases low in water are scanned as a drop between potassium bromide plates. If water-miscible phases have less than about 85% water, the samples are scanned between silver chloride plates or by ATR. It is often possible to subtract the spectrum of water and identify one or two major components. Liquid phases are dried at 80°C in a vacuum oven. By recording the weight loss on drying, the amount of volatile and nonvolatile material can be calculated. These quantities are very useful for ascertaining whether all of the major components of a sample have been determined, and if not, where the fault might lie. The nonvolatile portion is scanned separately. It is sometimes possible to scan a sample between salt plates as received, to dry the sample by separating the plates, and then to scan the nonvolatile portion. The spectrum of the volatile portion can then be obtained by subtraction.

The screening spectra from each of the separated phases are then interpreted using published reference spectra, especially those of commercial products, as a guide (Anderson et al., 1980). Although some spectra are, of course, complex, a surprising number of spectra are readily interpretable. Often there are only one, two, or three major components in a given phase and the experienced analyst can readily sort them out. Even if specific compounds cannot be identified, the identification of functional groups or compound types, such as phenolic compounds, can be very valuable. The presence of minor components, and the only occasional presence of a pure compound, make the use of computer-aided library searching rather unsuccessful, at least in our experience. The results from the interpretation of the screening spectra as well as the other screening results are then used to guide further analyses.

Using IR to screen samples can almost always reduce the amount of subsequent work. For example, if a sample is a saturated hydrocarbon, with perhaps some aromatic hydrocarbons, it is almost always sufficient to know that the sample is oil or other petroleum product and there is no need to identify the many components by GC/MS. If the sample contains a low percentge (1 to 5%) of a halogenated solvent, for example, this can be observed in the infrared spectrum. If the volatile components of a sample are toluene and the xylene isomers, this can be recognized in the infrared spectrum. GC/MS would not be necessary, and GC could be used if quantification were necessary. On the other hand, if a mixture of acetate esters and ketones is observed, GC with perhaps MS confirmation would be necessary.

INFRARED ANALYSES

Infrared has many uses during the analysis of hazardous wastes after the screening is complete, and this will be illustrated in a wide range of examples, but first a few generalizations. Infrared spectroscopy can be used to perform many analyses which are done by gas chromatography and mass spectroscopy. This is particularly true of Fourier transform IR when coupled with GC, and this application has been discussed earlier. Rather, the intention here is to highlight and illustrate applications which complement and extend the commonly used methods of hazardous waste analyses.

In several years experience at NEIC (Blackman et al., 1984) much less than half the organic content of hazardous waste samples can be gas chromatographed and thus identified by GC/MS. While often the nonchromatographable components are not the most hazardous, it is important to identify all major components if hazard is to be assessed and if futile efforts to identify additional compounds by GC/MS are to be avoided. The application of IR to the nonvolatile organic portion of hazardous waste samples is the best way to identify these components. The remainder of this paper presents examples of the analysis of selected hazardous waste samples encountered in this laboratory.

RAPID ANALYSES

The versatility and speed of IR analysis is well illustrated by a case in which several drums were found in a neighborhood concerned about purported high cancer rates. The three drums analyzed each contained a clear liquid and viscous, opaque, colored liquid as a separate phase. An automated Karl-Fischer titration quickly established that the clear liquid was water. IR scans of the viscous liquids, both as received and dried (dried on the KBr plates), showed the presence of styrene and an unsaturated polyester resin. The styrene was identified in the difference spectrum obtained by subtracting the spectrum of the dried material from the spectrum of the original material. This quick analysis was sufficient to allow disposal of the waste and assess the hazard.

A similar case illustrates some of the same virtues of IR. Several drums were found on a farm, the result of "midnight" dumping. A quick analysis was needed to guide removal and safety measures. An IR spectrum showed that the major component was a chlorinated aldehyde. After some searching of reference spectra, chloral (trichloroacetaldehyde), used as an intermediate in pesticide production, was identified. Quick identification was important since chloral reacts vigorously with water.

INORGANICS

Since almost all species with covalent bonds absorb in the midinfrared region, IR has much to offer for inorganic analysis. Infrared spectrometry can aid in inorganic compound speciation and is most useful for identifying inorganic moieties which contain polyatomic ions, such as chromate or sulfate. Infrared can easily identify some species such as carbonate and insoluble cyanides often missed by other methods. The usual analytical approach is to perform an elemental analysis, sometimes targeting priority pollutants and at other times the major elements. Often elemental analysis will not sufficiently characterize a waste. For instance, it may be necessary to trace a waste to its source or to accurately assess its potential dangers, and for this compound identification is valuable. Elemental composition may be enough to infer which compounds are present, but to establish identity, X-ray diffraction or infrared spectroscopy are necessary, with IR better suited to routine use.

An example is a waste recovered from a drum found to contain chromium.

The chromium pigment zinc yellow was identified in the infrared spectra, as was the mineral talc. The latter would have been impossible to identify from elemental analytical results which found magnesium and silicon as well as many other elements. Talc and zinc yellow are consistent with a paint waste, which was much better characterized as a result of the infrared analysis.

There are several specific types of samples where IR is especially useful. There are several anions which are not usually identified by ion chromatography or other routine methods which IR can identify fairly easily. One is carbonate, which is quite common as calcium carbonate in such wastes as those from paint or polymer formulations. A second is insoluble forms of cyanide, which are often missed by spot tests, and are present, for example, in plating bath and pigment wastes. For example, in a number of samples, a ferriferrocyanide pigment has been identified by IR after being missed by spot tests.

A common type of speciation by IR involves silicate minerals. An elemental analysis which finds silicon and other common elements is not very useful for speciation, but by identifying the mineral, the waste source may be identifiable. Using infrared it is possible to identify mica, talc, diatomaceous earth, quartz, etc. which are commonly used ingredients in industrial formulations.

NONCHROMATOGRAPHABLE ORGANICS

Well over half of the organic waste found in drums is not gas chromatographable and thus will not be identified by the usual GC/MS procedures. Furthermore, many of the organic compounds which find their way into groundwater are very polar or ionic at neutral pH and are not gas chromatographable. Infrared spectrometry has proven useful for such identifications and has the potential to do more in the future.

The contents of several drums found dumped in a stream in a small mountain town illustrate the utility of IR for nonchromatographable organics. One drum contained over 80% ethylene glycol. When the sample was analyzed by the standard GC/MS approach, the ethylene glycol was not detected. Because it is so polar, the chromatography showed only a broad hump which was not further investigated. Despite the presence of water, the IR spectrum easily revealed the presence of ethylene glycol. This points up another advantage of IR. The GC/MS data, which aren't expected to include everything, can't be readily evaluated in terms of the total mass fraction of the sample which has been determined. In contrast, IR allows the estimation of major components since the entire sample can be scanned as received, or at least with minimal modification.

This is illustrated by a second drum from the same site. GC/MS analysis found about 10% of the sample to be several solvents, with the remainder unaccounted for. IR easily revealed that the bulk of the sample was a polyalkoxylated phenol surfactant. IR established the identity of this component and allowed estimation of its approximate concentration.

A major use of infrared has been to identify polymers found in waste drums. A third drum found at the above site contained 10% water, about 1% of several metals, and the rest a copolymer of ethylene and propylene glycols. IR was the only way readily available to analyze this sample. Similarly, paint wastes are quite common and the only way to identify the organic resin present is by IR. Often this is an o-phthalate-based polyester, and GC/MS will only identify traces of several o-phthalate esters. Polymers are often major components of a waste stream and their identification is important for establishing the type of waste, even if they are not highly toxic. It is often important to positively identify all major components, and if polymers are present, IR must be used.

Recently, our laboratory has investigated several cases in which hazardous wastes from a factory were dumped into a sewer. Solvents were found in the sewer, but since common solvents such as toluene are widely used, the presence of solvents in the sewer did not tie the waste to a particular facility. However, a resin was found in the sewer, and a good infrared spectrum of the resin was obtained. Analysis of several waste drums from the factory yielded an identical resin spectrum. This was sufficient to tie the waste to a particular generator. In addition, the matching spectra were made into an easily understood court room exhibit.

As a final example of a nonchromatographable organic, the analysis of contaminated groundwater emanating from a dump site will be described. The total amount of organic carbon in monitoring wells just off the site was well over 0.1%, but only about 10% of the carbon was accounted for by summing the results of GC/MS analysis. In order to attempt to identify the remainder, a portion of the contaminated groundwater was evaporated to dryness and then extracted with ethanol. An IR spectrum of the extract suggested the presence of an aromatic sulfonate with little or no alkyl content. By comparison with the spectra of several possibilities, it was determined that parachlorobenzene sulfonate was present as a major component. This was confirmed by dissolving the dried ethanol extract in acid and extracting with methylene chloride. After evaporating the methylene chloride and then scanning the residue, a good match with the spectrum of the acid form was obtained. The parachlorobenzene sulfonic acid, as determined by liquid chromatography, accounted for slightly over 50% of the organic carbon in the contaminated groundwater. While the isolation technique can not be considered generally applicable to polar organics in water, this example does point out that IR can be used to identify compounds which will not be found by the usual GC/MS approach. In the future, it may be that liquid chromatography can be interfaced to infrared instruments in order to more fully characterize polar organics in highly contaminated groundwater.

COMPLEMENTING GC/MS

It is not the purpose of this section to show how IR can be used to duplicate the results of GC/MS analysis, but rather to show that there are many situations in which IR analysis can complement GC/MS analyses. As is often the case in difficult analyses, a second technique for confirming initial results is very beneficial. Surprisingly, frequently in hazardous waste drum samples, one or two compounds predominate, and IR can be used to confirm and check the identification and quantification.

IR has the attribute of looking at the entire sample, often allowing deductions about the original sample form. For example, consider an incident in which a contract laboratory found about 70% pentachlorophenol in a sample. A metals analysis showed the presence of a major amount of sodium, and this was difficult to reconcile with the original result. An IR spectrum of a KBr pellet prepared from the original sample revealed that a phenolate ion, not a phenol, was present. Comparison with published spectra showed that pentachlorophenolate was present, undoubtably as the sodium salt. The solid sample for mass spectrometry had first been extracted into an acidic aqueous solution and then into a nonpolar organic solvent, which was used for the GC/MS analysis. IR was able to examine the original sample, while the GC/MS analysis had examined a preparation in which the phenolate ion had been protonated.

A second example of IR complementing GC/MS analysis involves a sample containing major amounts of carbaryl. Another laboratory found about 25% carbaryl and 12% 2-naphthol by GC/MS analysis of a methanol extract. IR analysis showed carbaryl to be present at greater than 80% and little naphthol present. Further study showed that, in methanol. carbaryl would decompose to 1-naphthol under the conditions used for the GC/MS analysis. The naphthol isomer thus produced is 1-naphthol, not 2-naphthol, a difference hard to discern by MS. The quantitative result found by GC/MS was low because of sample decomposition. While a more experienced laboratory might not have made this mistake, a laboratory analyzing a variety of waste samples by a set scheme might easily do so.

In sum, infrared spectroscopy is a valuable adjunct to the usual techniques of hazardous waste analysis, especially if complete analyses must be done in an efficient manner.

REFERENCES

1. Anderson, D. G., Duffer, J. K., Julian, J. M., Scott, R. W., Sutliff, T. M., Vaickus, M. J., and Vandeberg, J. T. "An Infrared Spectroscopy Atlas for the Coatings Industry," Federation of Societies for Coatings Technology, Philadelphia, 1980.
2. Bellamy, L. J. *The Infrared Spectra of Complex Molecules*, Chapman and Hall, London, 1975.

3. Blackman, W. C., Garnas, R. L., Preston, J. E., and Swibas, C. M. "Chemical Composition of Drum Samples from Hazardous Waste Sites," in *The Proceedings of the Fifth National Conference on Management of Uncontrolled Hazardous Waste Sites,* Hazardous Materials Control Research Institute, Silver Springs, MD, 1984, pp. 39–44.
4. Colthup, N. B., Daly, L. H., and Wiberley, S E. *Introduction to Infrared and Raman Spectroscopy*, Academic Press, New York, 1975.
5. Griffiths, P. R. and DeHaseth, J. A. *Fourier Transform Infrared Spectrometry*, John Wiley & Sons, Inc., New York, 1986.
6. Gurka, D. F., Hiatt, M., and Titus, R. "Analysis of Hazardous Waste and Environmental Extracts by Capillary Gas Chromatography/Fourier Transform Infrared Spectrometry and Capillary Gas Chromatography/Mass Spectrometry," *Anal. Chem.* 56:1102-1110 (1984).
7. Gurka, D. F. and Pyle, S. M. "Qualitative and Quantitative Environmental Analysis by Capillary Column Gas Chromatography/Lightpipe Fourier Transform Infrared Spectrometry," *Environ. Sci. Technol.* 22:963–967 (1988).
8. Harris, D. C. and Bertolucci, M. D. *Symmetry and Spectroscopy*, Oxford University Press, New York, 1978, pp. 93–224.
9. Kendall, D. N. *Applied Infrared Spectroscopy*, Reinhold, New York, 1966.
10. Lambert, J. B., Shurvell, H. F., Verbit, L., Cooks, R. G., and Stout, G. H. *Organic Structural Analysis,* Macmillan Publishing Co., Inc., New York, 1976, pp. 151–314.
11. Miller, F. A. and Wilkins, C. H. "Infrared Spectra and Characteristic Frequencies of Inorganic Ions," *Anal. Chem.* 24:1253–1294 (1952).
12. Miller, R. G. J. and Stace, B. C. *Laboratory Methods in Infrared Spectroscopy*, Heyden and Son, London, 1972.
13. Nakanishi, K. and Solomon, P. H. *Infrared Absorption Spectroscopy*, Holden Day, San Francisco, 1977.
14. Nyquist, R. A. and Kagel, R. O. *Infrared Spectra of Inorganic Compounds*, Academic Press, New York, 1971.
15. Pouchert, C. J. *The Aldrich Library of FT-IR Spectra*, Aldrich Chemical Co., Milwaukee, 1985.
16. Reedy, G. T., Bourne, S., and Cunningham, P. T. *Anal. Chem.* 51:1535–1540 (1979).
17. Shafer, K. H., Hayes, T. L., Brasch, J. W., and Jakobsen, R. J. Analysis of Hazardous Waste by Fused Silica Capillary Gas Chromatography/Fourier Transform Infrared Spectrometry and Gas Chromatography/Mass Spectrometry, *Anal. Chem.* 56:237-240 (1984).
18. Silverstein, R. M., Bassler, G. C., and Morrill, T. C. *Spectrometric Identification of Organic Compounds*, John Wiley & Sons, Inc., New York, 1981, pp. 95–180.
19. Wilkins, C. L. "Linked Gas Chromatography Infrared Mass Spectrometry," *Anal. Chem.* 59:571A–581A (1987).
20. Woodruff, H. B. and Smith, G. M. "Computer Program for the Analysis of Infrared Spectra," *Anal. Chem.* 52:2321–2327 (1980).
21. Wurrey, C. J., Bourne, S., and Kleopfer, R. D. "Application of Gas Chromatography/Matrix Isolation/Fourier Transform Infrared Spectrometry to Dioxin Determinations," *Anal. Chem.* 58:482–483 (1986).
22. Ying, L., Levine, S. P., Tomellini, S. A., and Lowry, S. R. "Self-Training, Self-Optimizing Expert System for Interpretation of the Infrared Spectra of Environmental Mixtures," *Anal. Chem.* 59:2197–2203 (1987).

CHAPTER 9

X-Ray Fluorescence Field Method for Screening of Inorganic Contaminants at Hazardous Waste Sites

G. A. Raab, R. E. Enwall, W. H. Cole, III, M. L. Faber, and L. A. Eccles

INTRODUCTION

Two of the most severely constraining problems encountered in the cleanup cycle for hazardous waste sites are the high costs of chemical analysis of samples and the long analytical turn-around times. Figure 1 depicts the traditional progression from onsite collection of intrusive samples through packaging, shipping, chemical analysis, and dispatch of the analytical results to site personnel. This process typically takes 20 to 45 days per batch to complete, and direct analytical costs typically range from $200 to $400 per sample.

In an effort to reduce the impact of high analytical costs and long turn-around times, the U.S. Environmental Protection Agency (EPA) Environmental Monitoring Systems Laboratory, Las Vegas, NV, with principal assistance from Lockheed Engineering & Sciences Company (LESC), initiated a project to develop and evaluate field methods utilizing X-ray fluorescence (XRF) technology. In contrast to such traditional laboratory methods as atomic absorption spectroscopy (AAS) and inductively coupled argon plasma emission spectroscopy (ICP), a field-portable XRF (FPXRF) system can be utilized for immediate determination of the identities and concentrations of inorganic contaminants. Concentration data can be acquired, processed, and plotted to produce a map of concentration isopleths directly onsite, as shown in Figure 2. Direct analytical costs incurred by FPXRF for a single in situ determination are roughly estimated at a few pennies; turn-around time is as short as 15 sec.

Routine field procedures begin with development of a spatial sampling scheme designed to attain the required data quality objectives. A field crew makes in situ determinations for the analytes of interest at the specified sampling locations by using a field portable energy dispersive X-ray instrument. Concentration data are entered into computer disk files for storage, and are processed to produce planimetric maps showing sampling locations and

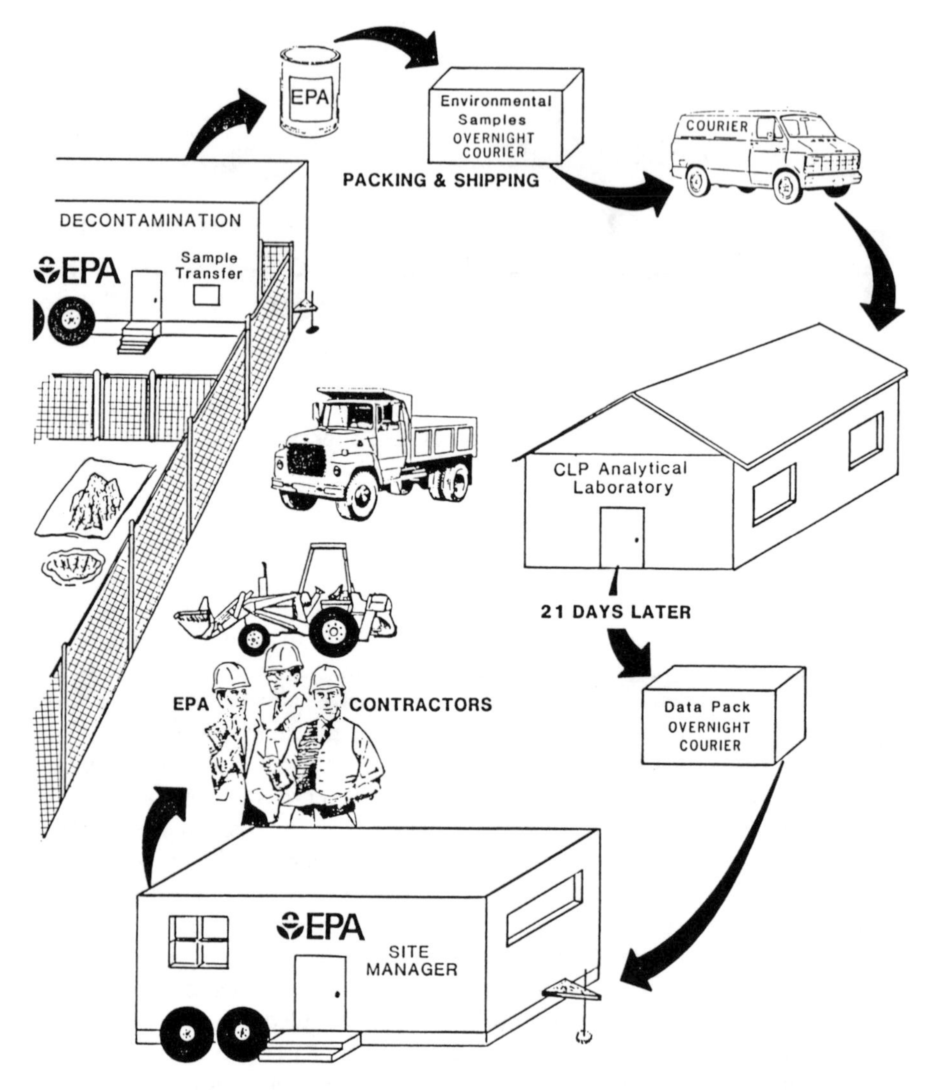

Figure 1. Traditional pathway for soil sampling and analysis at Superfund sites. Elapsed time from sample collection to receipt of data by site manager is approximately 21 days. (Graphics by Steve Garcia, LESC.)

concentration values (e.g., Figure 3) and concentration isopleth maps (e.g., Figure 4A). If a project involves remediation, the FPXRF system can be used to resample very rapidly after each phase of work to delineate areas requiring further remediation. Figures 4A through 4C depict the simulated progression of a site from its initial state (4A) to clean state (4C) established by resampling with an FPXRF system.

Data quality is of primary concern in the development and application of

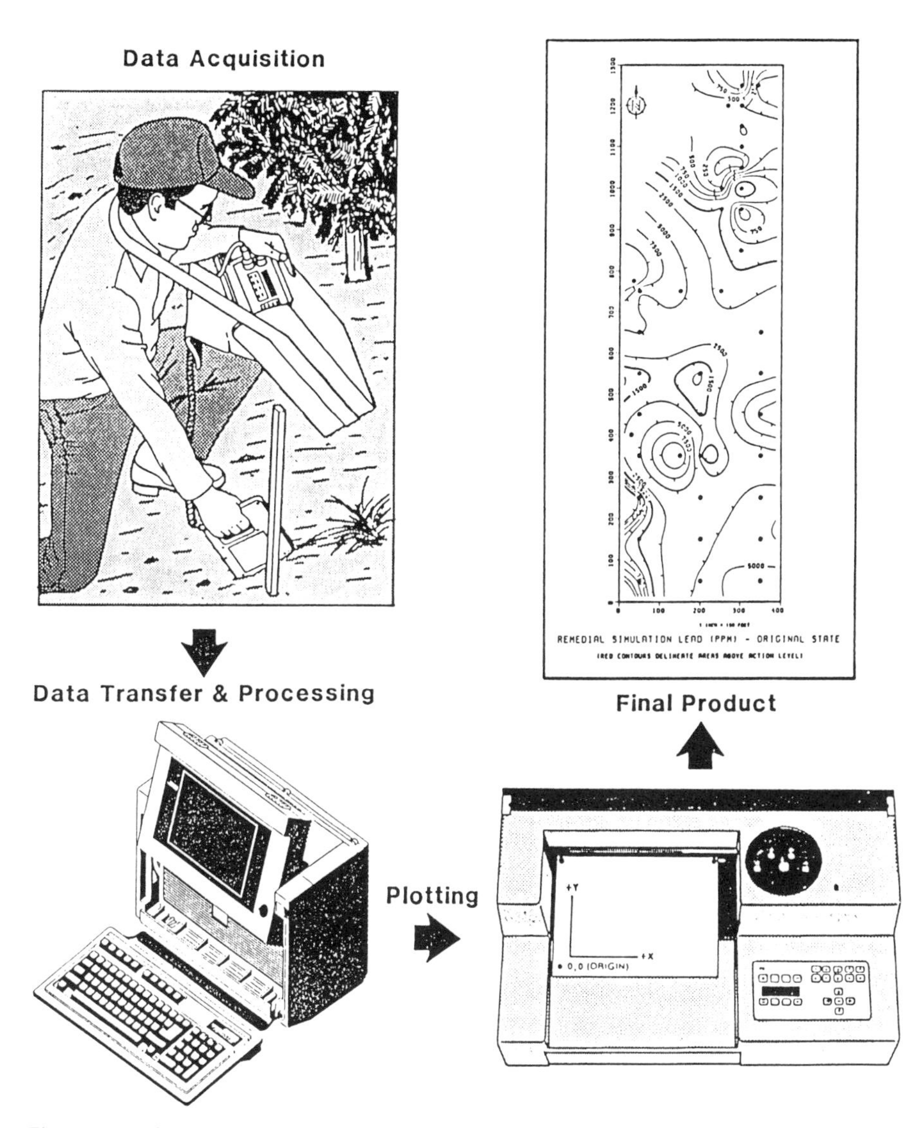

Figure 2. Hazardous waste site screening with FPXRF methodology provides data acquisition, transfer, processing, and plotting capability in real time, onsite. (Graphics by Steve Garcia, LESC.)

field procedures for environmental sampling and analysis. A recent EPA document (U.S. EPA, 1987a, p. 4/17) defines several parameters that serve as standard measures of data quality. These parameters are precision, accuracy (including detection limits), representativeness, completeness, and comparability, commonly referred to in the literature as PARCC. Of these, representativeness is ultimately the most important, because a sample must spatially

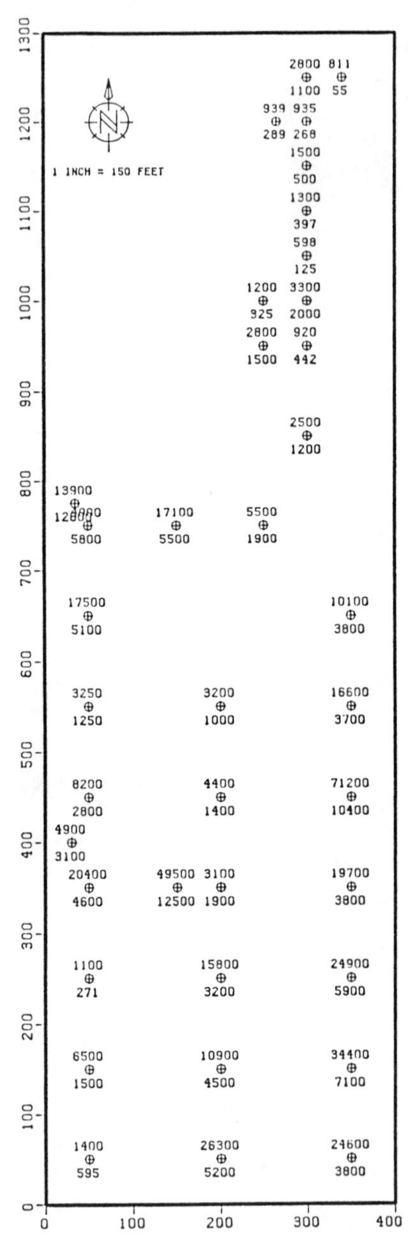

Figure 3. Simulation of a coordinate and concentration map generated by an idealized FPXRF instrument. Zinc values (in parts per million) are plotted above the sampling location symbols, and lead values (in parts per million) are plotted below the symbols. (Graphics by Steve Garcia, LESC.)

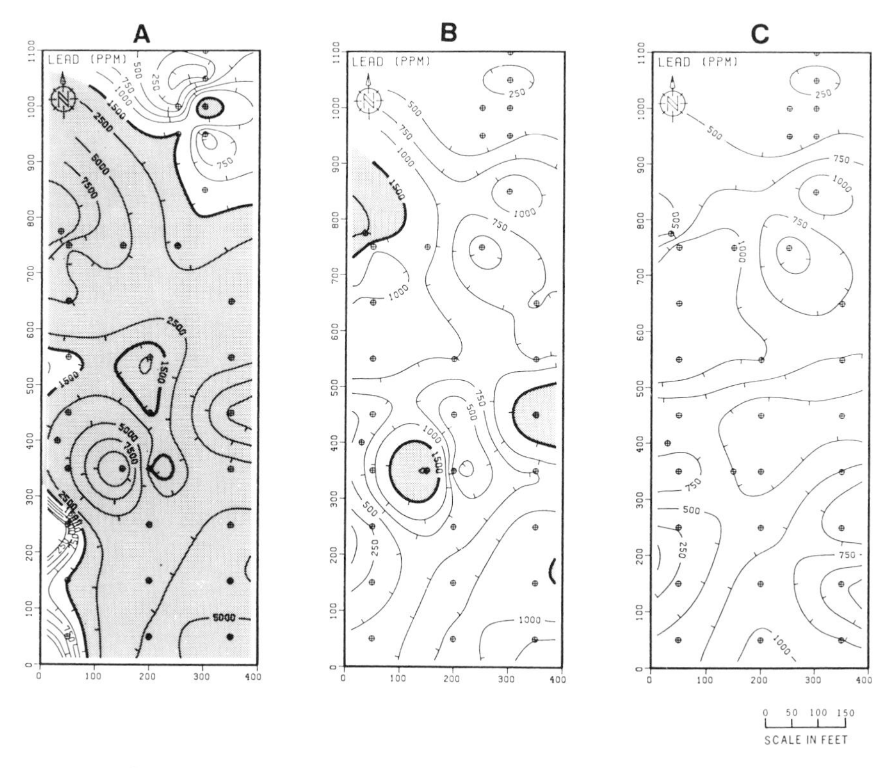

Figure 4. Simulation of concentration isopleth maps generated by the FPXRF instrument. (A) Original concentrations of lead (in parts per million), corresponding to lead values shown in Figure 3; (B) concentrations recorded after first-pass site cleanup; (C) concentrations recorded after second-pass site cleanup. Screened areas show concentrations above remedial action level. (Graphics by Steve Garcia, LESC.)

represent something besides itself to have any practical value. Representativeness is diminished by many sources of variability. Chemists and data quality auditors are chiefly interested in those sources of variability that affect analytical precision and accuracy. Hence, quality control procedures developed for environmental sampling and analysis focus on obtaining the highest possible analytical precision and accuracy. Site engineers and geoscientists, on the other hand, are concerned not only with the reliability of analytical results, but also with spatial variability and spatial representativeness, because contaminants are not dispersed uniformly throughout any given sampling medium. As end users of the data, scientists and engineers must infer the characteristics of target populations from sample data, and the variability of target populations has a spatial component. Therefore, the correct data quality objective is one that focuses on obtaining the most accurate spatial definition of contaminant concentrations that is possible for a given sampling situation.

Attaining accurate spatial definition requires that site personnel have specific knowledge of the spatial variability of contaminant concentrations within the site. Methods traditionally used to assess data quality are based on concepts and assumptions employed in classical statistics methodology, which is appropriately applied to sampling and analytical problems in the laboratory. Classical statistics, however, does not provide specific tools for analyzing and describing spatial variability. Geostatistics, on the other hand, does provide the tools necessary for defining and measuring spatial variability, and thereby provides the means for direct quantitative assessment of spatial representativeness.

Because we believe that application of geostatistical procedures is an integral part of the FPXRF system, we give geostatistics strong emphasis in this chapter. In many cases, the use of geostatistical procedures allows us to achieve a more accurate spatial definition with a field analytical method such as FPXRF than can be obtained with laboratory methods, in spite of the inherently lower analytical precision and accuracy that FPXRF provides. This assertion is based on the fact that spatial estimation error is reduced by the increased sampling density that is made possible by the lower sampling and analytical costs of FPXRF.

PRINCIPLES OF X-RAY FLUORESCENCE

X-ray fluorescence (XRF) is based on the principle that photons produced from an X-ray tube or radioactive source bombard the sample to generate fluorescence (Jenkins et al., 1981). The incident photons impinge on the electron cloud of the atom. Among other events, this process creates vacancies in one or more of the inner shells (e.g., Figure 5A), and the vacancies cause instability within the atom. As the outer electrons seek stability by filling the vacancies in the inner shells, the atom emits energies as X-ray photons (e.g., Figure 5B). The emitted energy (fluorescence) from a particular shell is characteristic of the atom in which it was produced and is equal to the difference in binding energy between the outer shell electron and the vacant shell. Most elements under the photon bombardment fluoresce simultaneously to produce a spectrum of characteristic radiation. It is this spectrum that the detector senses and counts.

There are two types of X-ray fluorescence spectrometers: energy dispersive and wavelength dispersive. The principal differences are in the method of detection of the fluorescent energies of the specimen and in the method of quantifying the analytes of interest. The FPXRF instrument discussed here (the X-MET 880) utilizes energy dispersive spectrometry.

The X-MET 880 is a field-portable, energy-dispersive X-ray fluorescence spectrometer marketed by Outokumpu Electronics, Inc., Austin, TX (e.g., Figure 2). The unit is self-contained, battery-powered, microprocessor-based, and weighs 8.5 kg. The surface analysis probe is specifically designed for field use. The X-MET 880 is hermetically sealed and can be decontaminated with soap and water. The probe includes one or two radioisotope sources, a gas

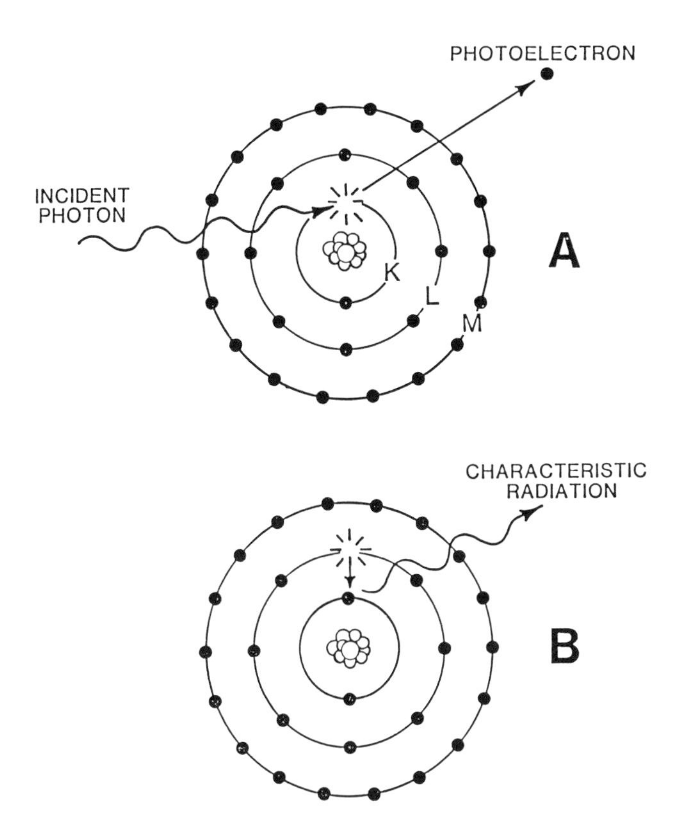

Figure 5. Excitation of a traditional Bohr atom to produce characteristic radiation. (A) Incident photon bombards sample, produces photoelectron, and creates vacancy in inner shell of atom; (B) outer electron seeks stability and falls into vacancy, producing characteristic radiation (fluorescence). (Graphics by Steve Garcia, LESC.)

proportional counter, and the associated electronics. The source is protected by an NRC-approved safety shutter.

The electronic unit has 32 calibration memories called "models." Each model can be independently calibrated for as many as six elements and can be used to measure elements from silicon to uranium, assuming the proper isotope source is available. The measured sample intensities are compared against the calibration curves to estimate concentrations.

DATA QUALITY CONSIDERATIONS

The suitability of data for different remedial investigation (RI) and feasibility study (FS) applications is of major concern. Suitability is determined by data quality, which in turn is expressed as a measure of the accuracy, precision, and the detection limits of the analytical method employed and by the objectives of the sampling program. Our goal is to supply site managers with data of known quality. To do so, we must define and minimize sources of error. Because we are functioning in a field environment, we cannot minimize those errors as efficiently as in the laboratory, but we can ensure that the errors (or variances) are quantifiable as explained below.

In XRF analysis of soils, the most pronounced sources of analytical error are introduced by heterogeneity, where the error is random, and by a wide range of particle size fractions, where the error is a negative bias in the specimen to be analyzed (Wheeler, 1988). For example, if a spoon was dipped into a jar of soil that had not been homogenized, the subsample in the spoon would be biased; that is, it would not represent the sample in the jar with respect to the particle size distribution. It is not possible to tell, a priori, if that bias is negative or positive; therefore, the error is random. Conversely, if a loose subsample of a well-homogenized, unpulverized soil sample is analyzed by XRF, then pulverized and reanalyzed by XRF, the result will be an increase in the analyte concentrations. Thus, an unpulverized soil sample, which represents a much wider range in particle sizes than a pulverized soil sample, displays negative bias in analyte concentration with respect to the unpulverized soil sample.

To minimize these sources of error in the laboratory, the sample is dried, pulverized, homogenized, subsampled, and then pressed to a pellet or fused into a glass disk, which becomes the specimen for analysis. Such sample preparation and analysis steps are generally written into laboratory standard operating procedures which ensure that every sample coming into the laboratory is handled in a uniform manner.

The crew operating from a field laboratory generally does not have the luxury of complying with a single set of standard operating procedures because they encounter a broad range of analytical objectives and sample handling situations from one site to another. At one site, the project manager may want only in situ measurements for preliminary assessment. At another site, the field crew may be asked to analyze samples in or out of a sample container with or without sample preparation. A third situation might require extensive sample preparation prior to specimen analysis.

In Situ Measurement

The simplest situation with regard to sample preparation is the in situ measurement, because there is no intrusive sample (i.e., a sample which is physically extracted from the ground [U.S. EPA, 1987a, p. C-3]), thus no sample handling error. To make an in situ measurement, it is necessary only to place the probe on the ground and pull the trigger, which activates the analysis mode. We generally chose a measurement time of 30 sec.

In situ measurements are the ideal implementation of FPXRF because they are a quick and cost-effective way to conduct preliminary assessments of the extent and distribution of the contaminants on a site. The preliminary assessment can be used to guide the decision of where to collect corroboratory samples to be sent to a certified laboratory for detailed analysis. The purpose is to reduce and more wisely select samples that must be sent to the laboratory thus cutting overall costs and data turn-around time.

The in situ mode of analysis is the simplest, but it also has the highest degree of potential error because there has been no effort to reduce the heterogeneity or the potentially wide range of particle size fractions. When these data are used in spatial estimation, however, the extension errors associated with spatial estimation far exceed accuracy and precision errors introduced by heterogeneity and the particle size effect.

An extension of the in situ measurement is the situation where numerous intrusive samples have been taken prior to the arrival of the FPXRF crew, and the project manager requires rapid screening of the samples to decrease the number of samples to be delivered to a certified laboratory. Where the sample container is large enough (e.g., a large resealable freezer storage bag), the XRF probe can be inserted into the container and a direct measurement can be made. Otherwise the sample must be poured onto a clean surface for direct measurement. Without further sample preparation, the degree of heterogeneity was fixed at the moment the sample was placed in the sample container. Several measurements (with the sample probe moved between each) reported as a mean value will reduce the variability (i.e., the effects of heterogeneity) of the measurements. This is a compositing technique using the ability of FPXRF to make rapid multiple analyses.

Much of the quality assessment and quality control protocol published in the past applies to analysis of aqueous samples and may not apply to XRF analysis of soils (U.S. EPA, 1987a, p. B-9). By running low concentration range check samples and midcalibration range check samples at regular intervals, we can show that gain change and baseline drift do not occur in XRF analysis with the regularity that they do in AAS and ICP analyses. Matrix spikes are unnecessary in FPXRF; enhancement and adsorption effects are well understood and corrected for by fundamental parameters software on the more sophisticated instruments and are minimized by the microprocessor in our field-portable instrument.

Analysis of Prepared Samples

If data of the level of quality suitable for preliminary assessment are unacceptable and the goal is to approach laboratory analytical results, then thorough sample preparation is necessary prior to the subsampling for a specimen for analysis. Ideally, the sample should be oven dried; sieved through a 2.0-mm (ASTM #10) mesh sieve; pulverized to pass a 0.053-mm or finer (ASTM #270) mesh sieve; homogenized, subsampled, and pressed to a pellet which is then analyzed. The specimen for analysis is now homogeneous, of a considerably reduced particle size, and at a maximum with respect to bulk density. If the portable laboratory is equipped with a transportable laboratory-grade XRF

Table 1. The Inorganic Target Analyte List (TAL)

Aluminum	Calcium	Magnesium	Silver
Antimony	Chromium	Manganese	Sodium
Arsenic	Cobalt	Mercury	Thallium
Barium	Copper	Nickel	Vanadium
Beryllium	Iron	Potassium	Zinc
Cadmium	Lead	Selenium	Cyanide

unit, specimen analysis will be extremely accurate and quite repeatable (precise). Furthermore, the specimen can be stored indefinitely for future reanalysis, confirmation, or litigation purposes.

As in all data acquisition programs, the end use of the data should guide the initial data quality objectives (DQOs). If preliminary assessment of the distribution of contaminants on a site is the end goal, then in situ measurements will suit the DQOs. If low-concentration samples are of extreme importance, as may be the case in looking for analyte migration from soil horizons to the water table, then thorough sample preparation prior to specimen analysis would be required to meet the DQOs. In either case, corroboratory samples should be sent to an offsite certified laboratory to ensure that the chosen mode of sampling and analysis is indeed fulfilling the DQOs.

It should be noted that EPA wet chemistry methods (U.S. EPA, 1989) for metals can produce very different analytical results than those of XRF. For example, a soil sample sent to an EPA-approved laboratory for analysis of extraction procedure (EP) toxicity is subjected to an acetic acid leach designed to obtain the immediately mobile analyte(s) of interest. If the analyte of interest was lead in the form of the mineral galena (PbS), then the weak extraction would probably recover little of the contaminant lead. If the lead was in the form of the minerals anglesite ($PbSO_4$) or cerrusite ($PbCO_3$), then the weak extraction would probably recover a great deal of the contaminant lead, as anglesite and cerrusite are the more soluble weathering/oxidation products of galena. XRF, on the other hand, would yield a total elemental analysis despite the phase of the analyte. A soil sample sent to an EPA-approved laboratory for routine analysis is subjected to a more active extraction (nitric acid and hydrogen peroxide), but not to a total soil digestion. Routine analysis for metals on the Target Analyte List (Table 1) will more closely approach the analytical results of XRF than EP toxicity analysis, but both EPA methods will often yield lower analyte results than XRF or a total soil digestion method. The authors are currently engaged in research aimed at quantifying the differences among several partial soil extraction methods, total soil digestion methods, and analysis by XRF.

METHOD AND QUALITY ASSURANCE

We prefer to adhere to three modes of analysis in applying FPXRF to characterizing hazardous waste sites: screening in situ (SIS), where no intrusive sample is taken and where calibration is based on site-typical standards; in situ

Table 2. Site-Typical vs. Site-Specific Calibration Standards

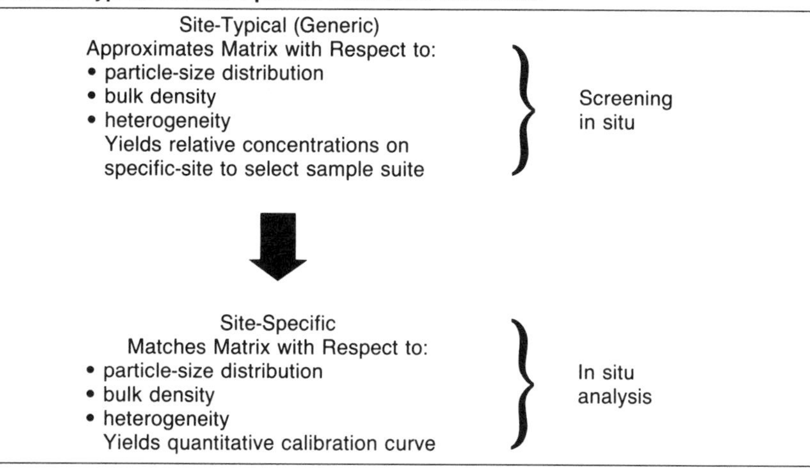

Standard	Use
Site-Typical (Generic) Approximates Matrix with Respect to: • particle-size distribution • bulk density • heterogeneity Yields relative concentrations on specific-site to select sample suite	Screening in situ
Site-Specific Matches Matrix with Respect to: • particle-size distribution • bulk density • heterogeneity Yields quantitative calibration curve	In situ analysis

analysis, where no intrusive sample is taken and calibration is based on site-specific standards; and field laboratory verification (FLV), where an intrusive sample is taken, prepared, and analyzed.

Two types of samples are suitable for FLV analysis: loose soil samples and pressed pellets. Loose soil samples can be prepared more quickly than pellets, but the more homogeneous pellets improve accuracy and confidence levels.

Instrument Calibration and Detection Limits

The X-MET 880 requires calibration curves, where the analytical values are regressed against the measured intensities. The software establishes a calibration curve for each element and stores the data for reference. It also automatically performs spectral deconvolutions and back-scatter corrections. The software processes the analytical data from each routine sample and quantifies the data by comparison to the appropriate calibration curve. Initial modeling of the curve allows the analyst to minimize matrix interferences such as absorption and enhancement.

For more sophisticated but less portable instruments (i.e., for field laboratory verification), a combination of calibration standards and/or fundamental parameter calculations are used to quantitate the analyses.

The X-MET 880 calibration curves are established on site-specific or site-typical calibration standards (Table 2). Site-specific calibration standards are collected from the specific site to match the soil matrix as closely as possible. Site-typical standards can be selected from commercially available characterized soils, or in our case, from standards previously characterized and used as site-specific standards. All calibration standards are unpulverized, loose soils in 31-mm diameter plastic cups. They are intended to more closely match the soil matrix than would a pulverized soil pressed to a pellet. We use site-typical standards to calibrate an instrument prior to going to a site on preliminary reconnaissance. Site-typical standards will yield only a *relative* range of analyte concentrations because they only approximate the soil matrix. The site reconnaissance is to select a suite of soils to be characterized and used as site-specific calibration standards.

Preparation of Site-Specific Standards

The site-specific calibration standards are selected from a suite of samples collected onsite to represent the range of the analyte of interest and to match the soil matrix as closely as possible. More than one model (calibration curve) may be necessary to maintain linearity over the concentration range of an analyte. One curve may range from 0.0 to 0.5 wt% of the analyte(s) of interest to better characterize low-concentration samples. A second curve may range from 0.0 to 5.0 wt% of the same analyte(s) to maintain linearity at higher concentrations at the expense of accuracy and precision below 0.5 wt%.

The samples selected as site-specific calibration standards should be subjected to total elemental analysis by using a total digestion procedure (we use aqua regia and hydrofluoric acid in a Parr bomb [Bernas, 1968; Buckley and Cranston, 1971]) and analysis by ICP and/or AAS. The analysis should conform with EPA quality assurance (U.S. EPA, 1989) and should include all the elements listed on the EPA Target Analyte List (Table 1). We include a blind audit sample (usually NIST Standard Reference Material or equivalent) with the samples sent for analysis to ensure traceability and comparability, as explained in the following sections on data quality objectives. Because the existing EPA sample preparation procedure is a *leaching* method (HNO_3 and H_2O_2), it does not necessarily yield a total elemental analysis, whereas XRF does. Because of this potential disparity in values, we strongly recommend careful review of the data when employing values obtained following EPA extraction procedures.

Detection and quantitation limits (DL and QL) are determined from the numerous low-concentration range check sample analyses (see section entitled "Precision"). They are calculated as follows:

$$\text{DL or QL} = k * \text{SD} \qquad [1]$$

where k = confidence factor (The American Chemical Society Committee on Environmental Quality, 1983, defines k as 3 for a detection limit and k as 10 for a quantitation limit.)

SD = standard deviation of a series of nonconsecutive low concentration level check samples.

Data Quality Objectives

Precision

Precision of the FPXRF instrument and the calibration curve is determined by repeated measurements of a low-concentration range check sample and a midcalibration range check sample (e.g., after every 10th routine analysis). If we are taking intrusive samples, then we split every 10th or 20th sample at the time of collection to obtain sampling precision. Routine samples are also split during the preparation procedure to enable the quantification of sample preparation precision. For strictly in situ analysis, where no intrusive samples are taken, sample splitting (field and preparation duplicates) becomes unnecessary.

Accuracy

In the laboratory, accuracy of XRF instruments is generally determined by repeatedly analyzing a reference sample (e.g., NIST, ASTM) or by spike recoveries. We have tried to find suitable certified standards, but have found that the soil standards are too thoroughly pulverized and that the analyte concentrations are too low to suit our needs. Spiked samples do not apply to in situ analysis, and their preparation in the field laboratory is sufficiently time consuming to defeat the purpose of *rapid* onsite analysis.

In the case of FPXRF, we select a soil that we have previously characterized and used as a site-specific calibration standard. This sample is chosen at a concentration midrange of the site-specific calibration curve to be used on the current site. This check sample can also double as the midrange sample used to determine instrument precision.

Comparability

We establish comparability and traceability through our site-specific calibration standards. When a set of samples is sent to the EPA-approved laboratory for metals analysis, we insert a blind standard reference material obtained from the National Institute for Standards Testing (NIST, formerly the National Bureau of Standards). The blind sample gives our site-specific calibration standards traceability to NIST and allows us to audit the quality of the data that we receive from the analyses. Analyzing routine samples on a calibration curve established on samples analyzed in an EPA-approved laboratory gives us comparability to that laboratory. This is further corroborated because routine samples are periodically sent to the EPA-approved laboratory for analysis, and these analyses are compared to the values obtained from FPXRF.

Completeness

Completeness is addressed on a site-specific basis; if samples or sample data are lost or misplaced, the completeness is less than 100%.

Representativeness

Samples, of and by themselves, have little direct value for site characterization and remediation because they constitute a very small fraction of an entire site. Their chief value lies in what they represent about the remainder of the site. Federal regulations that govern the management of hazardous wastes define representative samples as exhibiting average properties of the whole waste (U.S. EPA, SW-846, Volume II, 1986, p. NINE-5). SW-846 extends this definition by relating representativeness to the accuracy with which sample statistics serve as estimators of population parameters (p. NINE-6).

Because contaminants exhibit spatial variation over any given site, knowledge of the average properties of the site has very limited value for characterization and remediation. Knowing the average concentration of a particular contaminant, for example, provides a site engineer with no practical basis for making decisions concerning remediation. On the other hand, knowledge of the variability of concentration over the entire site enables the engineer to delineate areas and volumes that must undergo remediation. Sample representativeness, then, concerns how accurately samples represent properties of the waste at specific locations within the site. Stated explicitly, the representativeness of a sample is defined by the accuracy with which it represents concentration at a specific *unsampled* location. This accuracy is a direct function of the spatial variability of the contaminant.

A brief discussion of sample representativeness is presented in the EPA DQO document addressing remedial response activities (U.S. EPA, 1987a, p. 4/18). Qualitative means for satisfying the representativeness criterion consist of explicit specification and description of sampling rationale and procedures, and ensuring that a sufficient number of samples are collected. Collocated samples are suggested as a more quantitative means for assessing representativeness. According to this approach, collocated samples provide a measure of the spatial variability of concentration over very small distance increments, spatial variability being generally antithetical with representativeness. One problem with this approach is that the observed variability contains components related to sample handling, sample preparation, and chemical analysis, as well as to spatial variability. Some effort must therefore be expended in assessing the contributions of different components. Another more serious problem is that spatial variability is observed for only those distance increments specified by protocol for the collocated samples, thereby precluding assessment of representativeness throughout the site.

Measurement of spatial variability at different distance increments is clearly a critical requirement for assessing sample representativeness. The following section discusses the statistical concepts and models relating to sampling, spatial variability, and representativeness.

SPATIAL VARIABILITY AND SAMPLING

This section presents some essential concepts necessary for the proper application of statistics to spatial sampling and to the question of sample representativeness. Traditional concepts of classical sampling theory are presented first, along with discussion of some inherent shortcomings in spatial applications (classical statistics is loosely defined here as the statistics of random variables). Following this discussion is a section describing some of the concepts embodied in geostatistics (geostatistics is defined here as the statistics of regionalized variables). Both sections are limited to applications for spatial sampling and sampling design. Procedures for spatial estimation (e.g., kriging) and related processes involved in generating isopleth maps are left unaddressed in this chapter for the sake of brevity. The reader should consult some of the listed geostatistical references for further information concerning these subjects.

Three definitions are necessary to avoid any ambiguity in the subsequent discussions.

- Sample—Classical statisticians define a sample as a set of observations or measurements made on a variable of interest, such as the chromium concentration of a waste sludge. In this chapter, the word sample is used to mean an individual observation or measurement, more in keeping with the common usage of geoscientists and engineers.
- Sample Value—A sample value is the actual datum or number resulting from a sample measurement or observation.
- Spatial Field—A spatial field is a specifically defined area or volume of interest that constitutes the subject of sampling. It is analogous to the target population of nonspatial statistics. Examples might be a physically delimited waste site, an impoundment pond, or a contamination plume.

Classical Statistical Models

Two important theorems constitute the theoretical underpinnings of classical sampling models: (1) The Law of Large Numbers, and (2) the Central Limit Theorem. The former holds that the sample mean will approach the mean of the target population as the number of samples increases. Hence, if a sufficient number of samples are taken, the sample mean can be used as an estimate of the mean of the target population (Chou, 1975, p. 211). The Central Limit Theorem holds, in essence, that the distribution of the means of a large number of sample values is approximately normal (Chou, 1975, p. 213), thus permitting assumption of the normal distribution as a model to calculate the probability that the mean of the target population, which is unknown, meets or exceeds some arbitrarily selected value. Probabilistic confidence limits can thereby be established for the standard error of the mean.

Several assumptions are essential for appropriate application of these two theorems to sampling problems. One of the most important is the assumption of sample independence, i.e., that there is no dependent relationship between sample concentrations, regardless of their proximity or spatial geometry. Random drawing of numbered balls from a box is an appropriate example. The number on a selected ball would not depend in any way on the numbers on physically adjacent balls within the box prior to its drawing. Balls with extremely different numbers could exist side by side in completely random fashion.

The assumption of independence is only realized if there is no spatial component of contaminant variability, implying that the frequency distribution of sample values should be independent of sampling locations. Given that sample values are independent, the Law of Large Numbers says that the mean concentration of the target population can be approximated by the mean of N sample values, and the Central Limit Theorem allows calculation of the estimation variance of the mean based on a normal distribution. These results are expressed as follows (Provost, 1984, p. 83):

$$\sigma_E^2 = \sigma_S^2/N \quad [2]$$

where σ_E^2 = estimation variance of the mean concentration
σ_S^2 = variance of sample concentrations
N = number of samples

Equation 2 can be rearranged to calculate the number of samples required to achieve a preselected value for the estimation variance of the mean:

$$N = \sigma_S^2/\sigma_E^2 \quad [3]$$

Equation 3, or a modification thereof, is commonly utilized for field sampling design. Data available from a pilot study or from other previous sampling can be used to calculate the sample variance σ_S^2. An arbitrary value, representing an optimizing criterion, is selected for the estimation variance σ_E^2. Substitution of these values in Equation 3 yields the requisite number of samples to meet the desired value of σ_E^2. As an example, suppose that preliminary sampling of a contaminated site indicates arsenic variability with a standard deviation (σ_S) of 700 ppm As. The site manager wishes to estimate the mean arsenic concentration to within an error of plus or minus 70 ppm As (σ_E). Substitution of these values for σ_S and σ_E in Equation 3 yields a value of 100 for the number of samples required to achieve the desired level of estimation error. Coordinates for sample locations are then derived from random number tables. In this manner, an entire sampling scheme can be developed, based entirely on the assumption of sample independence.

The problem with the foregoing sampling model is that environmental variables, such as contaminant concentration, rarely warrant the assumption of independence. In most cases, concentrations exhibit some degree of spatial dependence or autocorrelation (Flatman et al., 1988, p. 74; Yfantis et al., 1987, p. 183). Hence, random sampling designs based on classical statistical models can lead to highly unreliable results. Consider the waste site depicted in Figure 6A. It consists of a zone, delimited by the dashed line, where contaminant concentrations are generally higher than those in the surrounding zone.

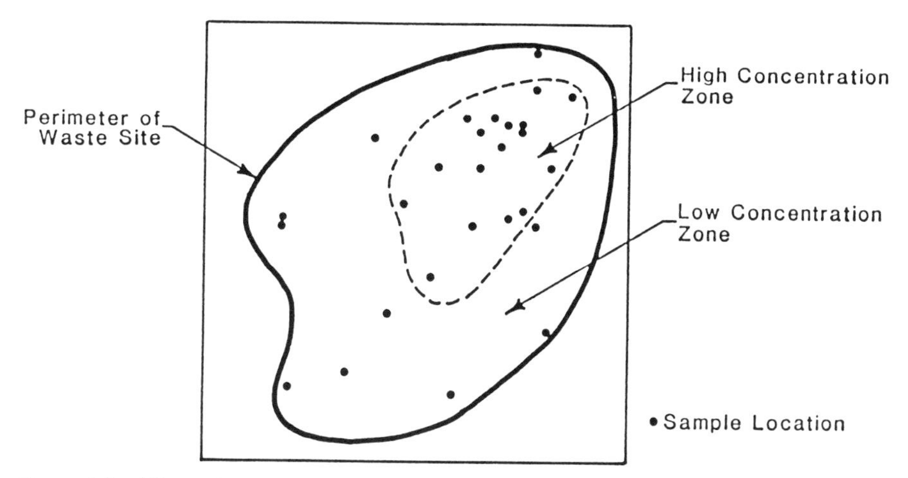

Figure 6A. Waste site where random sampling plan has been applied. (Graphics by Steve Garcia, LESC.)

This is a direct indication that the concentration is spatially autocorrelated. Random numbers were used to locate several sampling points. Purely by chance, a cluster of points occurs in the high-concentration zone (the tendency of random numbers to produce clusters is discussed by Borgman and Quimby, 1988, p. 33). It is clear that the sample mean would be more representative of high values because of the location of the clustering and would therefore be a poor estimate of the target mean for the entire site. If further samples were taken, with the greater portion falling in the high zone, the sample mean would be biased even higher. Conversely, clustering in the low concentration zone would produce an opposite bias.

Figure 6A illustrates the fact that the sample statistics and histogram of a spatially autocorrelated variable are partially determined by the relative proximities and spatial geometry of the sample locations. Thus the Law of Large Numbers cannot guarantee the representativeness of the sample mean. Authors such as Smith et al. (1988, p. 162) are categorically incorrect in their assertion that the representativeness of a set of samples is assured by random sampling from the target population, if the samples are autocorrelated. Quite to the contrary, a systematic sampling scheme, such as that depicted in Figure 6B, would yield a more representative estimate of the mean concentration for the entire site because the regular sample spacing provides, in effect, spatial weighting according to the relative areas or volumes comprising the zones. Corroboration is provided by the mathematical arguments of Ripley (1981, p. 25), who shows that systematic sampling schemes are superior to random schemes in the case of spatially autocorrelated variables, provided no unrecognized periodicity exists. Further corroboration is provided by the sampling simulation studies of Olea (1984, p. 21).

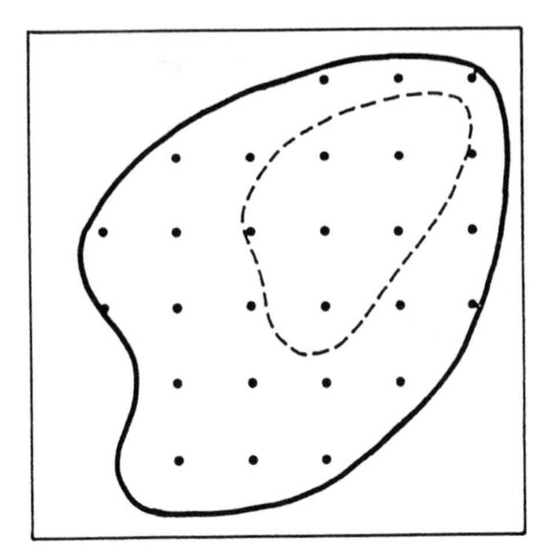

Figure 6B. Waste site where systematic sampling plan has been applied. (Graphics by Steve Garcia, LESC.)

Geostatistical Models

Spatial Variability

Whereas classical statistical models begin with the assumption of sample independence, geostatistical models begin with the opposite assumption and introduce the concept of regionalized variables. A regionalized variable is a random variable, in the classical statistical sense, with a spatially structured component of variability. It is a particular realization of an unknown random spatial function (Huijbregts, 1975, p. 39) that is continuous throughout the spatial field under consideration. A regionalized variable does not vary spatially with completely random abruptness, but rather in a more gradual manner such that the values of closely spaced samples tend to be more similar, and those of more widely spaced samples tend to be less similar. Thus, a sample possesses a certain degree of spatial representativeness (i.e., a volume or area of influence) in contrast to the independent sample of classical statistics. The spatial variation of a regionalized variable, such as contaminant concentration, is a direct result of the unique combination of physical and chemical processes that govern the dispersion of the contaminant within the spatial field, and it gives rise to coherent spatial patterns or entities such as plumes or zones.

Although a random function cannot be known *a priori*, and would be too complex to deal with directly if it could, some of its properties can be elucidated from samples. A very important property for sampling considerations is the spatial correlation between values taken by the variable at different locations. To minimize certain assumptions required about the random function, spatial variance rather than correlation or covariance is usually studied in geostatistics. The variance structure of the spatial function can be approximated by an experimental semivariogram in which the averages of squared differences between pairs of sample values are plotted against intersample distances for a given direction. It is expressed mathematically as follows (modified from David, 1977, p. 74):

$$\gamma(h) = \frac{1}{2n}\sum_{i}^{n} [Z(X_i) - Z(X_{i+h})]^2 \qquad [4]$$

where h = intersample distance in a given direction
$Z(x_i)$ = sample value at location X_i
$Z(x_{i+h})$ = sample value at location X_{i+h}
$\gamma(h)$ = semivariance for intersample distance h
n = number of sample pairs

Figure 7A shows a semivariogram for some hypothetical concentration data in units of parts per million. Each point in the diagram is obtained by the algorithm expressed in Equation 4. A structural pattern is exhibited by the tendency of the semivariance values to increase with the intersample distance. After reaching a specific distance known as the "range," the semivariance appears to be randomly scattered about a limiting value called the "sill," which represents the raw variance of the sample values. At distances less than the range, the semivariance is less than the sill because of the intercorrelation, or spatial dependence, of the sample values. When the distance reaches or exceeds the range, sample values are independent. Thus, the range defines the spatial limit of representativeness of a sample in the given direction, and the

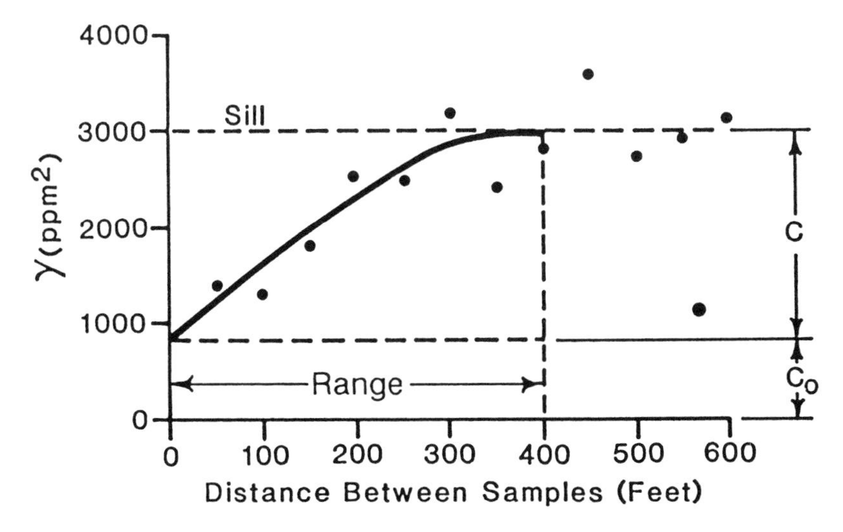

Figure 7A. Semivariogram for hypothetical concentration data. (Graphics by Steve Garcia, LESC.)

value of a sample provides some information about concentrations at unsampled locations within this range.

An analytical mathematical model has been fitted to the semivariogram points in Figure 7A and is shown as the heavy solid line. Note that the model does not intersect the origin when distance is zero. Rather, it tends toward a finite value of semivariance known as the "nugget variance." Nugget variance represents the residual uncertainty of the sample value itself, and originates from spatially independent sources of variability such as errors in sample handling, in preparation or analysis, and from spatially dependent variations of concentrations with ranges less than the sample spacing (also called "microstructure").

Figure 7B displays a semivariogram generated for the same set of hypothetical data as Figure 7A along a different direction. This semivariogram exhibits a much shorter range than the range in Figure 7A, exemplifying the fact that concentrations may possess a higher degree of variability in one direction than in another. Such anisotropic behavior is quite familiar in the case of a spatial entity such as a plume, in which variability is frequently observed to be least along its long dimension and greatest in the orthogonal direction. In addition to anisotropy, a semivariogram may display multiple ranges and sills, reflecting the influence of several physicochemical factors.

A different type of spatial structure is displayed by the semivariogram in Figure 8. Instead of a transition structure such as that shown in Figure 7A, this semivariogram indicates that variability simply increases linearly as a function of distance. There is no finite variance (i.e., no sill) and no finite range of

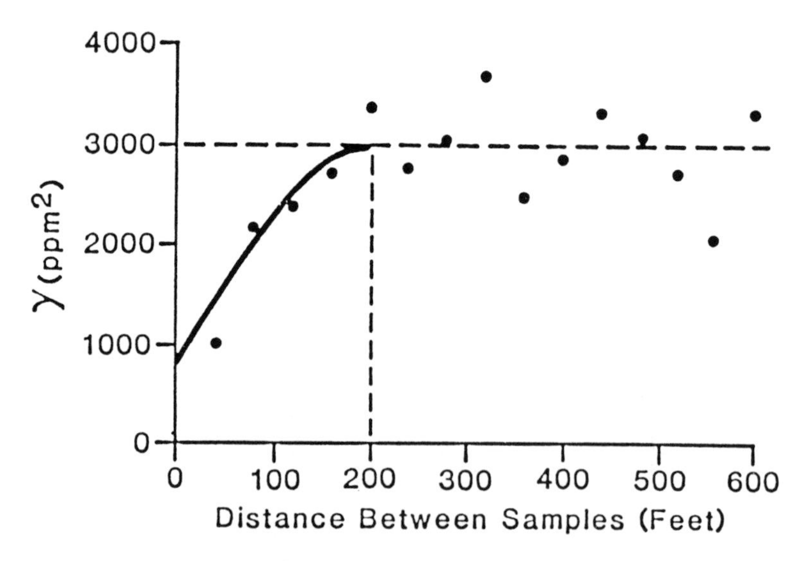

Figure 7B. Semivariogram for the same data set but along a different direction. (Graphics by Steve Garcia, LESC.)

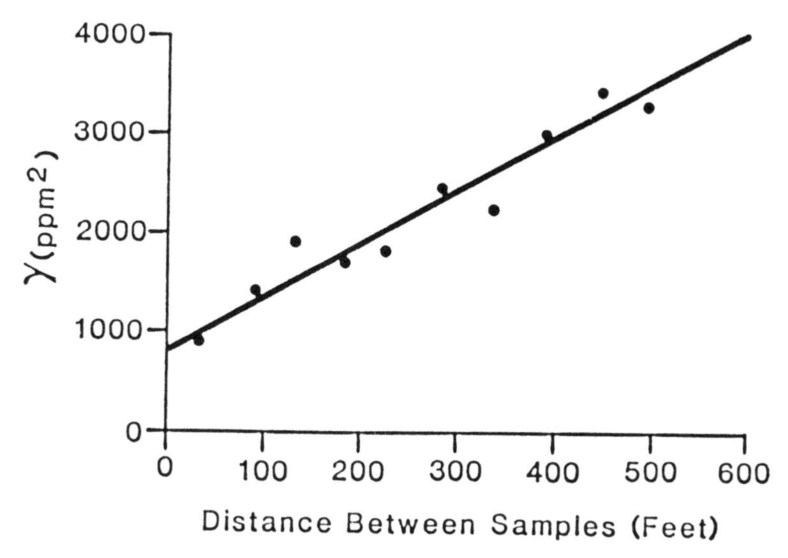

Figure 8. Spatial structure showing monotonic increase in variability as a function of distance. (Graphics by Steve Garcia, LESC.)

influence, at least within the range of intersample distances represented in the diagram.

A third principal type of semivariogram is shown in Figure 9. Semivariance values are seen to fluctuate randomly in a band around the sill. The absence of a gradually increasing trend from the origin indicates that sample values are

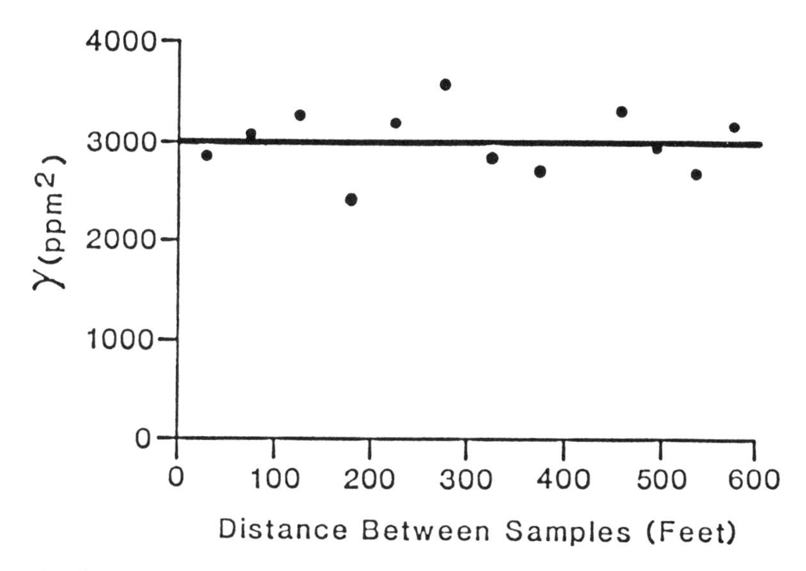

Figure 9. Spatial structure showing random variability around the sill. (Graphics by Steve Garcia, LESC.)

not correlated at any distance, thus representing the condition of true independence requisite for application of classical statistical procedures. However, this condition is rare because physicochemical processes that govern the spatial dispersion of contaminants are usually not completely random.

Sample Support

Contaminants dispersed in heterogeneous solid particulate media, such as soils or sludges, exhibit heterogeneity on physical scales that typically extend downward to the dimensions of individual particles. Hence, a given sample always consists of a mixture of contaminated and uncontaminated material. The relative proportions of these materials vary spatially so that the concentration of a sample is partially determined not only by its location, but also by its size, shape, and orientation within the spatial field. Support is a term that specifies these physical attributes quantitatively.

The concept of sample support (also referred to as data support) is very important because histograms, semivariograms, and statistics relating to sample variability are unique to any given support. For example, the variance of concentration values of a set of cubic samples, measuring 3 in. on an edge, would differ from the variance of a similar set of 3-ft long by 3-in. diameter vertical auger samples taken at the same locations. Semivariograms for these two types of support would also be different. It follows that data homogeneity, in the statistical sense, requires maintenance of constant support.

Constant support can rarely be maintained throughout an RI/FS study. Concentration data are typically obtained from many different types of support such as soil borings or drill cores of different lengths and diameters; more or less equidimensional intrusive samples of different volumes; large volume samples taken for bench chemistry investigations; or in situ XRF measurements representing very small volumes of investigation (typically on the order of 0.05 in.^{3}). The spatial variability of such different supports, expressed by semivariograms, can be related through a procedure known as regularization (e.g., Journel and Huijbregts, 1978, p. 77–94), thus allowing the use of data representing different supports for various geostatistical purposes.

Sample Representativeness and Estimation Error

Data QA documents frequently address the issue of sample representativeness in terms of a variety of errors that tend to diminish this parameter. Routine QA practice employs analysis of error variability, usually expressed as error variance, that relates to different aspects of sampling such as collection and handling, preparation, and chemical analysis (for example, see Taylor, 1988, p. 102). Sampling errors are usually uncorrelated so their variances can be summed to provide an estimate of the total error variance:

$$\sigma_S^2 = \sigma_C^2 + \sigma_P^2 + \sigma_A^2 \qquad [5]$$

where σ_S^2 = total variance of sampling errors
σ_C^2 = variance of sample collection and handling errors
σ_P^2 = variance of sample preparation errors
σ_A^2 = variance of analytical errors

Representativeness has an inverse relationship to the total sampling variance so that a higher value of σ_S^2 implies less representativeness, and conversely.

Although this approach is certainly valid, it is also insufficient because it fails to address spatial variability. In order to be spatially representative, it must be possible to estimate contaminant concentration at a given unsampled location from the known concentration of the sample(s). Spatial estimation incurs errors that arise from the variation of concentration within the distance intervals between the locations of samples and estimation targets. The variance of the estimation error, σ_E^2, is defined by the semivariogram (modified from Journel and Huijbregts, 1978, p. 55 and 151):

$$\sigma_E^2 = 2\,\gamma\,(h) = 2\,[\sigma_S^2 + \gamma_1\,(h)] \qquad [6]$$

where σ_E^2 = estimation variance
$\gamma(h)$ = total semivariance for separation distance h
σ_S^2 = total sampling error from equation
$\gamma_1(h)$ = semivariance due to spatial extension

σ_E^2 thus serves as a true, comprehensive measure of representativeness because it accounts for all sources of variability that diminish this parameter, and shows that the statement of Taylor (1988, p. 106), asserting the "virtual impossibility of demonstrating sample representativeness," is largely incorrect in the case of regionalized variables.

The process of estimation, or interpolation, is the most important step in the RI/FS process because it establishes the inferential link between samples and the spatial population they are supposed to represent. This link is typically expressed in the form of a spatial model that gives a quantitative description of contaminant concentration throughout a site, and it is upon this model that all subsequent decisions concerning remediation are based. The reliability of spatial estimation should therefore be a matter of utmost QA concern, particularly in view of the fact that errors incurred in the spatial extension of sample values are usually much larger than those arising from chemical analysis and sample preparation (Flatman et al., 1988, p. 73), provided these processes are in overall control. Published QA procedures concentrate strictly on sampling errors (i.e., the σ_S^2 components in Equation 6) and ignore the largest source of errors arising from spatial extensions. Current practices, therefore, fail to achieve an adequate assessment of sample representativeness, and should be modified to incorporate the tools for spatial data analysis afforded by geostatistics and regionalized variable theory.

Equation 6 is specifically valid when estimating the concentration of a volume with size, shape, and orientation equal to that of the sample used as the estimator (again, the concept of equivalent support). A more general case involves a change of support such as the estimation of a large volume (or of an area if the spatial field is restricted to two dimensions) from one or more small samples. A common example is depicted in Figure 10, where the variance for a large remediation block is estimated from vertical auger samples at the center of the block and at the centers of the four adjacent blocks. In this case, the estimation variance is as follows (modified from Journel and Huijbregts, 1978, p. 54):

$$\sigma_E^2 = 2\,\bar{\gamma}\,(s;B) - \bar{\gamma}\,(s;s) - \bar{\gamma}\,(B;B) \qquad [7]$$

where σ_E^2 = estimation variance
$\bar{\gamma}\,(s;B)$ = average semivariance of sample volume v within the rectangular block B
$\bar{\gamma}\,(s;s)$ = average variance of samples within the spatial field
$\bar{\gamma}\,(B;B)$ = average variance of blocks within the spatial field

All of the terms in Equations 6 and 7 are derived from semivariogram models which need only be supplied with distance values for solution. Thus, if reliable semivariogram models are available from prior sampling, Equations 6 and 7 can be used to compute σ_E^2 for subsequent sampling. It follows from regionalized variable theory that a unique value of σ_E^2 exists for any given sampling configuration, so σ_E^2 can be employed as an optimizing criterion for *a priori* sampling configuration design, or as a spatial representativeness criterion for QA evaluation of an existing sampling configuration.

To illustrate the relationship between σ_E^2 and sampling configuration, Figures 11A through 11C display blocks with arbitrarily selected dimensions of

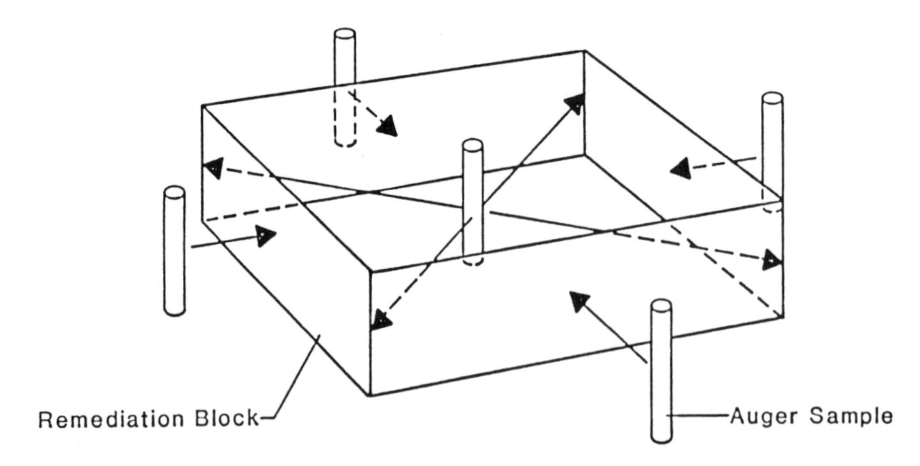

Figure 10. Concentration of a large remediation block is estimated from vertical auger samples at the center of the block and at the centers of the four adjacent blocks. (Graphics by Steve Garcia, LESC.)

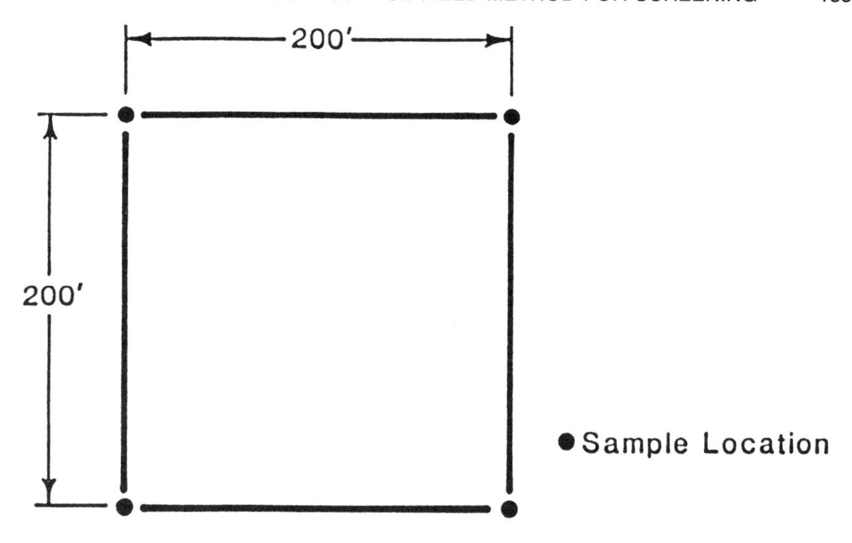

Figure 11A. Block in which samples are situated at its corners (σ_E^2 = 485.4 ppm^2). (Graphics by Steve Garcia, LESC.)

200 by 200 ft containing four samples arranged in different configurations. Estimation errors have been calculated for each pattern by using the semivariogram model given in Figure 7A. The largest error is obtained for Figure 11A, in which the samples are situated at the corners of the block, representing the largest distance from the block itself. Less error is obtained for the random configuration in Figure 11B, because the samples are situated within the block. Least error is obtained in Figure 11C, where the samples are arranged in symmetrical, equidistant locations within the block.

These diagrams exemplify two important concepts that follow from region-

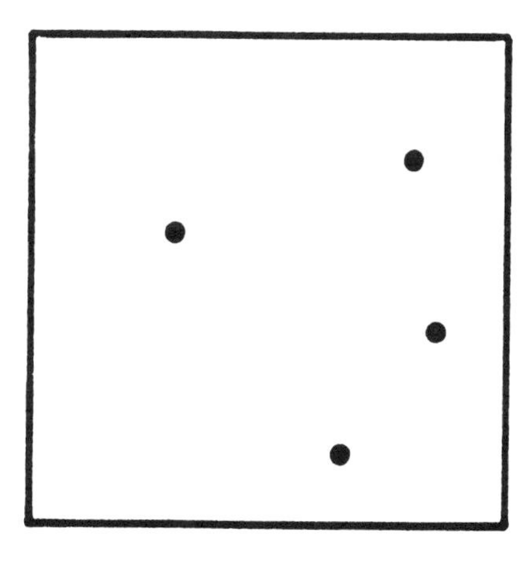

Figure 11B. Block with random samples (σ_E^2 = 301.9 ppm^2). (Graphics by Steve Garcia, LESC.)

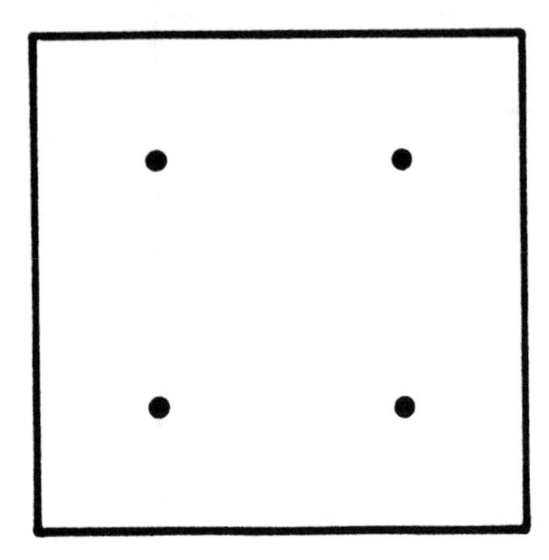

Figure 11C. Block in which samples are arranged in symmetrical, equidistant locations (σ_E^2 = 249.4 ppm^2). (Graphics by Steve Garcia, LESC.)

alized variable theory: (1) Estimation errors are smaller when samples are situated on a regularly spaced configuration, and (2) estimation errors diminish with decreasing distance between samples and estimation targets. The latter concept has particular importance for the application of FPXRF. Sampling density can be greatly increased because of the very low sampling and analytical costs incurred by this method, thereby reducing spatial extension errors which constitute the largest component of estimation error. In many cases, it may be possible to obtain a more reliable and representative spatial model by employing FPXRF, even with lower analytical precision and accuracy, than by employing CLP methods which achieve higher analytical precision and accuracy, but which permit much lower sampling densities because of the inherently greater expense. Thus, we believe that a combination of geostatistics for optimizing spatial sampling design and spatial estimation, and FPXRF for maximum sampling density, can fulfill data quality objectives in many site screening and RI/FS sampling situations.

Spatial Sampling Design

Geostatistical procedures are computationally intensive and require a computer with appropriate software for processing data. Software packages can be obtained commercially from several proprietary sources. Limited public-domain packages are also available from government agencies, including the EPA. Some of these packages include utility programs specifically for the design of spatial sampling configurations. The user specifies semivariogram parameters, dimensions of the target estimation block, and location of the samples relative to the block. The program returns a value for the estimation variance associated with the specified configuration. Acquisition of such a program is highly recommended because it allows the user to evaluate many different sampling configurations very rapidly and to make an optimal selection based on the estimation variance.

Optimizing constraints for spatial sampling design should include the minimization of sampling costs concurrently with achieving an established level of estimation variance based on an estimate of acceptable risk. All too often, sampling design is constrained to fit a preconceived budget or contract bid, with no concern for estimation variance and the implied spatial representativeness. Such false economy leaves parties responsible for sampling design at risk in subsequent technical or contract review and leaves responsible parties at risk for inadequate remediation and possible litigation. It is essential to recognize the importance of risk qualified estimation variance for establishing the defensibility of a spatial sampling design.

Specifications for an optimal design depend to some extent on the sample acquisition and analysis methods mandated for the study. One of the principal benefits of applying the FPXRF system is the low cost of sampling and analysis in comparison to other methods. More spatial representativeness can be achieved per unit cost, and multistage sampling becomes a routinely viable and cost-effective procedure. Multistage designs, exemplified in a paper by Myers and Bryan (1984), are highly recommended for any field situation involving spatial sampling because of conditional objectives that must be fulfilled in sequential order to achieve overall optimization. Two principal stages are defined for the purposes of structural analysis and spatial modeling, and are reviewed in the following discussions.

Structural Analysis Stage

Initial sampling in a field project should be conducted to determine the presence and character of spatial structure by means of semivariograms. Reliable definition of semivariograms can best be achieved by collecting samples on regular, closely spaced intervals, thus limiting the amount of area that can be covered at this stage. Achieving areal coverage is a proper objective for the subsequent spatial modeling stage.

Sampling configurations may consist of transects, grids, or combinations of both. Transects minimize the number of samples required and are recommended by McBratney et al. (1981, p. 332) and Webster (1984, p. 906). In many cases, they result in good quality semivariograms. Flatman (1988, p. 79), however, states that in some cases transects result in poor semivariograms, possibly due to erratic variation typical of pollution data. Another drawback is that transects are restricted to single directions and are therefore less useful than grids for investigating anisotropy in the spatial structure (Starks et al., 1986, p. 61). Flatman et al. (1988, p. 79) recommend use of a combination of several transects and a square grid. This design allows a preliminary determination of the maximum extent of contamination using the transects concurrently with detailed evaluation of possible anisotropy from directional semivariograms based on the grid data.

No firm rules exist for initial selection of sample spacing at this stage. If existing information concerning the spatial variability of the contaminants is unavailable, some arbitrary fraction of the maximum dimension of the site or exclusionary zone can be selected as a preliminary spacing. Semivariograms generated from the preliminary sampling are inspected for high nugget variance, or for highly irregular or absent structure. If any of these attributes is in evidence, the sample spacing should be decreased to some submultiple, such as one half or one fourth of the original spacing, and samples should be taken at the new locations. This process can be repeated until good semivariograms are obtained or until it becomes obvious that the point of diminishing returns has been reached. If high nugget variance is still in evidence, some type of compositing scheme might be considered (e.g., Flatman et al., 1988, p. 81). In some situations, time constraints may occasionally necessitate combining structural analysis sampling with spatial estimation sampling in a single stage. Such cases warrant a combination of grid and transect configurations, the latter having more closely spaced sample locations for the purpose of improving semivariogram definition in distance ranges corresponding to microstructure.

Incorporating existing information into the sampling design cannot be overstressed. The spatial structure represented by semivariograms is directly determined by the physical and chemical processes operating to disperse contaminants in different sampling media. Knowledge of the likely interactions between contaminant-bearing solutions and soil minerals, for instance, may give some indication of spatial variability and may provide a more objective basis for selecting a preliminary sample spacing. Factors such as prevailing winds or topography that influence contaminant transport indicate potential directions of structural anisotropy. Spatial variability is frequently (but not always) least in the direction of transport, so that knowledge of likely transport directions may assist in proper orientation of grids or transects. In the case of an elliptical plume, for example, one transect should be aligned parallel to its major axis and one aligned parallel to its minor axis for optimal semivariogram definition.

In some cases, data from prior site sampling are available. Such data can sometimes be used for generating semivariograms and for analyzing the spatial structure. If the semivariograms represent the same sample support that will be used for spatial estimation sampling, and they are of good quality, it may be possible to utilize them directly for the spatial estimation stage and to forego the structural analysis sampling. If a different support is represented, the semivariograms can be modified to represent the desired support by regularization. At the very least, poor quality semivariograms from prior site sampling may assist in the design of structural analysis sampling. Existing data should be evaluated for potential problems due to the use of different analytical laboratories or methods, different sampling protocols, or other sources of statistical inhomogeneity, and utilized accordingly.

Some words of caution are appropriate for the evaluation, interpretation, and application of semivariograms. Like the histograms of classical statistics, they are subject to sampling error. They are sensitive to extreme values, spatial outliers, and poor choice of distance classes during the semivariance calculation (Armstrong, 1984). Proper application requires the validity of certain assumptions about the stationarity of the underlying random function (Huijbregts, 1975, p. 39). The presence of multiple spatial structures may result in a very complex or "nested" semivariogram. In view of these potential problems, it is highly recommended that experienced individuals perform the semivariogram interpretation and structural analysis. Because the spatial structure depends on physical and chemical factors relating to the contaminants and to the sampling medium, it is extremely important that geologists, geochemists, mineralogists, or other individuals who have specific understanding of these factors are involved in the analysis. General guidelines and procedures for analyzing semivariograms and for performing structural analysis are discussed in Journel and Huijbregts (1978, pp. 195–235). Although presented in a mining context, most of the concepts are closely analogous to applications in environmental sampling.

Spatial Modeling Stage

The objective of this stage of sampling is to provide a base of samples from which spatial estimates can be made. Sometimes the target estimates represent points or blocks at specific isolated locations. Most often, the geoscientist or site engineer requires a spatial model, for each contaminant, that describes its concentration everywhere within the site. A spatial model enables the use of such graphical interpretive tools as contour maps and isometric plots to analyze contaminant distribution and to locate areas where concentration levels exceed critical values.

The first consideration in designing a scheme for spatial estimation sampling is whether the topical contaminant(s) display spatial structure. Direct evidence for the presence or absence of structure is provided by reliable semivariograms, made available either from structural analysis stage sampling or from existing data. Random fluctuation of semivariance values about a sill constitutes direct evidence for lack of structure, provided that the structural analysis samples were not taken on a spacing that exceeds the range of an unrecognized macrostructure.

If semivariograms are not available, indirect evidence can sometimes be used to indicate the likelihood of structure. If a contaminant is highly resistant to mobilization, transport, and dispersal because of its chemical or physical properties, there is a good possibility that it may not display coherent structure and that its spatial distribution is governed solely by its original deposition or placement. On the other hand, if its properties conceptually indicate the likelihood of transport and dispersal, or if existing data show that it tends to concentrate in definable zones or plumes, then the presence of structure is probable. If semivariograms and indirect evidence are both lacking, it is recommended that the presence of structure be assumed because this is most often the case. Complete randomness is rarely encountered in contaminated media.

If there is evidence directly indicating a lack of spatial structure, sample independence can be assumed and traditional procedures for sampling design based on classical statistics can be employed. Coordinates for sample locations can be derived from random numbers for either a simple random or a random-stratified pattern. Optimum sampling density is usually determined by selecting a maximum value for the estimation variance of the mean concentration and by solving an expression such as Equation 3 for the requisite number of samples. More specific discussions of such procedures can be found in Provost (1984), Triegel (1988), Barth and Mason (1984), and Bruner (1986). Pertinent EPA guidance documents include U.S. EPA (1987a,b) and U.S. EPA (1986). However, readers should be aware of the logical incongruity inherent in designing spatial sampling schemes using statistical procedures that assume spatial independence, and using these samples to develop spatial models and isopleth maps, all of which require the assumption of spatial dependence.

Classical statistical procedures can be modified for sampling design when spatial structure is assumed but semivariograms are not available. Samples should be taken on a systematic grid with randomized origin, or on a random-stratified grid if further randomization is desired. Sampling density can be determined by the same procedure described in the preceding paragraph. Application of the raw sample variance in Equation 3 ignores the spatial component of variance and may lead to overestimation of the required number of samples. However, this procedure can be used as a conservative approach in lieu of semivariograms.

Geostatistical procedures should be used for sampling design and data analysis when spatial structure is directly indicated by semivariograms. Sampling points should be located on a systematic grid because of spatial autocorrelation. Square or rectangular grid configurations are probably the most commonly utilized and are the simplest to lay out in the field. The efficiency of different configurations has been investigated by several authors. In a very comprehensive simulation, Olea (1984) studied the relative efficiencies of random, random-stratified, clustered, hexagonal, transect, triangular, and square configurations for local estimation assuming a linear semivariogram with no nugget variance in an isotropic spatial field. Greatest efficiency under these conditions was obtained with the hexagonal configuration, followed closely by the square and triangular configurations. Most inefficient and unreliable from the standpoint of estimation error are the clustered and random configurations. Sensitivity of the efficiency to changes in sampling density, semivariogram slope, and number of nearest neighbors was also studied. Yfantis et al. (1987) studied square, triangular, and hexagonal configurations using a spherical semivariogram in an isotropic spatial field. They also studied the sensitivity of the mean squared estimation error to changes in nugget variance and in sampling density. Under conditions of small nugget variance, the triangular configuration produces optimum results. Under conditions of high nugget variance and sparse sampling density, the hexagonal configuration produces optimum results. McBratney and Webster (1981) report the results of a similar study.

Grid orientation should be determined by structural anisotropy as indicated by directional semivariograms. Anisotropy is viewed as an ellipse with its major axis oriented along the direction of the semivariogram with the longest range and its minor axis oriented along the orthogonal direction. Lengths of the major and minor axes are equal to the ranges of the respective semivariograms. Sampling grid lines should be laid out parallel to the axes of the anisotropy ellipse. If randomization is desired, it is applied only to specification of the grid origin.

A commonly used rule of thumb for sample spacing is two-thirds of the length of the anisotropy axis in the given direction. This spacing allows some overlap of the area of influence of each sample so that, in effect, every point within the sampling domain is statistically represented. A more rigorous approach consists of generating plots of estimation error versus sample spacing for each pattern considered. Examples of such plots appear in the papers of Olea (1984), Yfantis et al. (1987), and McBratney and Webster (1981). Using this approach, the designer can determine the appropriate sampling density to achieve a preselected level of estimation error for a given sampling configuration. The sampling grid should extend somewhat beyond the known or suspected limits of contamination to avoid the necessity of extrapolation when applying the local estimator to the sample data (Flatman et al., 1988, p. 80).

Readers will realize that these discussions of spatial sampling design do not exhaustively cover this subject, nor should they be construed as a blueprint for sampling design. Rather, the strategies discussed here represent some general guidelines along with some of the most current published references. These guidelines are primarily applicable to screening and characterizing sites that will likely require delineation of subordinate areas or volumes for remediation. Ultimately, those designing spatial sampling schemes must approach each investigation as a unique case. Readers should keep in mind some general principles that will apply in all cases in which contaminant concentrations are regionalized:

1. Spatial variability can be expressed quantitatively by semivariograms.
2. Estimation variance depends strictly on the locations and relative spatial configurations of sampling points.
3. Semivariograms enable calculation of estimation variance and thereby the quantitative assessment of spatial representativeness for any given sampling configuration.
4. Lowest estimation variance is achieved with regularly spaced, symmetric sampling configurations.

SUMMARY

The FPXRF system is an effective, onsite complement to traditional wet chemistry methods such as ICP and AAS, as well as to laboratory XRF methods (Raab et al.,1989). The FPXRF method packages traditional XRF capability in an instrument that one person can carry and that a two-person team can operate in the field. The portability of the instrument has major advantages: during an initial (screening) visit to a hazardous waste site, project personnel can analyze soil samples in situ and review the computer-processed results immediately, identify and verify the spatial distribution of contamination, and determine whether contamination levels are high enough to warrant further investigation or remedial work. The application of such a field-portable system for hazardous waste site investigations can net an overall decrease in the time and cost of analyses for the inorganic constituents detected because it can provide data in real time, onsite.

During the initial screening of a hazardous waste site, we can process sample data produced by the FPXRF system immediately by computer. The field crew then can use the processed data to:

- identify the contaminated areas
- restrict the investigation to these areas
- determine approximate contaminant concentrations and choose samples on the basis of those determinations
- determine whether investigation or remediation beyond the screening level is necessary

For exposure assessment analysis, samples must undergo more rigorous sample preparation than is required for screening. If the onsite coordinator uses FPXRF during screening, he can identify the contaminated areas and areas that register as "background" (noncontaminated) before his field crew collects samples for the more stringent analyses. As a result, he can exclude samples collected in noncontaminated areas from the rigorous preparation required for offsite laboratory analysis. Consequently, the onsite coordinator can restrict and intensify the sampling effort to the known areas of contamination and thereby reduce the number of samples that must be sent to the laboratory for analysis.

ACKNOWLEDGMENTS

We are greatly indebted to Dr. R. Bryan of Geostat Systems International, Inc., Dr. C. Lytle of NEA, Inc., J. McLaughlin of CMC, and Dr. D. Lietzke of Lietzke Soil Science Services for their technical reviews. Our thanks also extend to Annalisa Haynie, for without her technical skills in word processing, this chapter would never have reached completion.

NOTICE

Although the information in this paper has been funded wholly, or in part, by the United States Environmental Protection Agency under contract number

68-03-3249 to Lockheed Engineering & Sciences Company, it has not been subject to Agency review. It therefore does not necessarily reflect the views of the Agency and no official endorsement should be inferred. The mention of trade names or commercial products does not constitute endorsement or recommendation for use.

REFERENCES

1. American Chemical Society Committee on Environmental Quality. "Principles of Environmental Analysis," *Anal. Chem.* 55:2210–2218 (1983).
2. Armstrong, M. "Common Problems Seen in Variograms," *Math. Geol.* 16:305–313 (1984).
3. Barth, D. S. and Mason, B. J. "Soil Sampling Quality Assurance and the Importance of an Exploratory Study," in *Environmental Sampling for Hazardous Wastes,* G. E. Schweitzer and J. A. Santolucito, Eds., American Chemical Society, Washington, D.C., 1984, pp. 97–104.
4. Bernas, B. "A New Method for Decomposition and Comprehensive Analysis of Silicates by Atomic Absorption Spectrometry," *Anal. Chem.* 40:1682–1686 (1968).
5. Borgman, L. E. and Quimby, W. F. "Sampling for Tests of Hypothesis When Data are Correlated in Space and Time," in *Principles of Environmental Sampling,* L. H. Keith, Ed., American Chemical Society, Washington, D.C., 1988, pp. 25–72.
6. Bruner, R. J. "A Review of Quality Control Considerations in Soil Sampling," in *Quality Control in Remedial Site Investigation: Hazardous and Industrial Solid Waste Testing,* C. L. Perket, Ed., ASTM Publication 925 (1986) pp. 35–42.
7. Buckley, D. E. and Cranston, R. E. "Atomic Absorption Analyses of 18 Elements from a Single Decomposition," *Chem. Geol.*7:273–284 (1971).
8. Chou, Y. L. *Statistical Analysis,* Holt, Reinhart & Winston, New York, 1975.
9. David, M. *Geostatistical Ore Reserve Estimation,* Elsevier Scientific Publishing Co., New York, 1977.
10. Flatman, G. T., Englund, E. J. and Yfantis, A. A. "Geostatistical Approaches to the Design of Sampling Regimes," in *Principles of Environmental Sampling,* L. H. Keith, Ed., American Chemical Society, Washington, D.C., 1988, pp. 73–84.
11. Huijbregts, C. J. "Regionalized Variables and Quantitative Analysis of Spatial Data," in *Display and Analysis of Spatial Data, Proceedings NATO Advanced Study Institute,* J.C. Davis and M.J. McCullagh, Eds., John Wiley & Sons, Inc., New York, 1975, pp. 38–53.
12. Jenkins, R., Gould, R. W. and Gedcke, D. *Quantitative X-ray Spectrometry,* Marcel Dekker, Inc., New York, 1981.
13. Journel, A. G. and Huijbregts, C. J. *Mining Geostatistics,* Academic Press, Inc., New York, 1978.
14. McBratney, A. B., Webster, R., and Burgess, T. M. "The Design of Optimal Sampling Schemes for Local Estimation and Mapping of Regionalized Variables. I. Theory and Method," *Comput. Geosci.* 7:331–334 (1981).
15. McBratney, A. B. and Webster, R. "The Design of Optimal Sampling Schemes for Local Estimation and Mapping of Regionalized Variables. II. Program and Examples," *Comput. and Geosci.* 7:335–365 (1981).
16. Myers, J. C. and Bryan, R. C. "Geostatistics Applied to Toxic Waste. . .A Case Study," in *Geostatistics for Natural Resources Characterization,* G. Verly, M.

David, A. G. Journel, and A. Marechal, Eds. D. Reidel Publishing Co., Dordrecht, Holland, 1984, pp. 893–901.

17. Olea, R.A. "Systematic Sampling of Spatial Functions," Series on Spatial Analysis No. 7, Kansas Geological Survey, University of Kansas, Lawrence, KS (1984).
18. Provost, L. P. "Statistical Methods in Environmental Sampling," in *Environmental Sampling for Hazardous Wastes,* G. E. Schweitzer and J. A. Santolucito, Eds., American Chemical Society, Washington, D.C., 1984, pp. 79–96.
19. Raab, G. A., Cardenas, D., and Simon, S. J. "Evaluation of a Prototype Field-Portable X-ray Fluorescence System for Hazardous Waste Screening," EPA/600/4–87/021, U.S. Environmental Protection Agency, Las Vegas, NV (1987).
20. Raab, G. A., Kuharic, C. A., Cole, W. H., Enwall, R. E., and Duggan, J. S. "The Use of Field-Portable X-ray Fluorescence Technology in the Hazardous Waste Industry," in *Advances in X-Ray Analysis, Proceedings of the Thirty-eighth Annual Denver Conference on Applications of X-ray Analysis,* Plenum Press, New York, 1990.
21. Ripley, B. D. *Spatial Statistics.* John Wiley & Sons, Inc., New York, 1981.
22. Smith, F., Kulkarni, S., Myers, L., and Messner, M. J. "Evaluating and Presenting Quality Assurance Sampling Data," in *Principles of Environmental Sampling,* L.H. Keith, Ed., American Chemical Society, Washington D.C., 1988, pp. 157–168.
23. Starks, T. H., Brown, K. W., and Fisher, N. J. "Preliminary Monitoring Design for Metal Pollution in Palmerton, Pennsylvania," in *Quality Control in Remedial Site Investigation: Hazardous and Industrial Solid Waste Testing,* C.L. Perket, Ed., ASTM Pub. 925 (1986), pp. 57–66.
24. Taylor, J. K. "Defining the Accuracy, Precision, and Confidence Limits of Sample Data," in *Principles of Environmental Sampling.,* L. H. Keith, Ed., American Chemical Society Washington, D.C., 1988, pp. 101–107.
25. Triegel, E. K. "Sampling Variability in Soils and Solid Waste," in *Principles of Environmental Sampling,* L.H. Keith, Ed., *American Chemical Society,* Washington, D.C., 1988, pp. 385–394.
26. U.S. EPA. "Test Methods for Evaluating Solid Waste. Volume II: Field Manual Physical/Chemical Methods," SW-846, U.S. Environmental Protection Agency, Washington, D.C., (1986).
27. U.S. EPA. "Data Quality Objectives for Remedial Response Activities: Development Process," EPA/540/G-87/003, U.S. Environmental Protection Agency, Washington, D.C. (1987a).
28. U.S. EPA. "Data Quality Objectives for Remedial Response Activities: Example Scenario," EPA 540/G-87/004, U.S. Environmental Protection Agency, Washington, D.C. (1987b).
29. U.S. EPA. "Contractor Laboratory Program Statement of Work for Inorganic Analyses," SOW No. 788, Attachment A, U.S. Environmental Protection Agency, Washington, D.C. (1989).
30. Webster, R. "Elucidation and Characterization of Spatial Variation in Soil Using Regionalized Variable Theory," in *Geostatistics for Natural Resources Characterization,* G. Verly, M. David, A. G. Journel, and A. Marechal, Eds., D. Reidel Publishing Co., Dordrecht, Holland, 1984, pp. 903–913.
31. Wheeler, B. D. "Accuracy in X-ray Spectrochemical Analysis as Related to Sample Preparation," *Spectroscopy* 3:24–33 (1988).
32. Yfantis, E. A., Flatman, G. T., and Behar, J. V. "Efficiency of Kriging Estimation for Square, Triangular, and Hexagonal Grids," *Math. Geol.*19:183–205 (1987).

CHAPTER 10

The Application of X-Ray Fluorescence Spectrometry to the Analysis of Hazardous Wastes

Douglas S. Kendall

X-ray fluorescence (XRF) spectrometry has many important applications to the analysis of hazardous wastes. This potential is still being developed, and there are many elemental analysis problems involving hazardous waste in which more consideration should be given to XRF.

XRF spectrometry has long had an important niche in industrial chemical analysis. Steel mills, foundries, nonferrous metal smelters, and related industries use XRF for rapid, precise, and accurate analyses. The mining industry uses XRF in all aspects of its work from prospecting to assaying the finished products. The petroleum industry uses XRF to accurately determine metals in petroleum products.

XRF could be almost as useful to those doing hazardous waste analyses, if it were properly understood and more widely available. Environmental analyses require great sensitivity and XRF does not always offer sufficient sensitivity, as detection limits are in the 1 to 100 ppm range. However, hazardous waste analysis is very different from measuring natural levels in the environment and is in many ways more akin to the analysis of industrial products. For those analyses in which significant concentrations of heavy metals or other target elements are present, conventional XRF instrumentation very often has sufficiently low detection limits.

Once it is established that XRF has sufficient sensitivity for a particular hazardous waste analysis, the many advantages of XRF determinations can be appreciated. Often very little sample preparation is needed. Solids or liquids can simply be placed in disposable cups with an X-ray-transparent window and analyzed directly. Viscous waste oils and paint sludges can be characterized without problematic digestions such as those required by AA or ICP. The easy sample preparation makes XRF spectrometry ideal for rapid qualitative analysis.

XRF instrumentation is usually quite trouble free and easily maintained. It is very stable and holds calibration well. The precision of repeated measurements is usually very good. It is multielement, in either a simultaneous or rapid sequential fashion. Computer-controlled spectrometers with sample changers can analyze many samples while unattended. A wide variety of quantitative methods are available. Often methods developed for industrial materials can be adapted to hazardous wastes. These factors combine to make XRF very useful for efficiently analyzing a large number of similar samples, such as determining toxic metals in a large number of soil samples or used oil samples.

The remainder of the chapter describes some applications of XRF spectrometry to hazardous waste analysis. The advantages of XRF and areas in which it is particularly useful are illustrated.

This chapter is divided into sections. First, the basic principles and the instrumentation are briefly reviewed. Then the use of XRF for the rapid, qualitative analysis (screening) of hazardous wastes is described. The analysis of soils contaminated with heavy metals by XRF is discussed next. The standardless analysis of hazardous wastes promises to greatly increase the utility of XRF. The potential of standardless analysis combined with easy sample presentation are described in the following section. A final part describes the analysis of used oil by XRF spectrometry.

This chapter will not cover advanced X-ray spectrometry techniques which have demonstrated enhanced performance compared with the instrumentation in widespread use. The current research areas of most promise include the use of synchrotron radiation (Iida et al., 1986), total reflection geometry (Pella and Dobbyn, 1988), and proton-induced X-ray emission (PIXE) (Clayton and Wooller, 1985). These techniques are still developing and are not readily available to an average laboratory and are not yet cost-effective for most hazardous waste analyses. Also not discussed here is the use of XRF in conjunction with electron microscopy. This combination can characterize small particles of airborne pollutants. XRF has also been used extensively to analyze particulates collected on air filters.

FUNDAMENTALS

X-rays are a form of radiation more energetic and of shorter wavelength than visible and ultraviolet light. The wavelengths of X-rays are traditionally measured in angstroms (1 Å = 10^{-10} m), although the more correct metric unit is nanometers (10 Å = 1 nm). The shorter the wavelength of an X-ray, the higher the energy. X-ray energies are measured in kiloelectron volts (keV). Dividing 12.4 by an energy in keV gives the wavelength in angstroms. Of course, dividing 12.4 by a wavelength in angstroms gives the energy of the X-ray in keV. Analytically useful X-rays are in the range from about 0.2 to 20 Å or from about 0.6 to 60 keV. For example, the K alpha emission of sulfur is at

5.373 Å (2.31 keV), the K alpha emission of cadmium is at 0.54 Å (23.0 keV), and the L beta emission of lead is at 0.98 Å (12.7 keV).

X-rays can be emitted by an atom when there is a vacancy in an inner electron shell. If an outer-shell electron moves into the vacant inner-shell orbital, the excess energy can be removed from the atom by the emission of an X-ray. X-rays of analytical interest derive from vacancies in the K, L, and M shells, and are called the K, L, and M series of X-rays. Since inner-shell electrons are involved in X-ray fluorescence, the emission spectrum is little affected by oxidation state or bonding. The X-ray emission spectrum of an element is relatively simple compared to ultraviolet and visible emission spectra (which are used in ICP). A regular pattern is followed. Within a given series, such as the K lines, the emission energy increases with increasing atomic number (the wavelength gets shorter).

When X-rays irradiate a material, they can be either absorbed, scattered, or transmitted. Absorbed X-rays lead to fluorescence or emission of characteristic X-rays. Scattered X-rays contribute to the background. The amount and type of scattering is a source of information about a sample. Absorption and scattering coefficients are well-known for almost all elements. It is possible to use these values in calculations of matrix effects which greatly reduce the need for matrix matching of samples and standards. Such calculations are known as fundamental parameters methods and are described more completely in the section on standardless analysis.

There are two types of X-ray fluorescence spectrometers: energy dispersive and wavelength dispersive. Wavelength dispersive spectrometers have been in analytical laboratories for a longer time. Figure 1 is a simplified diagram of a wavelength dispersive spectrometer, which uses a high-powered X-ray tube to irradiate a sample with X-rays. The fluorescent X-rays, which are characteristic of the elements present in the sample, are directed by a collimator onto the analyzing crystal, where Bragg diffraction separates X-rays by wavelength. In

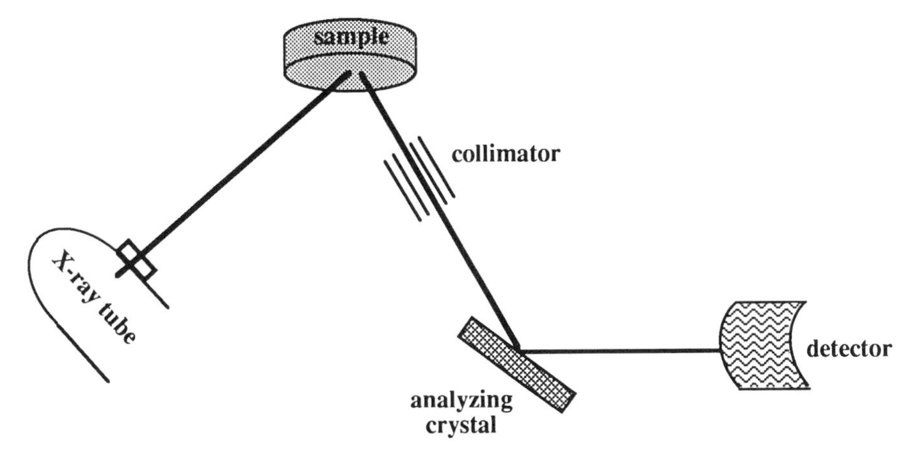

Figure 1. Basic configuration of a wavelength dispersive X-ray fluorescence spectrometer.

a sequential spectrometer, the detector and analyzing crystal are moved in concert so that each wavelength is individually focused on the detector. The detector, either a scintillation or a proportional counter, counts all the X-rays which reach it. In a simultaneous spectrometer, there is a separate crystal and detector for each element of interest. A typical sequential spectrometer will cost about $150,000 and a simultaneous spectrometer over $200,000.

One product of research on radioactivity and transuranic elements is the energy dispersive detector, a lithium-drifted silicon semiconductor (the SiLi detector). This detector produces a current proportional to the energy of an X-ray which strikes it. The resolution of a typical detector, operating at liquid nitrogen temperatures, is about 160 ev. It can resolve the characteristic X-rays of the elements and is used in energy dispersive spectrometers as both the dispersive element and the detector. As in wavelength dispersive instruments, the sample is irradiated by X-rays produced by an X-ray tube. The SiLi detector is placed close to the sample and receives X-rays of all energies. X-rays are counted and sorted by energy. A multichannel analyzer accumulates the counts to produce a spectrum. An energy-dispersive spectrometer is naturally multi-element, although to collect a complete spectrum several sets of excitation conditions may be used. An energy dispersive XRF spectrometer suitable for laboratory use costs $50,000 to $100,000.

A useful addition to an energy-dispersive spectrometer are secondary targets. By placing a specific target between the X-ray tube and a sample, it is possible to irradiate the sample with monochromatic radiation. If the target, usually a pure metal or metallic oxide, is well-chosen, excellent sensitivity for a particular analyte and a low background can be achieved. Monochromatic excitation is advantageous when using the fundamental parameters approach to calibration and calculating matrix absorption effects.

A number of books are available on X-ray fluorescence spectrometry, including an introductory text (Jenkins, 1988) and a complete treatise (Bertin, 1975). A book reviewing quantitative methods is available (Jenkins et al., 1981). The reviews in *Analytical Chemistry* should be consulted for recent work.

SCREENING

An energy dispersive X-ray fluorescence spectrometer can produce spectra covering all elements from sodium to uranium in only a few minutes. This permits a rapid qualitative analysis of an unknown with detection limits for most elements of less than 100 ppm. A rapid, multielement qualitative analysis such as this can be of great utility. In our laboratory, qualitative XRF analysis is used as a screening technique. Screening means initial, preliminary analyses which serve to direct subsequent analyses. If no toxic or priority pollutant metals are found in a sample, then it may be possible to eliminate additional metals determinations.

In the EP Toxicity test, solid samples and samples which cannot be filtered are diluted 20 times during the extraction. Even if the target metal is completely soluble, the original solid sample must have at least 20 times the regulatory limit if the EP extract is to contain a concentration greater than the regulatory limit. One implication of this is that XRF is suitable for screening nonfilterable samples prior to performing the complete EP Toxicity test, which is time consuming and labor intensive. XRF spectrometry can detect concentrations of EP metals which have the potential to cause such a sample to characterize as an EP Toxicity hazardous waste. This is especially valuable at an enforcement laboratory which can screen a number of samples and pick out those with the highest possibility of being violations.

Screening can be valuable in other situations as well. Our laboratory often receives samples of quite pure compounds which were found at manufacturing sites or former manufacturing sites. XRF is used to identify which elements are present. This may allow speciation of the compound when combined with prior knowledge. In any case, subsequent analyses can then be planned to most easily and quickly confirm the preliminary results. Examples include a red pigment consisting of elemental selenium and a yellow pigment containing cadmium sulfide and barium sulfate. A paint with an iron oxide pigment does not need any additional metal analyses. Difficult samples such as sludges, paint wastes, and used oils can be checked for the presence of toxic metals. If none are found, further work is not necessary.

CONTAMINATED SOILS

The contamination of soils by toxic metals is a serious environmental problem. Often large areas are involved. Since cleanup usually requires removal of the soil, it is important that the contaminated area be well-defined. After remedial actions, careful measurements must be made in order to certify the success of the abatement actions. X-ray fluorescence spectrometry is an effective way to meet the analytical challenges posed by this type of problem.

A more common approach, especially at Superfund sites bound by Contract Laboratory Program (CLP) protocol, is atomic absorption or inductively coupled plasma measurements. These techniques require extensive sample preparation, must be done in a laboratory, and are expensive if several metals must be determined. Graphite furnace atomic absorption spectroscopy, while being very sensitive, requires a skilled operator. ICP spectrometry has the advantage of easy multielement measurements, but suffers from the problem of interferences. This is especially true at sites which are highly contaminated with heavy metals. Determinations made from digestates with high concentrations of major elements can cause interference problems. Calibrations designed for dilute solutions of drinking water often can not cope well with digestates containing high concentrations of major elements and significant amounts of heavy metals. Our laboratory has observed problems in the determination of metals such as cadmium and arsenic in soils by the standard protocol of the Superfund Contract Laboratory Program.

Consider the alternative of analysis using X-ray fluorescence spectrometry. XRF has long been used to accurately determine major and minor elements in rocks. The analysis of soil is a very similar problem; a few examples are cited (Bain et al., 1986; Matsumoto and Fuwa, 1979; Paveley et al., 1988; Zsolnay et al., 1984).

Soil or rock samples can be analyzed as ground powders, first drying and then grinding them to a uniform size. This, of course, should be done no matter what analytical technique is applied. A 5- to 10- g portion of the sample is then placed in a disposable plastic cup with a thin mylar or polypropylene window, which is X-ray transparent. The XRF analysis is done on this sample, non-destructively, and can easily be repeated.

Prior to laboratory measurements, another use of XRF has great potential. This is the use of portable XRF equipment to perform a preliminary survey of a site. The main sampling can be better planned and executed if some basic facts about the site are known. A preliminary survey with a portable XRF spectrometer can set the boundaries of the contaminated area, find hot spots, and establish gradients. This type of information will greatly increase the likelihood that the principal sampling will be proper and valid. This brief mention of portable XRF units should indicate their potential, which is discussed in another chapter of this book.

Several studies done at the National Enforcement Investigations Center of the EPA have compared the results of soil analyses by ICP and XRF. An early study showed good agreement between XRF and ICP for ten elements (Kendall et al., 1984). The results of a more recent study will be described here. The samples came from a Superfund site which had been the location for numerous mines and mills. Parts of the site were heavily contaminated with lead and arsenic. Some samples even contained several hundred parts per million of mercury, a remnant of gold refining, which was used to amalgamate the gold. Concentrations of arsenic reached almost 1%, and there was several times as much lead as arsenic.

The samples were dried and then ground to -100 mesh. Finer grinding would have been preferable. The XRF analysis was done on this ground material placed in a disposable cup. The ICP analysis was done on digestions prepared by KOH fusions using 0.25 g of sample. This total digestion can handle silicates and the organic material in soils, and is generally useful for hazardous wastes as well.

Obtaining suitable standards is the main difficulty with XRF measurements of contaminated soils. The most accurate calibrations are done with standards which closely match the samples. Ideally, the standards should span the concentration range of the analytes and the major elements in the standards should match those in the samples. The parameters in the calibration equations, which may contain interelement corrections, are adjusted by least squares techniques to fit the standards. In order to accurately determine the parameters, there should be more standards than parameters. Several standards should be left out of the calibration and then used to test the final fit.

The approach used for this study was to take samples from a previous project as calibration standards. They had been prepared and analyzed by several techniques and gave self-consistent calibrations. Another possibility is to add known amounts of an element, as a solution, to a base material. This is followed by drying and mixing. A third approach is to use certified reference materials such as those from NBS or the Canadian Certified Reference Material Project (Canadian Centre for Mineral and Energy Technology, Ottawa, Ontario).

The need for matrix matching is reduced by using scattered radiation as an internal standard. As discussed previously, the composition of the matrix determines the scattering of X-rays. By measuring the scattering for all standards and samples and by correcting for differences in scattering, it is possible to significantly improve calibration. This technique can compensate for differences between samples and standards. Andermann and Kemp (1958) were some of the earliest investigators to use scattered X-rays as internal standards. Reynolds (1963) was an early user of Compton scattering as an internal standard. Since absorption is roughly inversely proportional to scattering, these investigators found that by ratioing the intensities of characteristic lines to scattered intensities, they could compensate for varying absorption by different matrices. A common choice was the Compton peak of one of the characteristic lines of the X-ray tube. Scattering is used as an equivalent to an internal standard, attempting to compensate for inadequacies in empirical calibrations. More recent applications of this technique have included trace elements in geological materials (Feather and Willis, 1976, Giauque et al., 1977, and Croudace and Gilligan, 1990).

The needs of the study, the availability of standards, and the time available all affect the calibration. Usually, environmental studies can be done with a basic set of standards with scattering corrections used to compensate for matrix matching problems. Care in XRF calibrations is well worth the time and effort. XRF instruments are very stable and hold calibration well. Once calibrated, analyses proceed quickly.

Results of this study for lead and arsenic are shown in Figures 2 and 3, respectively. For both elements, the agreement between ICP results and XRF results is very good. Some divergence at higher concentrations is explained by the lack of good standards at these concentrations. The excellent agreement between the two types of measurements means that XRF is of satisfactory accuracy for these measurements and that a representative portion of the sample is observed by the spectrometer. This latter point is important because of concern that XRF is not suited to trace metal analyses in soils since the spectrometer does not observe a representative subsample. This may be true for very low background levels, but in our experience is not true for soils contaminated with heavy metals.

One of the few interference problems of any consequence in XRF spectrom-

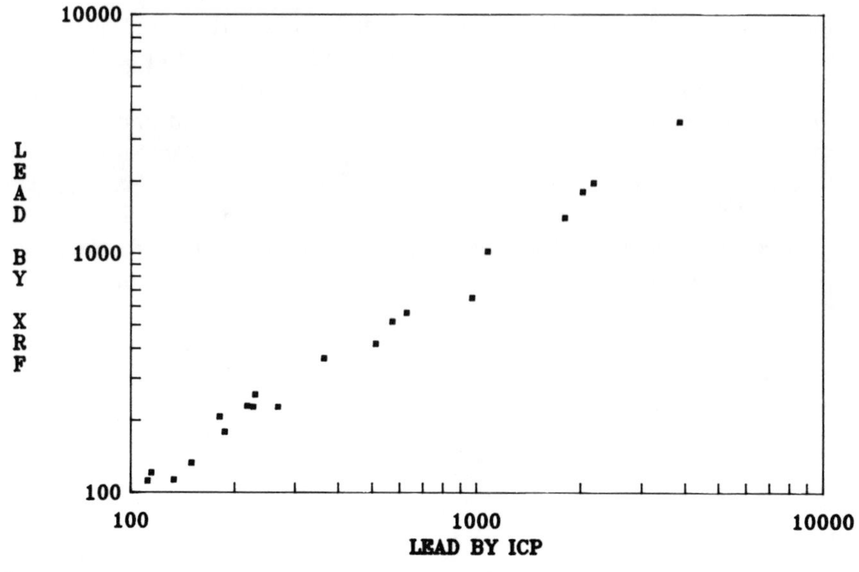

mg/kg

Figure 2. Lead as determined by XRF versus determination by ICP.

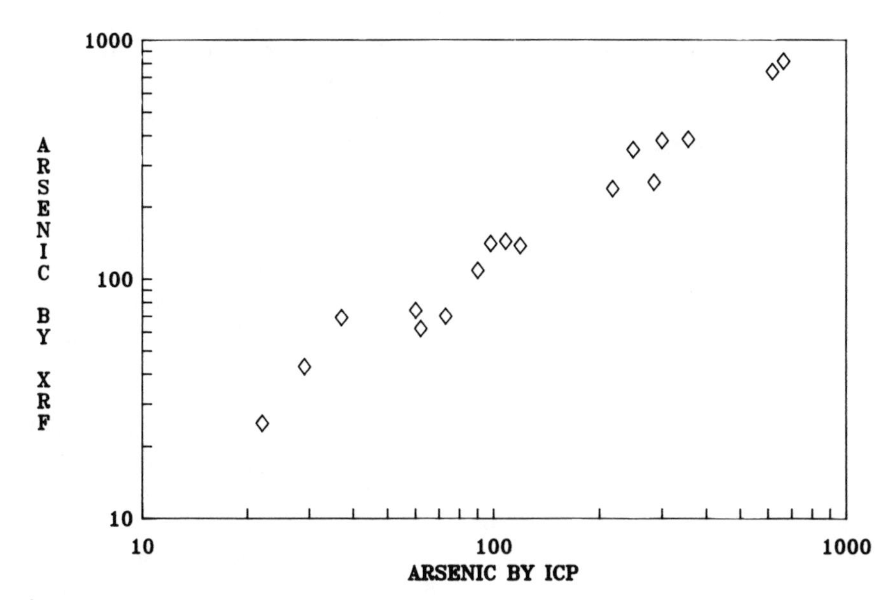

mg/kg

Figure 3. Arsenic as determined by XRF versus determination by ICP.

etry is the almost exact coincidence of the lead L alpha line and the arsenic K alpha line. Lead can be measured just as well using the L beta line, but the alternative isn't as good for arsenic. To avoid the interference from lead, arsenic can be measured using the K beta line, which is not as intense as the K alpha line. Using the K beta line of arsenic raises the detection limit by a factor of four or five. An alternative suitable for some situations is to measure the combined line from lead and arsenic. The presence of either or both is a concern. When assessing a cleanup, both must be removed and this can be checked with the combined line. A potential disadvantage can be turned to good use.

STANDARDLESS ANALYSIS

The principal difficulty with the quantitative analysis of hazardous waste by XRF spectrometry is calibration. Although the best XRF work is done with an empirical calibration which uses standards which closely match the samples, this type of approach is not possible with most hazardous wastes. A laboratory at a treatment and disposal facility, an enforcement laboratory, or a laboratory at a remedial action site does not usually know what will be in the next waste it encounters. While XRF is ideal for qualitatively analyzing such wastes, as described earlier, quantitative analysis is not so straightforward. However, it is now possible to analyze samples by XRF spectrometry with only minimal standardization. A few standards are used to calibrate the sensitivity of the spectrometer, but there is no need for close matrix matching of standards to unknown wastes.

Although empirical calibrations are still the most accurate and precise, great progress is being made in standardless analyses. These methods offer the possibility of quantitative analysis of hazardous wastes with acceptable accuracy. For most wastes, sampling is the source of the largest error. This is particularly true for solids, soils, and viscous liquids. If the analytical error is kept under 20%, then the sampling error will be the dominant contribution to the overall error. Accuracies of 20% or better are possible with standardless calibration procedures.

Methods which rely on few standards, but instead are based on the fundamentals of X-ray physics, are called fundamental parameters methods. These methods take advantage of the well-known behavior of X-rays and their interactions with matter. Using a few parameters such as the X-ray tube voltage, current, and anode material, it is possible to predict the output of the tube. For each X-ray photon of a particular energy which strikes a particular element in the sample, the probabilities of scattering or absorption are well-known. The fluorescent yield of those atoms excited by the incident X-rays can also be predicted. Absorption or scattering of the emitted X-rays must also be considered. By integrating these factors for the whole sample, the complete emission spectrum of any material can be predicted by calculation. It is more usual that the emission spectrum is measured and the composition of the sample is sought. This too can be done. With a given set of experimental conditions and an observed spectrum, the sample composition can be varied until the predicted spectrum matches the observed spectrum. Fundamental parameters methods rely on accurate data for fluorescent yields, absorption cross-sections, etc. and on only a few experimental parameters. These include geometric factors for instrumental design and some measure of detector sensitivity.

The basis for these methods has been understood for a long time. Development of practical fundamental parameters procedures has required the availability of extensive computer power and the talent of many researchers (Jenkins et al., 1981). One of the more influential computer programs implementing the fundamental parameters approach was developed at the Naval Research laboratory (Criss et al., 1978). This program required a mainframe computer, but similar programs have been written for microcomputers. Programs such as these generally require that one measurement for each element of interest be made from standards, which may be pure elements or compounds. They require that at least one emission line be measured for each element in the sample (except that one element can be determined by difference) and allow specification of a particular compound for a given element (such as an oxide). Fundamental parameters programs have been shown to work well on such materials as metal alloys and the major elements in rocks.

There is one major limitation preventing the application of the fundamental parameters approach described here to the XRF analysis of hazardous wastes. Wastes often contain significant amounts of organic material, which, because of its light element nature, is not directly observed by conventional XRF spectrometers. The composition of the light element part of the sample matrix must be known if the fundamental parameters approach to quantification is to be successful. Fortunately approaches have been developed which allow estimation of the light element nature of a sample.

When X-ray photons impinge upon a sample, they are either transmitted, absorbed, or scattered. There are two types of scattered radiation. Rayleigh (or coherent) scattering does not cause a change in wavelength. Compton (or incoherent) scattering produces scattered radiation which is of longer wavelength or less energy than the incident radiation. Each element is characterized by a different ratio between scattering and absorption and by a different ratio between Compton and Rayleigh scattering. Light elements scatter the most, and heavier elements absorb a higher proportion of incident X-rays. Thus, the scattered X-rays contain information about the composition of the whole matrix, including the light element content.

A combination of these approaches promises to provide standardless analysis for any sample. Characteristic line spectra are used to identify and quantify the heavier elements (sodium and above), and the scattered radiation is used to determine the nature of the light elements. The light elements are determined by comparing the measured scattered radiation with the calculated contribution to the scattering of the heavier elements. A fundamental parameters approach is used to adjust the sample composition until the calculated spectrum matches the experimental spectrum, including the scattered radiation. Since basic physical constants are used to account for absorption, scattering, and fluorescence of each element, there is no need for standards which closely match the sample. A few standards, perhaps pure elements or standard reference materials, are used to calibrate geometric factors and detector sensitivities.

Fundamental parameters programs which include scattering calculations to account for the light elements have been implemented by several investigators. Nielson (1977 and 1983) used the ratio of the incoherent scatter to the coherent scatter to characterize the light elements of the sample. This is done by using the scatter ratio to select two light elements whose concentrations are adjusted to give the appropriate light element atomic number, absorption, and scattering. The program iteratively adjusts the sample composition, both light and heavy elements, until self-consistency is achieved and the calculated spectrum matches the observed spectrum. The spectrometer was calibrated with multielement thin films. The procedure was used to analyze coal, NBS orchard leaves, a soil, and a ground rock with excellent results. Measured values were within two standard deviations of the reference values for most elements, which included major and trace elements. This type of procedure seems ideal for hazardous waste analysis because it does not require similar standards.

A similar approach has been implemented by Van Grieken and colleagues (Van Grieken et al., 1979; Van Dyck and Van Grieken, 1980). The ratio of coherent to incoherent scattered radiation was used to characterize the light elements. For a number of standard reference materials, the absorption coefficients determined by this method were within 5% of the values calculated from the known compositions. Good results were obtained in the analysis of several reference materials. Calibration of the spectrometer was done with commercially available thin metal or compound films. There was no need for soil, rock, or organic calibration standards to match the reference materials analyzed.

The preceding section has outlined the potential of fundamental parameters methods to provide virtually standardless analysis of a wide variety of materials. This potential has yet to be realized in the analysis of hazardous waste. However, the basis exists for the development of XRF methods for the quantitative analysis of hazardous wastes. In addition to programs developed by independent investigators, such as those mentioned here, some instrument vendors have developed similar programs. With the necessary computer programs becoming increasingly available, the tools for the development of standardless XRF analysis of hazardous waste are available. It should soon be possible to place a small subsample of any hazardous waste in an XRF spectrometer and within a few minutes know its composition with an accuracy of 10 to 20% with limits of detection of about 10 ppm.

USED OIL

A very large amount of used oil is produced every year from vehicles and industrial sources. Much of it is burned as fuel, including a significant amount burned in small commercial and residential boilers. This large amount of used oil in commerce and its use as a fuel make it a convenient receptacle for hazardous waste. Spent halogenated solvents, often from degreasing, can be present in used oil. Sometimes this is inadvertent and sometimes intentional. The EPA assumes that if more than 1000 ppm chlorine is present in used oil, then halogenated solvents have been added to the used oil and the mixture is a hazardous waste. The mixture must be treated as a hazardous waste fuel. This presumption can be rebutted if the generator can show that the chlorine is not from chlorinated solvents. Another concern is the presence of toxic metals in used oil. Lead is often found in used automotive oil, although the amount has decreased over the last several years as the amount of lead in regular gasoline has decreased. If used oil contains more than 100 ppm lead or 4000 ppm chlorine, it is off specification used oil and is subject to regulation. Burning oil high in chlorine will produce hydrochloric acid, which can cause corrosion in boilers. Lead in used oil which is burned can lead to hazardous emissions.

Thus, there is a need to determine lead and total chlorine in used oil. Measuring total chlorine should be much easier than measuring individual chlorinated solvents by gas chromatography/mass spectrometry. The reference method for determining chlorine in oil should be combustion of the oil in an oxygen bomb followed by ion chromatographic analysis of the chloride in the combustate. This technique is slow and requires a skilled operator.

XRF analysis of used oil for lead and chlorine has much to recommend it. Little sample preparation is needed; the sample need only be placed in a disposable plastic cup. X-ray fluorescence spectrometry is an ideal way to determine chlorine in new oil and is perhaps the method of choice because of its accuracy and precision. Indeed, ASTM has a XRF method for sulfur in oil (ASTM D-2622), and sulfur is similar to chlorine in XRF analyses. Used oils present a much tougher problem. Two principal reasons are sediment and water. Sediment is often present in oil, sometimes as a product of combustion or as particulates from machining. Since almost all XRF spectrometers irradiate the bottom of the sample, sediment which settles to the bottom of the sample is a problem. If it contains the analyte, it will increase the reported concentration. In any event, the sediment is not representative of the oil matrix and it will absorb X-rays emitted by chlorine in the oil matrix. Water can also be a problem. Used oil tanks often have considerable amounts of very dirty water on the bottom. Also, some cutting oils are water-based with water miscible oils. Water can be a problem because it presents a substantially different matrix from the oil matrix used in calibration standards. An analyst may unknowingly attempt to analyze a very dirty water sample or a two-phase sample with water on the bottom. Another problem is the often messy nature of used oils. The viscosity may reach that of a sludge. Three phases may be

Table 1. Determination of Total Chlorine in Used Oil

Sample	Expected (mg/kg)	NAA (mg/kg)	XRF (mg/kg)
1	480	425	435
2	920	805	861
3	1,527	1,355	1,300
4	3,045	2,925	2,995
5	268	125	110
6	29,540	28,050	27,500
7	320	280	287
8	1,498	1,440	1,425
9	3,029	3,115	2,850

Notes: Expected = Value based on sample preparation.
NAA = Neutron Activation Analysis.
XRF = X-ray Fluorescence Spectrometry.

present: oil, water, and sediment. Large amounts of sediment may be present, including some that settles very slowly. All these factors make oil analysis a problem. However, these factors complicate all analytical methods for used oil. XRF, if used properly, still has much to recommend it.

The most suitable methodology for the determination of chlorine in used oil by XRF is not yet completely defined, but the following describes possible techniques. Calibration should be done using a suitably spiked oil matrix. Mineral spirits, mineral oil, or a similar chlorine-free material should be used. A readily available chlorinated hydrocarbon, such as 1-chlorodecane (available from Aldrich), is added. In the preparation of standards, volatile compounds should be avoided in order to produce stable standards. Standards which include the range from 1000 to 4000 mg/kg should produce a linear calibration curve. Samples high in chlorine should be diluted to the linear portion of the curve. Of course, the K alpha peak of chlorine would be measured and suitably background corrected.

The main area of concern is whether the sample matrix matches the matrix of the standards. When it does, XRF will give excellent results. The analyst should check this by preparing spiked samples. A spiking solution can be prepared in the same way as the calibration standards. If spike recoveries are not at least 85%, then the results for that sample should not be accepted. It is sometimes possible to improve results by diluting samples with mineral spirits. Dilution reduces the influence of X-ray absorbing sample constituents. In our laboratory used oil samples are often diluted by a factor of five. Until experience is acquired, it is best to treat each sample or type of sample separately and spike them all.

Methods for used oil analysis are still being developed and improved. The EPA has sponsored studies of methods for determining chlorine in used oil, which have been conducted by Alvia Gaskill of the Research Triangle Institute. X-ray fluorescence spectroscopy was one of the methods studied. The NEIC laboratory of the EPA participated in the study and the results are shown in Table 1. The other investigators who employed XRF spectrometry achieved similar results. In Table 1, the expected values are based on weighings done as part of sample preparation. The samples were prepared by adding known amounts of chlorinated compounds to used crankcase oil, used oil in fuel oil, and metalworking fluids. The neutron activation analysis (NAA) was done as a reference test method. Sample 5 was not stable, but the XRF results agree with the NAA results. The samples were analyzed at NEIC using procedures described here and the agreement is quite good. All samples were diluted and also spiked. The results show that XRF analysis is suitable for the determination of chlorine in at least some types of used oil.

XRF analysis is even more suitable for the determination of lead. This is because the X-rays emitted by lead are more energetic and penetrating than chlorine K alpha emissions, and thus less affected by the sample matrix. XRF is widely used to determine a wide variety of metals in petroleum products. XRF spectroscopy has been used for the determination of wear metals in used lubricating oil (Liu et al., 1986). Determining lead in all types of used oil is a similar measurement. The determination of lead can be done similarly to the XRF analysis of oil for chlorine. Standards containing lead in oil are available from Conostan (Conoco, Ponca City, OK). Detection limits for lead in oil are on the order of 10 ppm, so at the regulatory limit of 100 ppm, lead can be acurately determined. Our laboratory has had success using XRF to determine lead in used oil. Since preparation of oil samples for AA or ICP is a difficult task, additional development of XRF methods is well warranted.

CONCLUSION

The analysis of hazardous wastes is a complex and difficult task. Hazardous wastes are of widely varying chemical composition and physical characteristics. Improved accuracy and increased efficiency in their analyses are needed. Elemental analyses by X-ray fluorescence spectrometry have much to offer and should be considered as an alternative.

REFERENCES

1. Andermann, G. and Kemp, J. W. "Scattered X-rays as Internal Standards in X-ray Emission Spectroscopy," *Anal. Chem.* 30:1306–1309 (1958).
2. Bain, D. C., Berrow, M. L., McHardy, W. J., Paterson, E., Russell, J. D., Sharp, B. L., Ure, A. M., and West, T. S. "Optical, Electron and X-ray Spectrometry in Soil Analysis," *Anal. Chim. Acta* 180:163–185 (1986).
3. Bertin, E. P. *Principles and Practice of X-Ray Spectrometric Analysis,* Plenum Press, New York, (1975).

4. Clayton, E. and Wooller, K. K. "Sample Preparation and System Calibration for Proton-Induced X-ray Emission Analysis of Hair from Occupationally Exposed Workers," *Anal. Chem.* 57:1075–1079 (1985).
5. Criss, J. W., Birks, L. S., and Gilfrich, J. V. "Versatile X-ray Analysis Program Combining Fundamental Parameters and Empirical Coefficients," *Anal. Chem.* 50:33–37 (1978).
5a. Croudace, I. W. and Gilligan, J. M. "Versatile and Accurate Trace Element Determinations in Iron-Rich and Other Geological Samples Using X-Ray Fluorescence Analysis," *X-ray Spectrometry* 19:117–123 (1990).
6. Feather, C. E. and Willis, J. P. "A Simple Method for Background and Matrix Correction of Spectral Peaks in Trace Element Determination by X-ray Fluorescence Spectrometry," *X-Ray Spectrom* 5:41–48 (1976).
7. Giauque, R. D., Garrett, R. B., and Goda, L. Y. "Energy Dispersive X-ray Fluorescence Spectroscopy for Determination of Twenty-Six Trace and Two Major Elements in Geochemical Specimens," *Anal. Chem.* 49:62–67 (1979).
8. Iida, A., Yoshinaga, A., Sakurai, K., and Gohshi, Y. "Synchrotron Radiation Excited X-ray Fluorescence Analysis Using Total Reflection of X-rays," *Anal. Chem.* 58:394–397 (1986).
9. Jenkins, R. *X-ray Fluorescence Spectrometry,* John Wiley & Sons, Inc., New York, 1988.
10. Jenkins, R., Gould, R. W., and Gedcke, D. *Quantitative X-Ray Spectrometry.* Marcel Dekker, Inc., New York, 1981.
11. Liu, Y., Harding, A. R., and Leyden, D. E. "Determination of Wear Metals in Oil Using Energy Dispersive X-ray Spectometry," *Anal. Chim. Acta* 180:349–355 (1986).
12. Kendall, D. S., Lowry, J. H., Bour, E. L., and Meszaros, T. M. "Comparison of Trace Metal Determinations in Contaminated Soils by XRF and ICAP Spectroscopies," in *Advances in X-ray Analysis,* Vol. 27. J. B. Cohen, J. C. Russ, D. E. Leyden, C. S. Barrett, and P. K. Predecki, Eds., Plenum Press, New York, 1984, pp. 467–473.
13. Matsumoto, K. and Fuwa, K. "Major and Trace Elements Determination in Geological and Biological Samples by Energy-Dispersive X-ray Fluorescence Spectrometry," *Anal. Chem.* 51:2355–2358 (1979).
14. Nielson, K. K. "Matrix Corrections for Energy Dispersive X-ray Fluorescence Analysis of Environmental Samples with Coherent/Incoherent Scattered X-rays," *Anal. Chem.* 49:641–648 (1977).
15. Nielson, K. K. and Sanders, R. W. "Multielement Analysis of Unweighed Biological and Geological Samples Using Backscatter and Fundamental Parameters," in *Advances in X-ray Analysis,* Vol. 26, C. R. Hubbard, C. S. Barrett, P. K. Predecki, and D. E. Leyden, Eds., Plenum Press, New York, 1983, pp. 385–390.
16. Paveley, C. F., Davies, B. E., and Jones, K. "Comparison of Results Obtained by X-ray Fluorescence of the Total Soil and the Atomic Absorption Spectrometry Assay of an Acid Digest in the Routine Determination of Lead and Zinc in Soils," *Commun. Soil Sci. Plant Anal.* 19:107–116 (1988).
17. Pella, P. A. and Dobbyn, R. C. "Total Reflection Energy-Dispersive X-ray Fluorescence Spectrometry Using Monochromatic Synchrotron Radiation: Application to Selenium in Blood Serum," *Anal. Chem.* 60:684–687 (1988).
18. Reynolds, R. C. "Matrix Corrections in Trace Element Analysis by X-ray

Fluorescence: Estimation of the Mass Absorption Coefficient by Compton Scattering," *Am. Mineral* 48:1133–1143 (1963).

19. Van Dyck, P. M. and Van Grieken, R. E. "Absorption Correction via Scattered Radiation in Energy-Dispersive X-ray Fluorescence Analysis for Samples of Variable Composition and Thickness," *Anal. Chem.* 52:1859–1864 (1980).
20. Van Grieken, R., Van't Dack, L., Dantas, C. C., and Dantas, H. D. S. "Soil Analysis by Thin-Film Energy-Dispersive X-ray Fluorescence," *Anal. Chim. Acta* 180:93–101 (1979).
21. Zsolnay, I. M., Brauer, J. M., and Sojka, S. A. "X-ray Fluorescence Determination of Trace Elements in Soil," *Anal. Chim. Acta* 162:423–426 (1984).

SECTION IV

TOXICITY SCREENING METHODS FOR HAZARDOUS WASTE

CHAPTER 11

Characterizing the Genotoxicity of Hazardous Industrial Wastes and Effluents Using Short-Term Bioassays

Virginia Stewart Houk and David M. DeMarini

INTRODUCTION

According to estimates by the U.S. Environmental Protection Agency (EPA), more than 260 million metric tons of hazardous wastes are produced annually in the United States by approximately 14,000 generators. Chief producers include industry (e.g., chemical industries and petroleum refineries), the federal government (principally the Departments of Energy and Defense), agriculture, and institutions such as laboratories and hospitals.

Under the Resource Conservation and Recovery Act (RCRA) of 1976, the EPA was mandated to provide a definition of hazardous waste for regulatory purposes. The EPA developed two mechanisms by which a waste could be classified as hazardous: characteristics and listings. Characteristic wastes exhibit one or more of a number of hazardous properties, including ignitability, corrosivity, reactivity, and toxicity. The listing process plays a complementary role, identifying wastes that might escape detection through characteristic review. Included in this category are commercial chemical products and wastes from specific industries or industrial processes.

Provisions for identifying a waste's genotoxic potential *per se* have not been promulgated. Carcinogenesis, mutagenesis, and teratogenesis were candidate characteristics in an Advance Notice of Proposed Rulemaking published in 1978 (Federal Register, 1978), but final Hazardous Waste and Consolidated Permit Regulations (Federal Register, 1980) omitted these properties. Instead, the EPA chose to promulgate an all-encompassing toxicity characteristic to act as a regulatory umbrella to cover all of the toxic potentials exhibited by a waste. Called Extraction Procedure (EP) toxicity (and soon to be changed to Toxicity Characteristic Leaching Procedure, or TCLP), this toxicity characteristic was designed to identify wastes likely to generate a leachate containing hazardous concentrations of specified toxic chemicals. However, the list of toxic chemicals is comprised of only eight metals and six pesticides. (Under the proposed expanded toxicity characteristic, an additional 38 organic compounds will be added to this list.) Consequently, because the toxicity characteristic is determined by such a restricted set of compounds, wastes with the potential to elicit genetic damage are most commonly identified on the basis of the listing process outlined in Sections 261.31 through 261.33 and Appendices VII and VIII of RCRA. Wastes and substances listed there have been selected in part because of available scientific evidence indicating their mutagenic, carcinogenic, and teratogenic potential.

Both the characteristic and listing criteria for identifying a hazardous waste rely upon knowledge of the chemical constituents present in a waste. However, chemical characterizations are limited in the amount and type of information they convey. The presence of a known set of narrowly defined compounds is unlikely to reflect the overall chemical composition of most wastes. Furthermore, chemical analyses do not take into account the possibility of chemical interactions (antagonisms, synergisms, etc.) or the production of toxic metabolites resulting from various degradative pathways, both chemical and biological. Perhaps most importantly, biological effects cannot be reliably predicted from a knowledge of chemical composition alone.

To provide a more comprehensive and conclusive evaluation of a potential hazard, the toxicological impact of a waste should be assessed using biological test parameters. Included among the available test systems are epidemiologic methods (using human populations), long-term animal studies (using the whole animal model), and short-term bioassays (using the whole animal or, more commonly, simple microbial or cell culture assays). Short-term tests are most often chosen to measure biological endpoints such as genotoxicity, aquatic toxicity, etc. because of their relative simplicity and sensitivity, their low cost, and the speed with which results can be obtained.

More than 200 short-term tests utilizing microorganisms, insects, plants, and mammals have been developed over the last 10 to 15 years to aid in the identification of agents that pose a genetic hazard to humans (Waters et al., 1988). Genotoxic compounds, or compounds that damage DNA, are important for two principal reasons. They may damage the DNA residing in germ cells, resulting in an increased incidence of heritable genetic disease, or they may damage the DNA in somatic cells, leading to increased cancer incidence. Most short-term genotoxicity assays serve as indicators of carcinogenic potential. Employing endpoints other than neoplasia, they are based on some of the cellular and subcellular mechanisms underlying the carcinogenic process itself. Initial screening for genotoxicity using short-term bioassays costs between $500 and $10,000 per sample tested (depending on the test selected) and can be performed in 2 or 3 days to several weeks. By using different test systems, various genetic endpoints can be examined to ascertain the different types of damage that can be caused by a sample.

The short-term genotoxicity assays that have been used to test hazardous wastes can be separated into major groups on the basis of the genetic endpoint detected and the test system employed. Included are prokaryotic, eukaryotic, and mammalian cell assays designed to detect gene mutations, DNA damage and repair, and chromosomal damage such as chromosome aberrations and sister chromatid exchanges. The selection of an assay or battery of assays depends upon several factors, including the validity of the test system and its accumulated database, the appropriateness of the assay for complex environmental samples in general as well as for the particular sample to be tested, the purpose for which the data are intended, and relative cost and simplicity.

Often a waste is too toxic in its crude state to be tested directly in a bioassay. Furthermore, the waste may not be compatible with the test system selected due to its physical matrix (e.g., sludges, tars, oils, emulsions, etc.). Thus, the crude mixture is often filtered, extracted, concentrated, fractionated, or otherwise altered in order to make it suitable for delivery to one or more of the test systems. The preparation of crude wastes for bioassay will be discussed in this paper, as will the advantages and disadvantages of the various processing techniques.

A discussion of the genotoxicity of hazardous wastes also will be presented. Waste types will be categorized according to industrial source (e.g., petrochemical industries versus pulp and paper mills). Test results and their implications will be examined. The advantages and disadvantages of using short-term bioassays to study hazardous wastes will be addressed, along with an analysis of the ways in which bioassay results can aid in monitoring, cleanup, and treatment technologies.

SHORT-TERM BIOASSAYS USED TO EVALUATE THE GENOTOXICITY OF HAZARDOUS INDUSTRIAL WASTES

As mentioned previously, there is a large number of short-term bioassays for detecting genetic toxicity. These assays utilize a wide range of organisms and cell types and measure a variety of different genetic changes. Because of the diversity of biological systems and genetic endpoints, there is no single test that adequately detects the types of genetic damage that may be induced by all chemical classes of genotoxic compounds. However, only a few short-term bioassays have been used extensively with complex chemical mixtures and/or hazardous industrial wastes. These tend to be those assays that are the simplest to perform, the least expensive, and that are generally compatible with various types of complex mixtures. Of these short-term tests, the Salmonella assay (Ames test) is the one that has been used most widely for evaluating the genotoxicity of various effluents, wastes, and discharges.

The Salmonella assay was developed by B. N. Ames, P. E. Hartman, and coworkers in the early 1970s. The assay uses a set of histidine-requiring mutants of the bacterium *Salmonella typhimurium*. It is called a reverse-mutation assay because the bacteria are already mutated (in one of the genes of the histidine biosynthetic pathway) and thus require histidine, which is a common amino acid, in order to grow. A chemical is declared mutagenic if it introduces a mutation that corrects or compensates for the original mutation and restores the cells to histidine independence. Mutagenic potency is measured by the dose-related increase in numbers of histidine-independent colonies of cells (revertants) that occur upon exposure to the chemical.

In addition to the histidine mutation, selected tester strains have been genetically engineered to increase their susceptibility to mutagens. Modifications include a cell wall mutation, *rfa*, that permits large molecules to penetrate through the cell wall, and a DNA repair mutation, *uvrB*, that affects the nucleotide excision-repair system. When intact, this repair system removes incorrect nucleotides (the building blocks of DNA molecules) from damaged DNA. However, because the repair system is defective, the bacteria can no longer correctly repair the damage, and sensitivity to certain classes of genotoxic compounds is increased. In addition, some strains of the bacteria contain an additional piece of DNA called a plasmid (pKM101) that codes for DNA repair enzymes that are error-prone. Thus, strains containing this plasmid are frequently more sensitive to mutagenesis than nonplasmid-containing strains because the enzymes incorrectly repair certain types of DNA damage.

The initial set of histidine mutants introduced by Ames and co-workers consisted of four strains; eventually, additional strains were introduced. The idea of using several strains, which constitute a tester set, is based on the fact that different classes of chemicals induce different types of mutations, and no single strain is mutated by all chemical classes. Thus, some strains detect chemical mutagens that delete or add nucleotides to the DNA molecule (called frameshift mutagens); other strains detect chemicals that change or alter a single nucleotide in the DNA molecule (called base-pair substitution mutagens). During the past 15 years, two strains (TA98 and TA100) have emerged as the most sensitive and useful for the purpose of screening complex mixtures or wide varieties of chemicals. Thus, nearly all of the data on the mutagenicity of hazardous wastes has been generated in TA98 and TA100.

One limitation of a bacterial assay such as the Salmonella assay is that bacteria lack many of the metabolic enzymes present in mammals. These enzymes can activate chemicals into electrophilic compounds that can bind covalently to DNA, resulting in a potential mutation. This metabolic activation can be supplied exogenously to the assay by adding a portion of liver homogenate, called S9, from rodents. S9 can be added to a variety of test organisms, including bacteria, yeast, and mammalian cells in culture.

The genetic damage detected by the Salmonella assay represents a class of DNA damage called point mutations. In addition to the Salmonella assay, other bacterial systems, including *Escherichia coli* WP2, have been used to detect point mutations. Fungal systems such as *Aspergillus nidulans* in the haploid state (i.e., containing one set of chromosomes) and the yeast *Saccharomyces cerevisiae* also have been used to detect point mutations or small deletions.

A broader range of mutations may be detected by DNA repair assays such as the "rec" assay using various strains of the bacterium *Bacillus subtilis*. This assay uses DNA repair-deficient and repair-proficient strains of cells. If a chemical causes DNA damage, the DNA repair-deficient strain should be less able to repair the damage than the repair-proficient strain and should be more sensitive to killing by the chemical. Thus, a differential response between the two strains indicates that the chemical has caused DNA damage.

Another type of assay that detects a broad range of DNA damage is one that measures the induction of prophage λ in *E. coli*. A prophage is the DNA of a bacterial virus that is integrated into the bacterial chromosome. Damage to the bacterial chromosome induces a set of responses (called the SOS response) that includes the production of DNA-repair enzymes. In addition, DNA damage induces enzymes that cause the prophage to replicate, form intact virus particles, and lyse the cell in order to release the virus particles. Exposure of a prophage-containing bacterium (lysogen) to a chemical that induces point mutations, deletions/insertions, and/or DNA breaks will elicit the SOS response. DNA damage is measured by an increase in the number of virus particles produced, which can be quantitated as colonies of virus (plaques) on a background lawn of sensitive indicator bacteria. The genotoxicities of a number of industrial wastes and effluents have been evaluated for their ability to induce prophage.

Another major class of DNA damage is called chromosomal mutations, which include large deletions, insertions, or rearrangements of the DNA. Chromosomal mutations induced by hazardous industrial wastes have generally been determined by observing the chromosomes of plant cells (such as onion) or cultured mammalian cells (such as Chinese hamster ovary cells) with a light microscope in order to detect rearrangements of the chromosomes or changes in the number of chromosomes per cell. Chromosomal aberrations include inversions, insertions, translocations, and other types of visible rearrangements. Sister chromatid exchange (SCE) is another type of chromosomal change that may be visualized under the light microscope.

In addition to visual inspection of the chromosomes, genetic tests can be performed that indicate the induction of chromosomal mutations. Such assays have been developed using fungi such as *A. nidulans* in the diploid state (i.e., containing two sets of chromosomes). Some mammalian cell mutation assays, such as the mouse lymphoma L5178Y/TK +/- assay, measure mutation at a heterozygous, autosomal gene. Thus, such assays can detect both point mutations (small mutations within the gene) as well as chromosomal mutations (multiple-gene mutations involving genes adjacent to the target gene).

Mammalian cell transformation assays have been used to a limited extent with complex hazardous wastes. These assays measure morphological changes in the colonies of cultured mammalian cells. Such changes are frequently indicators of preneoplastic changes; thus, mammalian cell transformation by a hazardous waste would suggest that the waste may have some cancer-causing potential.

Recently, the development of the ^{32}P-postlabeling method for detection of

DNA adducts has permitted the determination of DNA adducts in cells or organisms exposed to complex mixtures. DNA adducts are xenobiotic molecules that are bound covalently to DNA. Such molecules may lead to mutations, and their presence is an indication of DNA damage. The ^{32}P-postlabeling method does not require knowledge of the type of chemical that is bound to the DNA; instead, it can detect a range of large, aromatic molecules. Although new, this technique has been used to a limited extent to detect the formation of DNA adducts in rodents exposed to hazardous industrial wastes.

PREPARATION OF CRUDE WASTES FOR BIOASSAY

A crude hazardous waste is a complex of hundreds or thousands of different chemical compounds that manifest diverse chemical, physical, and toxicological properties. To analyze this complex mixture for genotoxic potential, one of two approaches may be used. Selected constituents or fractions of the mixture may be individually bioassayed, or the mixture may be treated as a single entity and tested in its crude state. The advantages and disadvantages associated with each approach are discussed below.

Method-1

To analyze individual components of a waste, several alternative approaches are possible. A detailed chemical analysis of the waste may be performed to identify its constituents, and the genotoxic potential of each identified compound may then be assessed. A literature search may provide some information on the genotoxicity of the individual chemical compounds; alternatively, the chemicals may be evaluated in appropriate bioassays by the investigator. Several reports in the literature discuss the genotoxicity of specific compounds identified in, for example, pulp and paper mill effluents (Carlberg et al., 1980b; Nestmann et al., 1980), energy-related effluents or products (Epler et al., 1978), and chemical maufacturing effluents (Spanggord et al., 1982; Sundvall et al., 1984). However, this approach requires the testing of a multitude of identified chemicals and thus is both time-consuming and costly. Furthermore, this approach is insufficient in specifying which compounds in the sample are responsible for what proportion of the genotoxicity.

Even when a relatively detailed chemical analysis is performed, many mutagens and carcinogens present in a waste may remain unidentified or undetected due to the insensitivity of the analytical technique, the relatively small amount of the chemical present in the mixture (even though the chemical may be a potent genotoxicant), or the lack of an appropriate standard. For example, Jungclaus et al. (1978) performed a detailed chemical analysis of the wastewater and the receiving waters and sediments of a chemical manufacturing plant. Their analysis resulted in the accumulation of thousands of spectra. However, following an extensive and exhaustive evaluation using several analytical techniques, including vapor stripping of volatile organic compounds, direct aqueous injection gas chromatography, computerized gas chromatography and mass spectrometry (GC/MS), high pressure liquid chromatography (HPLC), and high resolution mass spectrometry (HRMS), only 123 compounds could be identified conclusively. Further evidence of the problems associated with chemical analysis was provided by Ellis et al. (1982), who tested secondary effluents from 10 municipal and industrial wastewater treatment plants. Although mutagenic activity had been detected in every effluent, only two of the 243 identified organic compounds were carcinogens or mutagens; thus some genotoxic compounds remained unidentified. Finally, Donnelly et al. (1987a) concluded from their analysis of a wood-preserving bottom sediment that the chemical characterization failed to identify the major mutagenic constituents detected in the waste because of matrix interferences or the detection limit of the analytical instrument.

Another problem associated with chemical analysis is that chemical interactions between chemical species (e.g., synergisms, antagonisms) and degradative or transformation potentials are ignored by this approach. Furthermore, an appropriate mathematical model for the determination of total genotoxic activity must be selected. For example, should final estimates of total activity be based on the simple summation of the activities of the individual constituents or should a more complex model be employed?

A second approach to analyzing the individual constituents of a complex mixture such as hazardous waste is to characterize chemical species by fractionating the mixture into chemically defined parts (such as acid, base, and neutral fractions) and then testing each fraction for genotoxicity in an attempt to determine the active chemical class. The goal of this approach is to isolate active components from inactive or toxic constituents and to identify the hazardous compounds when possible. This approach is often combined with an extraction/concentration step prior to fractionation to provide a sample with sufficient (detectable) activity. Extraction/concentration can be performed using organic or aqueous solvents or by a resin extraction procedure (e.g., XAD-2 or XAD-8 columns). Following extraction/concentration, the sample is subdivided by a combination of solvent partition steps into chemical classes based on relative acidic-basic properties or relative polarities. Keleti et al. (1982), for example, used increasingly polar organic solvents to serially extract coal liquefaction wastewater treatment sludges and solids to determine the most active fraction(s) and to provide tentative identification of the chemical class(es) associated with each fraction.

Through iterative chemical fractionation and biological testing, subfractions with relatively high concentrations of genotoxic compounds may be obtained for a final assessment of chemical constituents. This technique, called bioassay-directed fractionation, allows the biological effect to direct the fractionation scheme toward the ultimate identification of the biologically active constituent(s). Analytical techniques (such as HPLC and GC/MS) may be used at this point in an attempt to identify important constituents. An example of bioassay-directed fractionation is provided by Donnelly et al. (1987a), who performed a chemical analysis of the mutagenically active fractions of a wood-preserving waste to provide information on the potential source of genotoxic activity in each fraction. Disadvantages of using this approach include the limited amount of sample available for bioassay following processing and the probable loss or modification of components during any one of the processing steps (extraction, concentration, fractionation). For example, Chen et al. (1983) suggested that the lack of mutagenic potential in the various fractions of an above-ground retort process water (compared to the high mutagenic potential of the whole sample) may have been due to a loss of components during the extraction procedure.

Method-2

The second approach to testing hazardous waste for genotoxicity is to treat the mixture as a single entity. That is, test the whole (crude) sample instead of the individual parts. Houk and DeMarini (1988), DeMarini et al. (1987b, 1989), and Simmons et al. (1988) administered crude waste samples to several biological systems, including a bacterial assay for mutagenicity; a bacterial assay for DNA damage; a short-term rodent assay for lethality, hepatotoxicity, and urinary mutagenicity; and an assay for the detection of DNA adducts formed in tissues of rats exposed to the wastes. The obvious advantages of this approach include the relevancy of the tested sample to its environmental counterpart, the absence of artifacts generated by the extraction/concentration or fractionation processes, and elimination of the need for modeling interactive effects. Disadvantages include the possible incompatibility of the complex (and often toxic) sample with the biological test system, the inability to positively identify the constituent(s) of the mixture responsible for the genotoxicity, bacterial contamination (although this rarely poses a problem due to the toxicity of the sample), the potential for small amounts of contaminants to go undetected without concentration of the sample, and the possibility of constituent activity to be masked by chemical interactions within the whole sample. Cunningham et al. (1985), for example, demonstrated that the neutral fraction of a condensate tar from a coal conversion process had a much higher genotoxic potential than the whole tar sample, suggesting that compounds in the whole sample interacted in an inhibitory manner. Thus, although this testing stratagem seems more relevant for use with environmental samples, difficulties in testing may be encountered, and problems with interpretation of results may be introduced.

THE GENOTOXICITY OF HAZARDOUS WASTES

Many different classes of hazardous and industrial waste mixtures have been tested for genotoxicity, but it is often difficult to ascertain from published reports the regulatory classification of the tested sample. Therefore, we have included in this review RCRA-classified wastes, wastes of unspecified or unknown classification, and wastes that are not regulated by RCRA but which may be regulated under Section 402 of the Clean Water Act as point source discharges. Of the wastes that are RCRA-regulated, several were derived from specific sources, such as the bottom sediment from wood-preserving processes (K001) and bottoms from an acetonitrile purification column (K014); others originated from nonspecific sources, e.g., the organic still bottoms from a solvent recovery process (F005).

For purposes of this discussion, we have included effluents, discharges, sludges, and wastewaters from industrial sources (for example, pulp and paper mills, textile and dye manufacturers, steel mills and foundries, pharmaceutical supply companies, pesticide manufacturers, and industrial and hazardous waste treatment facilities). Omitted from this discussion are combustion emissions, airborne pollutants released from industries or wastesites, pure chemical compounds that would be classified as hazardous waste under RCRA, municipal wastewater treatment effluents and sludges, and sediments, surfacewaters, and groundwaters suspected of being contaminated with industrial wastes.

Petrochemical Wastes

The relatively recent emergence of petroleum as a feedstock for the synthesis of organic chemicals revolutionized the entire chemical industry. Hydrocarbons could be polymerized to form plastics. Coal tars could be converted to dyes, and solvents and degreasing agents could be produced from the more volatile fractions. Furthermore, the addition of halogen groups to hydrocarbons provided the means to manufacture complex organic compounds resistant to degradation, including chlorinated pesticides, polychlorinated biphenyls, and polyvinyl chloride. When tested, a large number of petrochemicals were found to be potent carcinogens, mutagens, and/or teratogens. Thus, the wastes generated by these industries represent an extremely important category for study.

Textiles and Dyes

The effluents from four different dye-related industries (manufacture of dyes and epoxy resins, textile dyeing and finishing, paper dyeing, and manufacture of hair dyes and herbicides) were investigated by McGeorge et al. (1985). All effluents were mutagenic in one or more strains of the Salmonella assay. In fact, of the 33 different industrial discharges studied, the dye-producing industries generated some of the most mutagenic wastes tested, with the dye and epoxy resin facility effluent being one of the most potent.

In another extensive study of industrial effluents, Sanchez et al. (1988) examined the genotoxicity of 82 industrial wastewaters from an urban industrial complex located in Greater São Paulo, Brazil. The majority of the effluents were discharged into surface waters that were later used for human consumption and agricultural and recreational activities. Three microbial mutagenicity assays were used to test the effluents: *S. typhimurium, E. coli* WP_2, and *S. cerevisiae*. Of the waste categories tested, the textile industries contributed the highest percentage of mutagenic effluents. (Additional results from this study are presented elsewhere in this paper.)

Somashekar (1987) described various chromosomal abnormalities in the flower buds of *Chlorophytum amaniense* that were induced by dye industry wastewater discharged into a sewage canal in India. Physicochemical characteristics indicated that the effluents were acidic and contained considerable amounts of heavy metals such as chromium, copper, and zinc. The buds were dipped for various periods of time into different concentrations of the wastewater. Following treatment, the buds were fixed and stained. The dose-dependent production of a high percentage of chromosomal abnormalities was observed and was linked to the presence of the heavy metals.

In a similar study, Somashekar et al. (1985) examined the induction by dye manufacturing effluents of chromosome abnormalities in onion (*Allium cepa*) root tip cells. The effluents were acidic (pH 3.4) due to the presence of Azored, Acid Violet, and Triphenyl methane, and extremely high levels of chromium were measured. Various chromosome abnormalities were induced by the effluents, including chromosome breaks, binucleate cells, anucleation, laggards, etc. Some of the abnormalities observed in this study were also observed by Ravindran and Ravindran (1978) in their analysis of textile factory effluents. In this study, river waters heavily polluted by discharges from a nearby rayon factory were used to irrigate two plant species: *Ornithogalum virens* and *Allium cepa*. Cytological analysis revealed that chromosome breakage, dicentrics, lagging, and stickiness were common, and the frequency of abnormalities increased from ~10% in control plants to ~36% in plants treated with undiluted river water. No chemical analysis was performed; however, the waters were slightly acidic (pH 5.8).

An unextracted ink pigment sludge was evaluated by Houk and Claxton (1986) in an assay that combines physical separation of chemical constituents by thin layer chromatography (TLC) with an in situ analysis of mutagenicity using the Salmonella assay. The primary constituents of this sample were identified as naphthalene, phenol, ethylbenzene, toluene, and high concentrations of lead. In addition, several mutagens/carcinogens (benzo(a)pyrene, fluoranthene, and phenanthrene/anthracene) were present at low concentrations. However, test results indicated no evidence for the presence of mutagenic frameshift constituents and results for base-pair substitution mutagens were inconclusive.

Møller et al. (1984) detected the presence of potent, direct-acting frameshift mutagens (using strain TA98 in the Salmonella assay) in an organic extract of an effluent from a textile plant. Mutagenic activity reportedly decreased by 50% upon addition of exogenous metabolic activation. Furthermore, the acid-neutral fraction of the extract contained significantly more mutagenic material than the basic fraction. The extract was then separated by TLC and tested in situ with the Salmonella assay. Three distinct clusters of mutagenic activity were observed on the plate, indicating the presence of at least three different compounds acting as frameshift mutagens.

In summary, dye-related industries appear to produce consistently genotoxic (and relatively potent) effluents. Genetic damage induced by the discharges from this industrial category include mutations and chromosome abnormalities. In addition to the dyes themselves, the presence of heavy metals in these wastes may also be a possible source of genotoxic activity.

Rubber and Miscellaneous Plastics Products

Houk and DeMarini (1988) tested a black tar produced during plastics manufacturing. Chemical characterization (limited to specific organics and/or metals identified in Appendix VIII of RCRA) showed that the tar contained high levels of aniline (550 mg/g). The tar was tested for its ability to induce prophage in *E. coli*, an endpoint that is indicative of DNA damage repair. The waste was positive when tested with or without metabolic activation in this assay. Moreover, among 14 wastes tested, it ranked third in potency behind a petrochemical and a pharmaceutical waste. This same waste was tested for mutagenicity in Salmonella (DeMarini et al., 1987b; DeMarini and Houk, 1988). The crude waste as well as organic (dichloromethane) extracts of the waste were evaluated and were found to be mutagenic. In addition, the crude waste was administered by gavage to rats, and 24-hr urine samples were collected and tested for urinary mutagenicity in Salmonella. The urines derived from exposure of rats to the crude waste were mutagenic, suggesting that mutagenic constituents of the waste (or mutagenic metabolites converted by the rat in vivo) were ultimately eliminated in the urine. Of 10 industrial wastes tested for urinary mutagenicity, this sample was the most potent.

McGeorge et al. (1985) used the Salmonella mutagenicity assay to test four resin manufacturing waste effluents (two polyvinyl chloride samples, a rosin-derived sample, and an ion exchange sample) and an effluent from a chemical industry that produced plasticizers. Only the rosin-derived and plasticizer samples were mutagenic. In a ranking of the mutagenic potencies of 15 industrial waste samples, the rosin-derived effluent ranked 14th, and the plasticizers effluent ranked 6th (behind two samples from an organic chemicals plant and three dye effluents).

Finally, two by-products from vinyl-chloride production—an EDC tar and an EDC spent caustic (water portion)—were tested for mutagenicity in Salmonella. The name EDC is derived from one of the main components of the waste, ethylene dichloride. The tar consisted primarily of chlorinated aliphatic hydrocarbons (Rannug and Ramel, 1977). The spent caustic was mostly ethylene dichloride (4.5 mg/g), but many other chlorinated compounds were also detected (Houk, 1984). Due to its poor solubility in water, Rannug and Ramel dissolved the tar in three different solvents prior to administration to Salmonella strain TA1535. Base-pair substitution mutagens were detected in all three samples, and the mutagenic effect was greatly enhanced by the metabolic activation system. Houk and Claxton (1986) administered the crude caustic effluent to a TLC plate for separation of constituents and performed mutagenicity testing directly on the developed plate. They detected the presence of a fluorescent, direct-acting, relatively nonpolar, base-pair substitution mutagen.

In summary, the results obtained with the effluents from rubber and plastics manufacturers, in contrast to the results from the textile and dye industries, were highly variable. Response appears to depend upon the production process or industry sampled. The tars from this category, both of which were derived from plastics manufacturing, were genotoxic in several test systems and caused a variety of genetic damage. Moreover, they were quite potent. However, other waste samples from this category were not particularly genotoxic.

Miscellaneous Chemical Plants

Commoner and his colleagues (1977, 1978) undertook a study to detect and isolate mutagens present in the effluents of various petrochemical plants situated along the Houston (Texas) Ship Channel, which received the discharges. Limited mutagenicity testing was performed (Salmonella strain TA1538 only, limited doses). Of 12 petrochemical plant effluents tested, two were weakly mutagenic and one was strongly mutagenic with metabolic activation, indicating the presence of indirect-acting frameshift mutagens.

In a similar study by Somani et al. (1980), effluents from various petrochemical industries were sampled after the last treatment process prior to discharge into the Illinois River. The Salmonella mutagenicity assay was performed using strains TA98, TA100, and TA1537 with exogenous metabolic activation. Chemical characterization was performed using GC/MS. Results from the chemical analysis suggested that aliphatic hydrocarbons and their derivatives were primary components of petrochemical wastes; polynuclear aromatics also were found. Many of these classes of compounds include carcinogens and/or mutagens. However, marked toxicity, which was exhibited by most of the wastes in the mutagenicity assay, together with the fact that there were limited amounts of sample to test, precluded the detection of mutagenic activity.

Sanchez et al. (1988) assessed a number of chemical industry wastewaters in their study of urban discharges into the rivers of Greater São Paulo (see *Textiles and Dyes*). They examined both acute toxicity and mutagenicity using three short-term microbial assays. Within the group of very toxic effluents, most (37%) came from the chemical industries. These discharges were also mutagenic, but were not among the most potent industrial waste types tested.

Sundvall et al. (1984) used the Salmonella assay to test extracts obtained from the wastewater of a chemical process industry that produces primarily nitrobenzoic acids and nitrotoluenes. Gas chromatography was used to analyze for the presence of nitrobenzenes and nitrotoluenes; HPLC was used for the analysis of benzoic acids and cresols. Four nitroreductase-proficient strains of Salmonella were used (TA98, TA100, TA1535, and TA1538), and nitroreductase-deficient strains were included to demonstrate the contribution of nitroaromatic compounds to the mutagenicity of the effluent. Both base-pair substitution and frameshift mutagens were detected, and results from the nitroreductase-deficient strains demonstrated that the mutagens were largely nitroaromatic compounds. Only 30–40% of the mutagenic activity could be related to the identified nitroaromatics; one compound (3,5-dinitrobenzoic acid) was responsible for more than 80% of this activity. Activity decreased in the presence of S9 activation.

The team of Donnelly and Brown and their colleagues at Texas A & M University has performed extensive research in the analysis of hazardous industrial wastes. Brown and Donnelly (1984) studied the mutagenic and DNA-damaging potential of a bottom stream from an acetonitrile purification column (RCRA-classified waste K014). Chemical analysis of the waste indicated that it was approximately 95% water, 3.46% acetonitrile, 0.92% acetamide (a carcinogen), and a small percentage of heavy ends. Acid, base, and neutral fractions of an organic extract of the waste were tested for mutagenicity in the Salmonella assay and for lethal DNA damage in six strains of *B. subtilis*. Results from the mutagenicity test indicated that activity could be detected in all three fractions, with the basic fraction being the most potent. Primarily indirect-acting mutagens were present. However, the activity could not be associated completely with the carcinogenic constituent (acetamide). Instead, their data indicated that small concentrations of other unidentified mutagens or promoters had to be present to account for the total observed activity. None of the fractions induced increased lethal DNA damage, and it was suggested that this was possibly due to the absence of metabolic activation in this assay, since activation enhanced detection in the Salmonella assay.

In another study, Donnelly et al. (1987a,b) looked at the bottom sediment produced by the wood-preserving industry (RCRA-classified waste K001). Chemical constituents in this waste were believed to include phenols, cresols, PAHs, pentachlorophenol, and other chlorinated hydrocarbons. Waste extracts were partitioned into acid, base, and neutral fractions and tested in two prokaryotic bioassays (for mutagenicity in Salmonella and lethal DNA damage in *B. subtilis*) and in a eukaryotic bioassay (for mutagenicity and chromosomal damage in the haploid and diploid forms, respectively, of *Aspergillis nidulans*). All three fractions as well as the crude waste were mutagenic in Salmonella with metabolic activation. Furthermore, the acid and base fractions induced increased lethal DNA damage with activation. Additional testing in *A. nidulans* for mutagenicity and chromosomal damage revealed that all three waste fractions induced increased mutation frequencies, both with and without metabolic activation, and that each fraction preferentially induced a different type of chromosome abnormality, indicating the presence of several different clastogens (i.e., chemicals that induce chromosomal damage) in the waste. A comparison of the chemical (GC/MS) and biological analyses of this waste led the authors to conclude that the chemical analysis was unable to identify the major mutagenic constituent(s) of the waste.

Houk and Claxton (1986) used the TLC/Salmonella assay to detect the mutagenic constituents in two crude wastes from the chemical industries (an organic sludge from an industrial solvent recovery process and a latex paint manufacturing waste) and an ethanol extract of organic still bottoms. The waste samples were analyzed for inorganics by emission spectroscopy and for organics by GC/MS (Houk, 1984). Calcium and iron as well as a phthalate ester, 2-naphthol, and many unknowns were identified in the organic sludge. Results from the TLC/Salmonella assay indicated the presence of a relatively nonpolar, fluorescent, base-pair substitution mutagen that was not identified by the chemical analysis of the waste. The latex paint waste contained high levels of ethylbenzene, zinc, copper, and mercury, as well as smaller amounts of some carcinogenic/mutagenic metals such as arsenic and selenium. However, mutagenicity results from this sample were negative. (In general, the Salmonella assay does not detect genotoxic metals without some modification of the bioassay, so negative results are not surprising.) The organic still bottoms were shown to contain milligram per gram levels of several chlorinated organics, including tetrachloroethylene and ethylene dichloride, and high levels of copper. Low levels of a mutagen/carcinogen (chrysene/benzo(a) anthracene, 2.3 μg/g) were also detected. However, there was no evidence of mutagenic activity in this sample. These results illustrate the importance of combined chemical and biological testing. Wastes exhibiting no detectable traces of mutagens/carcinogens were mutagenic when biological screening was applied, possibly due to the presence of unidentified or undetected mutagens. Conversely, wastes that did contain detectable (albeit low) concentrations of known mutagens did not demonstrate activity in this particular assay. Because crude, unconcentrated samples of these wastes were tested, it is possible that the mutagens were present at levels too dilute to be detected by biological means.

In other investigations conducted by this group, a petrochemical waste oil was evaluated for its potential to cause mutations in the Salmonella assay (DeMarini et al., 1987b) and to induce DNA damage/repair in an *E. coli* prophage-induction assay (Houk and DeMarini, 1988). This waste oil was chemically analyzed for a number of organic chemicals and metals identified in Appendix VIII of RCRA, and high levels of toluene, carbon tetrachloride, and selenium were identified. In the prophage-induction assay for DNA damage, the waste was extremely potent relative to a group of 11 genotoxic industrial effluents (second only to a pharmaceutical effluent). The waste was also mutagenic in Salmonella, apparently containing base-pair substitution mutagen(s), and was the most potent of 15 industrial wastes tested. The chemical analysis performed on the waste was unable to suggest the remarkable genotoxic potencies observed. Once again, these results illustrate the importance of biological indicators of hazard.

In these same studies, chemical wastes of unspecified origin were collected from four commercial hazardous waste incineration facilities that composited wastes obtained from various industrial sources. Wastes were segregated prior to incineration according to their aqueous or organic properties. Liquids, semisolids, and tars were included, and chemical analyses detected the presence of cresols, chlordane, solvents (phenol, benzene, toluene, trichloroethylene, among others), and lead. Of 11 crude waste samples tested for mutagenicity in Salmonella, five were genetically active. In addition, when seven of the wastes were administered to rats by gavage, two of the wastes produced mutagenic urine (DeMarini et al., 1987b). When this same set of 11 crude wastes was tested in *E. coli* for the induction of prophage, eight were genotoxic, indicating DNA damage (Houk and DeMarini, 1988). It is probable that the genotoxic solvents and metals present in the waste were preferentially detected by the prophage assay due to the relative insensitivity of the Salmonella assay to these classes of compounds. A summary of this comparison was presented by DeMarini and Houk (1988).

Finally, McGeorge et al. (1985) examined a variety of organic chemicals industries and seven of nine effluents were mutagenic in the Salmonella assay. Of almost 30 waste types tested, industries manufacturing organic chemicals had some of the most genotoxically potent discharges.

In conclusion, chemical industry wastewaters are often quite toxic. Investigators should take this into consideration when devising test strategies; fractionating the waste prior to testing is a reasonable way to segregate the toxic constituents from the genetically active components. Chemical analyses performed on chemical manufacturing wastes have identified aliphatic and aromatic hydrocarbons (including chlorinated organics) with known carcinogenic/mutagenic potential, nitroaromatics, solvents, and carcinogenic and noncarcinogenic metals. The genotoxic potency of wastes from this industrial category obviously depends upon the chemicals manufactured and the process waste tested, but in general this group of wastes is significantly genotoxic. Because many of the samples reported here were collected just prior to discharge into rivers and other surfacewaters, wastes from this category could pose a substantial threat to human health and the environment.

Pesticides

Surprisingly few studies of pesticide wastes were found in the literature. An herbicide manufacturing acid was tested for mutagenicity by Houk and Claxton (1986) in the TLC/Salmonella assay. A fluorescent base-pair substitution mutagen that required metabolic activation for expression of activity was detected. An herbicide manufacturing waste (acetone/water sample) was tested by DeMarini et al. (1987a) and was found to be mutagenic in both the Salmonella and the mouse lymphoma mutagenicity assays, but did not induce cytogenetic effects or cell transformation in cultured mammalian cells.

Pulp and Paper Mills

The biological testing and analysis of wastes generated by the pulp and paper mill industries account for perhaps the greatest body of data in the literature on industrial effluents. The majority of the research has been conducted by Scandinavian countries (primarily Sweden and Norway) and Canada, where the forest products industry represents one of the leading industrial sectors.

In the manufacture of paper from wood, lignin is removed from the wood by a pulping process that entails "cooking" the wood at very high temperatures. Either a kraft (sulfate) plant or a sulfite plant performs this process. However, neither process is able to completely remove the lignin, so removal of residual lignin and various other organic constituents is achieved by a multistage bleaching process involving chlorination. Spent chlorination stage and alkali extraction liquors are formed as waste products and are often discharged directly into receiving waters.

The effluents from pulp and paper mills have been shown to contain dissolved lignin, cellulose degradation products, other wood extractives, and chlorinated compounds derived from the bleaching process. About 300 chemical compounds have been identified as constituents of the effluents, including three organic compounds with known carcinogenic properties (safrole, chloroform, and carbon tetrachloride) and 16 compounds listed in the NIOSH (National Institute for Occupational Safety and Health) subfile of "suspected carcinogens" (Lee et al., 1978). Because the receiving waters for these effluents are often used as sources of drinking water, there is great concern regarding the potential of these effluents to cause human health effects.

The genotoxic potential of the waste liquors and effluents produced by pulp and paper mills was first examined over a decade ago. Positive findings prompted researchers to try to isolate and identify those constituents responsible for the observed genotoxic activity. Since then, investigations have focused on understanding the processes leading to the formation of the genotoxic constituents, understanding their environmental fate, determining which process wastes are most hazardous, and evaluating effective treatment methodologies.

Due to the enormous body of literature within this particular industrial category, findings will not be reviewed on a paper-by-paper basis; rather, a more generalized summary of conclusions will be presented. An analysis of the hazardous constituents identified in pulp and paper mill effluents will be presented first, followed by a discussion of findings from the testing of the effluents themselves. In addition, the various treatment methodologies devised to reduce the genotoxicity of pulp and paper mill effluents will be discussed here rather than in the section on hazardous waste treatment.

Constituents

Many of the 300 chemical constituents identified in pulp and paper mill effluents have been examined for their potential to cause genotoxic effects. Those compounds found to induce genetic damage are referenced in Table 1. Many of the compounds responsible for the genotoxic activity found in the spent bleach liquors from sulfate pulp plants have been found to be responsible for the activity in effluents from sulfite pulp plants as well (Kringstad et al., 1981; Kringstad and Lindstrom, 1984; Carlberg et al., 1986); however, the concentration of the identified mutagens is lower in the effluent from the sulfite plants (McKague et al., 1981). The chemicals responsible for this activity are almost exclusively low molecular weight, nonpolar compounds (Stockman, et al., 1980; Rannug, 1980; Rannug et al., 1981; Møller et al., 1986). Many are relatively unstable and will decompose to genetically inactive products. However, others are not readily degradable and may persist in the environment and/or bioaccumulate. Kringstad et al. (1984), for example, showed that a small portion of the mutagenic effects caused by spent chlorination liquors could be ascribed to lipophilic compounds with propensity for bioaccumulation.

Effluents

Although effluents from both kraft and sulfite plants have been tested for genotoxic activity, the vast majority of analyses have been performed on the effluents from kraft pulp mills. Furthermore, of all the process streams studied, the chlorination stage liquors have been the most extensively examined, presumably because they are the most genotoxic of the effluents discharged by the pulp and paper mill industries.

In the first study of the genotoxicity of pulp and paper mill effluents (Ander et al., 1977), unconcentrated effluents obtained from each of the processes involved in the conventional kraft pulp bleaching sequence were tested for mutagenicity in the Salmonella assay. Effluents from the chlorination stages (C-stage), the alkaline extraction stages (E), the hypochlorite stage (H), and the chlorine dioxide stages (D) were tested. None of the unconcentrated effluents, except the chlorination stage effluent, exhibited significant mutagenic activity. The hypochlorite stage effluent demonstrated weak activity, but only when concentrated.

After these results were reported, studies of pulp mill effluents intensified. It was soon shown that the bleaching of sulfite pulps also produced mutagenic effluents, although the activity was not as high as that observed with kraft pulp effluents (Ericksson et al., 1979; Bjørseth et al., 1979; Carlberg et al., 1980a; Rannug, 1980; Møller et al., 1986). In addition, it was discovered that the bleaching of softwood pulp produced a more mutagenic effluent than the bleaching of hardwood pulp (Ericksson et al., 1979; Rannug, 1980). Straw pulp was also examined, and results indicated that the sulfite bleaching efflu-

Table 1. Genotoxic Compounds Identified in Pulp and Paper Mill Effluents

Compound	Genotoxic Endpoint	System	Reference
Acetovanillone	M	*S. cerevisiae*	Nestmann et al. (1983)
Benzyl chloride	M	*S. typhimurium*	McCann et al. (1975)
Bromocymene	M	*S. typhimurium*	Bjørseth et al. (1979)
Bromodichloromethane	M	*S. typhimurium*	Simmon et al. (1977)
	M (weak)	*S. cerevisiae*	Nestmann and Lee (1985)
	GC (weak)	*S. cerevisiae*	Nestmann and Lee (1985)
3-Chloro-4-dichloromethyl-5-hydroxy-2(5H)-furanone	M	*S. typhimurium*	Holmbolm et al. (1981) Ishiguro et al. (1987)
Chloromuconic acid	M	*S. cerevisiae*	Nestmann et al. (1983)
2-Chloropropenal	M	*S. typhimurium*	Kringstad et al. (1981, 1983)
Dibromochloromethane	M	*S. typhimurium*	Simmon et al. (1977)
	GC	*S. cerevisiae*	Nestmann and Lee (1985)
1,3-Dichloroacetone	M	*S. typhimurium*	Rapson et al. (1980) Stockman et al. (1980) Kringstad et al. (1981)
	M (weak)	*S. cerevisiae*	Nestmann and Lee (1985)
	GC	*S. cerevisiae*	Nestmann and Lee (1985)
Dichlorocatechol	M	*S. cerevisiae*	Nestmann et al. (1983)
Dichloro-p-cymene	M	*S. typhimurium*	Bjørseth et al. (1979)
Dichloroethane	M	*S. typhimurium*	Brem et al. (1974) Nestmann et al. (1980)
Dichloroguaiacol	M	*S. cerevisiae*	Nestmann et al. (1983)
Dichloromethane	M	*S. typhimurium*	Simmon et al. (1977) Nestmann et al. (1980)
Hexachloroacetone	M	*S. typhimurium*	Rapson et al. (1980) McKague et al. (1981) Kringstad et al. (1981) Douglas et al. (1985)
	GC	*S. cerevisiae*	Nestmann and Lee (1985)
Monochloroacetaldehyde	M	*S. typhimurium*	McCann et al. (1975) Kringstad et al. (1981)
Neoabietic acid	M	*S. typhimurium*	Nestmann et al. (1979)
	M	*S. cerevisiae*	Nestmann et al. (1983)
1-Oxa-6,7,10,10-Tetra-chlorospiro(4.5)dec-6-en-8,9-dione	M	*S. typhimurium*	Douglas et al. (1985)
1-Oxa-6,10,10-Trichloro-spiro(4.5)dec-6-en-8,9-dione	M	*S. typhimurium*	Douglas et al. (1985)
7-Oxydehydroabietic acid	M	*S. cerevisiae*	Nestmann et al. (1983)
Pentachloroacetone	M	*S. typhimurium*	McKague et al. (1981) Douglas et al. (1985)
	GC	*S. cerevisiae*	Nestmann and Lee (1985)
Pentachloropropene	M	*S. typhimurium*	Nestmann et al. (1980)
	M	*S. cerevisiae*	Nestmann et al. (1983)
	CA	CHO cells	Douglas et al. (1983)
	SCE	CHO cells	Douglas et al. (1983)
1,1,3,3-Tetrachloroacetone	M	*S. typhimurium*	Rapson et al. (1980) McKague et al. (1981) Douglas et al. (1985)
	M (weak)	*S. cerevisiae*	Nestmann and Lee (1985)
	GC	*S. cerevisiae*	Nestmann and Lee (1985)
1,1,2,2-Tetrachloroethane	M	*S. typhimurium*	Brem et al. (1974)
Tetrachloroethylene	M	*S. typhimurium*	Cerna and Kypenova (1977)
Tetrachloropropene	M	*S. typhimurium*	Nestmann et al. (1980)
	M	*S. cerevisiae*	Nestmann et al. (1983)
	M	*E. coli*	Ellenton et al. (1981)

Table 1 continued

Compound	Genotoxic Endpoint	System	Reference
Tetrachloropropene	CA	CHO cells	Ellenton et al. (1981)
			Douglas et al. (1983)
	SCE	CHO cells	Ellenton et al. (1981)
			Douglas et al. (1983)
Trichloroacetaldehyde	M	*S. cerevisiae*	Nestmann and Lee (1985)
1,1,3-Trichloroacetone	M	*S. typhimurium*	Douglas et al. (1985)
	M (weak)	*S. cerevisiae*	Nestmann and Lee (1985)
	GC	*S. cerevisiae*	Nestmann and Lee (1985)
Trichloroethane	M	*S. typhimurium*	Simmon et al. (1977)
			Nestmann et al. (1980)
Trichloroethylene	M	*S. typhimurium*	Cerna and Kypenova (1977)
			Simmon et al. (1977)
			Kringstad et al. (1981)
1,2,3-Trihydroxybenzene	M	*S. typhimurium*	Carlberg et al. (1980b)

Notes: M = mutation.
GC = gene conversion.
CA = chromosome aberrations.
SCE = sister chromatic exchange.

ent from wheat and rye straw pulp was comparable in mutagenicity to the bleaching effluents from kraft pulp (Lindgaard-Jorgensen and Folke, 1985). It should be noted that one group of investigators has shown that the effluent obtained from a thermomechanical pulping plant does not cause mutations in Salmonella, nor does it cause chromosome aberrations in CHO cells (Douglas et al., 1980; Lee et al., 1981).

As stated previously, the chlorination stage effluents are consistently the most genotoxic of all the process streams. Not only do they cause mutations in Salmonella (Ander et al., 1979; Bjørseth et al., 1979; Eriksson et al., 1979; Hoglund et al., 1979; Douglas et al., 1980, 1983; Rannug, 1980; Lee et al., 1981; Holmbolm et al., 1981, 1984; Kamra et al., 1983; Monarca et al., 1984), but they also cause mutations in mammalian (CHO) cells (Rannug et al., 1981), chromosome aberrations in mammalian (CHO) cells (Douglas et al., 1980, 1983, 1985; Lee et al., 1981), gene conversion and mitotic recombination in yeast (Douglas et al., 1983, 1985; Kamra et al., 1983; Nestmann, 1985), and sister chromatid exchanges in CHO cells (Douglas et al., 1983, 1985). The mutagenic effect of the effluent is dramatically reduced when an exogenous metabolic activation system is introduced to the test system (Hoglund et al., 1979; Bjørseth et al., 1979; Rannug et al., 1981; Kamra et al., 1983; Douglas et al., 1983), suggesting that in vivo metabolism may influence response.

Efforts to isolate and identify the compounds responsible for the genotoxicity of the chlorination stage effluents have been highly successful. The genetically active compounds have been found to originate from the residual lignin isolated from the unbleached pulp (Stockman et al., 1980; Nazar and Rapson, 1980). Three compounds have been identified as particularly potent mutagens in the kraft pulp effluent: 3-chloro-4-dichloromethyl-5-hydroxy-

2(5H)-furanone, 1,3-dichloro-acetone (a chlorinated derivative of the chlorofuranone), and 2-chloropropenal. It has been reported that the chlorofuranone may account for as much as 30 to 50% of the total mutagenicity observed in kraft chlorination effluent (Holmbolm et al., 1981). Furthermore, 2-chloropropenal has been shown to be responsible for as much as 45 to 55% of the activity observed in spent chlorination liquor (Kringstad et al., 1983). Thus, these two compounds alone may be responsible for almost all of the mutagenicity demonstrated by the chlorination stage waste. The spent chlorination liquors from sulfite pulp are not as thoroughly characterized; only a few weak mutagens have been identified, including 1,2,3-trihydroxybenzene and bromo-and dichloro-p-cymene.

In contrast to the genotoxicity demonstrated by the C-stage effluent, effluents from the alkaline extraction, hypochlorite, and chlorine dioxide stages generate little or no mutagenic effects (Ander et al., 1977; Bjørseth et al., 1979; Møller et al., 1986). Furthermore, untreated whole mill effluents are nonmutagenic or only weakly mutagenic (Hoglund et al., 1979; Rannug, 1980; Douglas et al., 1980; Lee et al., 1981). Presumably this is due to the fact that effluents from the different bleaching stages are combined at the plant prior to discharge, diluting the effects of the more potent C-stage effluent.

Treatment

There are several ways to reduce or eliminate the genotoxicity (and simultaneously the aquatic toxicity) of pulp and paper mill effluents: biological treatment, increasing the pH of the effluent to the alkaline level, high chlorine dioxide substitution for chlorine in the bleaching process, treatment of the effluent with sulfur dioxide, and ion exchange. Each of these treatment methods is discussed briefly here.

When pulp mill effluents are treated with biological sludge in an aerated lagoon, a considerable reduction in mutagenic activity is observed (Eriksson et al., 1979; Hoglund et al., 1979; Stockman et al., 1980; Douglas et al., 1980; Lee et al., 1981; Nestmann et al., 1984). Several mechanisms could be involved in the reduction, including biological degradation, physical adsorption of the mutagenic compounds on the biomass, and chemical instability of the compounds.

Treatment of the C-stage effluent under alkaline conditions (addition of either NaOH or $Ca(OH)_2$) is the most effective way to reduce or remove the mutagenicity of the effluent, provided that sufficient alkali is added to raise the pH to ~ 10 (Ander et al., 1977; Eriksson et al., 1979, 1982; Stockman et al., 1980; Rapson et al., 1980; Carlberg et al., 1980a; Rannug, 1980; Nazar and Rapson, 1980, 1982; Douglas et al., 1980, 1983; Lee et al., 1981; Møller et al., 1986). These results indicate that the mutagenic compounds in the effluent are alkali sensitive. Studies undertaken by Lee et al. (1981) have suggested that chlorinated organics undergo some amount of decay under alkaline conditions. Carlberg et al. (1986) later showed that the alkali treatment destroyed the highly potent mutagen 2-chloropropenal and most of the chlorinated acetones.

Replacing chlorine with chlorine dioxide in the first bleaching stage also reduces the mutagenicity of pulp and paper mill effluents (Eriksson et al., 1979; Stockman et al., 1980; Nazar and Rapson, 1980; Møller et al., 1986). Mutagenicity decreases with increasing chlorine dioxide/chlorine ratio; and with pure chlorine dioxide, the mutagenic activity returns almost to background levels. Sequential chlorination yielded higher activity than simultaneous or premixed addition (Donnini, 1983).

The "replacement" of chlorine with SO_2 resulted in a decrease in the mutagenic activity of the spent chlorination effluent, indicating that some or all of the mutagenic compounds present in the effluent react with the SO_2 (Eriksson et al., 1979, 1982; Douglas et al., 1980; Lee et al., 1981; Donnini, 1981, 1983; Møller et al., 1986). Although reduction was significant, the methodology may not be sufficiently practical to be used on a large-scale basis.

Ion exchange, which effectively removes colored materials from the bleaching effluent in the first alkaline extraction phase, also has been shown to remove a major portion of the mutagenicity (Eriksson et al., 1979; Stockman et al., 1980). It also has been demonstrated that oxygen/ozone bleaching prior to chlorination reduced the concentration of chlorinated organics in the final effluent, but did not reduce the potential for formation of mutagenic compounds during subsequent chlorination (Carlberg et al., 1980a).

Primary Metal Industries

This major group includes industries engaged in smelting and refining metals and in converting and alloying metals. Blast furnaces (including coke ovens) also fall under this category. As pointed out earlier, the genotoxicity of airborne contaminants emitted from these industries will not be reviewed.

Metallurgy

Sanchez et al. (1988) tested wastewater effluents from the metallurgical industries of an industrial complex in São Paulo, Brazil, for acute toxicity and mutagenicity in three short-term microbial bioassays. As a group, the metallurgical industries contributed a large percentage of highly toxic effluents (second only to the chemical industries). Furthermore, of the metallurgical wastes tested, a very high percentage (60%) were mutagenic. Mutagenic metallurgical (foundry) effluents have also been reported by Somani et al. (1980) and McGeorge et al. (1985).

Electroplating Industries

Somashekar and Arekal (1983) found that pollutants present in electroplating wastewater induced chromosome aberrations in plants (onion bulbs) grown in varying concentrations of the wastewater. They concluded that heavy metal complex ions and cyanides were responsible for the observed effects. However, when McGeorge et al. (1985) tested an electroplating waste for mutagenicity in the Salmonella assay, results were negative. If metals are indeed responsible for the cytogenetic effects observed in electroplating wastes as suggested by Somashekar and Arekal, such wastes would not necessarily be expected to be mutagenic in Salmonella because most carcinogenic metals are not detected by this assay.

Steel Mills

Commoner et al. (1978) studied the mutagenicity of an effluent from a steel mill by means of the Salmonella assay. The sample was only slightly mutagenic (in strain TA1538 only) before toxicity interfered with response. Consequently, the sample was applied to TLC plates to partition it into a series of consecutive zones, which were then eluted from the plates and tested in the usual manner in the Salmonella assay. Testing of the zones revealed the presence of both bacteriostatic and mutagenic material; the toxic constituent of the effluent had been successfully separated from the mutagenic constituent(s), allowing the genotoxic potential of the waste to be unmasked.

Coke Ovens

In a study designed to locate the source of mutagens present in the water of the River Meuse, which serves as a source of drinking for regions of Belgium and the Netherlands, Van Hoof and Verheyden (1981) identified two coke oven effluents that contained high levels of frameshift mutagens detected by the Salmonella assay. Fractionations were performed to isolate and identify the mutagens, and activity was shown to reside in subfractions containing polynuclear aromatic hydrocarbons and azaarenes (Van Hoof and Manteleers, 1983).

Wastewaters from a coke plant were extracted with DCM and tested for mutagenicity in the Salmonella assay (Schaeffer and Kerster, 1985). Both the extracts and the raw unfiltered effluent demonstrated mutagenicity.

A coke plant waste (liquid with suspended solids) was tested for mutagenicity in the standard Salmonella assay (Andon et al., 1985) and in the TLC/Salmonella assay (Houk and Claxton, 1986), for mutation at the *tk* locus in mouse lymphoma cells, for SCE induction and chromosome aberrations in CHO cells, and for cell transformation in BALB/C-3T3 cells (DeMarini et al., 1987a). The sample was extremely potent in the conventional Salmonella assay, especially in TA98 with metabolic activation (for example, the ethanol extract of this waste generated a model slope value of 30 revertants/μg in TA98, +S9). Furthermore, the ethanol and DMSO extracts of this waste, as well as the crude waste itself, produced a mutagenic response when tested in the TLC/Salmonella assay. A DCM extract of this waste was mutagenic in the

Salmonella and mouse lymphoma assays and induced cytogenetic effects (both SCEs and chromosome aberrations) in mammalian cells. Verification of the genotoxic potential of the waste came from the GC/MS analysis (Battelle, 1981), which revealed the presence of the mutagens/carcinogens benzo(a)pyrene (1.12 μg/g), benzo(b)-and benzo(k)fluoranthene (0.69 μg/g), and chrysene/benzo(a)anthracene (0.5 μg/g).

As a group, therefore, effluents from the metals industries are substantially genotoxic. Cytotoxic responses were also commonly observed. Chemical analyses allowed investigators to associate the observed effects with the presence of heavy metals, carcinogenic PAHs, and azaarenes.

Miscellaneous Industries

Pharmaceuticals

Pharmaceutical wastes, in general, do not seem to be highly mutagenic, although they may be genotoxic by other mechanisms. A black, oily liquid obtained from pharmaceutical manufacturers was tested in its crude state in the Salmonella assay and the *E. coli* phage-induction assay and was fed by gavage to rats whose urines were then collected and tested for mutagenic activity (DeMarini et al., 1987b; Houk and DeMarini, 1988; DeMarini and Houk, 1988). A chemical analysis of this waste revealed high concentrations of phenylisocyanate and dichlorobenzenes, as well as carbon tetrachloride and trichloroethylene. The waste did not induce mutations in Salmonella, nor did the urine contain mutagenic constituents. However, it did induce prophage in *E. coli,* suggesting that some type of DNA damage had occurred. The preferential detection of genotoxicity by the phage assay may be due to the presence of the solvents or chlorinated compounds identified.

An ethanol extract of a pharmaceutical sludge was tested for mutagenicity by combining thin layer chromatography for the separation of constituents with an in situ Ames test (Houk, 1984; Møller et al., 1985). No mutagenic activity was detected.

Sanchez et al. (1988) reported that effluents from pharmaceutical industries discharging into the rivers of Greater São Paulo were not mutagenic. However, of two pharmaceutical effluents tested by McGeorge et al. (1985), one was mutagenic. Furthermore, when its mutagenic potency was ranked on an activity-per-milliliter basis (i.e., by volume), the pharmaceutical waste ranked fifth relative to 15 other effluent types due to its high residue concentration.

Soaps and Detergents

The mutagenicity of two wastes from the soaps and detergents industry (surfactants and cleaning compounds industries) was investigated by McGeorge et al. (1985) using the Salmonella assay. The surfactants effluent was not mutagenic, and the waste from the cleaning compounds industry gave inconclusive results. Somani et al. (1980) also tested an effluent from an industry producing surfactants, detergents, and plastic moldings in the Salmonella assay; their sample was not mutagenic at the concentrations tested.

Munitions

The mutagenicity of 36 individual constituents present in munitions-production wastewaters was examined by Spanggord et al. (1982) using the Salmonella assay. In addition, these authors synthesized a "condensate wastewater" from the 36 constituents to represent an average discharge from such a plant. The synthesized wastewater was mutagenic in five strains of Salmonella, both with and without metabolic activation. The most potent compounds appeared as trace components of the discharge: 2,4,5-trinitrotoluene, 3,5-dinitroaniline, and 1,3,5-trinitrobenzene contributed 20% of the total mutagenicity of the blend even though their total concentration represented only 0.6% of the condensate.

An effluent obtained from an explosives manufacturer was tested by McGeorge et al. (1985) for mutagenicity using the Salmonella assay. The potential activity of this sample was masked by its extreme cytotoxicity; the exposed bacteria were unable to survive at the doses tested, and an insufficient sample was available to perform additional testing.

Hazardous Waste Treatment Processes and Facilities

Various treatment methods are available for the processing, conversion, or destruction of hazardous wastes. Included are physical treatments (e.g., carbon filtration), chemical degradation (oxidation, reduction, dechlorination, etc.), and biological treatment (e.g., activated sludges, aerated lagoons, and soil degradation). Several investigators have examined the effectiveness of some of these processes by testing the genotoxic potential of a waste both prior and subsequent to treatment.

The physical treatment of the wastewaters discharged from a hair dye facility was reported by McGeorge et al. (1985). Effluents were collected prior and subsequent to granular activated carbon filtration, and results indicated that the carbon treatment substantially reduced the level of mutagenic substances in the secondary wastewaters.

In another study of physical treatment, Brown and Donnelly (1982) compared the genotoxic effects of water samples collected before and after the dredging of a retention pond containing oily runoff water and waste oil. The Salmonella assay was used to measure mutagenic potential, and a DNA repair assay was used to measure increased lethal damage in repair-deficient strains of *Bacillus subtilis* compared to repair-proficient strains. Water samples were concentrated on a nonpolar resin and tested in four strains of Salmonella and six strains of *B. subtilis*. A chemical analysis was performed using GC/MS, which indicated that the "before" sample contained both hexachlorobenzene and trace quantities of polychlorinated biphenyls, whereas the "after" sample contained neither. Results from samples collected prior to dredging were inconclusive in the Salmonella assay due to bacteriocidal (toxic) effects. However, the presence of a DNA-damaging agent was clearly demonstrated using the DNA repair assay. The water collected after the dreding operation was also toxic to Salmonella, but a slight increase in mutation frequency (in TA98 only) was observed at lower, nontoxic doses. Contrary to the sample collected prior to dredging, the "after" sample did not induce a great deal of DNA damage. Chemical analysis suggested that the reduction in DNA damage may have been due to the removal of chlorinated hydrocarbons upon dredging.

Chemical destruction of some pure compounds found in wastes has been investigated by Lunn and Sansone (1987) and Castegnaro et al. (1985). The successful reductive destruction of compounds containing nitrogen-nitrogen bonds (e.g., nitrosamines) led Lunn and Sansone to report on the development of a method to safely degrade pharmaceutical wastes. Several antineoplastic and antidepressant drugs mutagenic in the Salmonella assay were subjected to reductive destruction and were completely degraded, producing only nonmutagenic reaction mixtures when tested in Salmonella. Castegnaro and coworkers demonstrated that wastes or spills containing mutagenic aromatic amines could be successfully decontaminated by oxidation with potassium permanganate in the presence of sulfuric acid. In their study, the degradation products of nine aromatic amines were tested for mutagenicity in the Salmonella assay. None of the final reaction mixtures were mutagenic.

Both chemical and biological treatment methods were examined by de Raat et al. (1985) in their evaluation of the discharge from a nitrofuran-producing factory in The Netherlands that disposed of its wastewater into the North Sea. Because the wastewater contained mother liquors from the synthesis of two nitrofurans that are generally mutagenic (furazolidone and nitrofurfural diacetate), the untreated wastewater effluent was tested for mutagenicity in the Salmonella assay and for the induction of chromosome abnormalities (sister chromatid exchanges) in mammalian cell culture. The wastewater was extremely mutagenic in strain TA100, indicating the presence of base-pair substitution mutagens; however, the addition of metabolic activation decreased its mutagenic potency. The wastewater also induced sister chromatid exchanges, although induction was not very great. The untreated mother liquors and wastewaters from two production processes at the plant were also tested in the Salmonella assay. All were clearly mutagenic. Experiments were then performed to test the efficacy of various detoxification treatments, including photochemical degradation, biodegradation by means of a chemostat, and detoxification by heating and/or increasing pH upon addition of calcium hydroxide. Light irradiation dramatically decreased the mutagenic activity of the wastewater. Passage of the wastewater through a chemostat decreased the concentration of one highly mutagenic nitrofuran (furazolidone) by almost 50%; surprisingly, however, only a slight decrease in mutagenicity was noted. After the wastewaters had been subjected to detoxification by heating and treatment with alkali, there was a decrease in the mutagenic activity by at least two orders of magnitude. The authors concluded that the wastewater contained mutagens that were not readily biodegradable; however, the photodegradation that would naturally occur subsequent to discharge would reduce mutagenicity. Chemical detoxification by heat and alkali treatment proved to be very effective in reducing the mutagenic potency of the discharge.

The biological treatment of hazardous waste has been investigated by several researchers. Goettems et al. (1987) evaluated the effectiveness of tertiary lagoons for use on effluents generated by a petrochemical complex in Brazil. Several species of resident fish obtained from a series of eight stabilization lagoons were examined for malformations; fin and eye deformities and abnormal jaw development were commonly detected. Incidences were inversely proportional to hydraulic detention of the effluent in the tertiary system. The mutagenicity of the water and sediment from the first and last of the eight lagoons was analyzed using the Salmonella assay. Of the sediment samples, 33% showed mutagenicity (in TA98 only), but results from Lagoon 1 did not differ significantly from Lagoon 8. Water collected from both lagoons was nonmutagenic. The authors concluded that some of the chemicals responsible for the fish malformations were deposited on the bottom sediment of the lagoons, and, because the waters were nonmutagenic, the stabilization lagoons played an important role in the removal from the water of chemicals discharged by the petrochemical industries.

In another study of biological treatment, researchers evaluated the mutagenic potential of runoff water and leachate from sludge-amended soils using the Salmonella mutagenicity assay and the *B. subtilis* DNA repair assay (Brown and Donnelly, 1984). Bottom sludges taken from a petroleum refinery and petrochemical plant were land treated using four different soil types. Tests conducted previously on both sludge samples revealed that the acid, base, and neutral fractions were mutagenic (Donnelly and Brown, 1981). Mutagenicity tests on waters that leached through the soil treated with the sludge indicated that migration of the refinery waste through the soil was initially very rapid, but repeated waste applications caused retention of the waste in the upper layer of the soil. High levels of mutagenic activity were detected in the runoff water, although activity did decrease with time. The petrochemical sludge was retained in the soil such that appreciable mutagenic activity was not detected in the leachate until six months following the second waste application. Greater amounts of mutagenic activity were detected in the runoff. The three-year study showed that the activity of both leachates eventually returned to background levels, presumably due to biological or chemical degradation or transformation.

In a similar experiment, Donnelly et al. (1983) monitored the migration and degradation potential of three hazardous wastes applied to soil using several microbial assays: *S. typhimurium* to detect mutagenicity, *A. nidulans* to detect mutagenicity and chromosome aberrations, and *B. subtilis* to detect DNA damage. Each of the wastes (including the petrochemical sludge described earlier, a wood-preserving bottom sediment, and the liquid stream from an acetonitrile purification column) was incorporated into two different soil types, and leachate and runoff samples were collected and analyzed. The genotoxicity of each crude waste was determined prior to land treatment, and it was found that mutagens were present in all three wastes. Furthermore, the wood-preserving sediment contained agents with the potential to induce DNA damage. Subsequent to land treatment, the leachate and runoff waters were collected in order to determine the potential for migration of mutagenic compounds. Results indicated that although low levels of mutagenic activity were present in the leachates, substantial amounts of activity were detected in the runoff. Activity returned to background levels within five months following the second application of the waste.

A more thorough analysis of the degradation of wastes by land treatment was reported by Donnelly et al. (1987c), who examined soil-amended wood-preserving waste. Bottom sediment was obtained from a sediment pond at a plant using both pentachlorophenol and creosote to treat wood. Mutagenicity tests were performed on the crude sediment as well as extracts periodically collected from two soils into which the waste was incorporated. Tests were performed using a prokaryotic *(S. typhimurium)* and a eukaryotic *(A. nidulans)* bioassay to detect mutations. Soils were extracted at various times over approximately a three-year period, and a rainfall simulator was used to mimic natural rainfall. Extract residues were partitioned into acid, base, and neutral fractions and analyzed by GC/MS. Results indicated that the quantity of extracted organic compounds was significantly reduced with time due to degradation and/or transformation of compounds. However, a highly mutagenic residue was detected in the soil up to 540 days after waste application. Major residual constituents were tentatively identified and included pentachlorophenol, fluoranthene, pyrene, and other mutagens/carcinogens.

Wastes obtained from plants designed to treat industrial wastes also have been evaluated for genotoxicity. Commoner et al. (1977) collected and analyzed wastes from two industrial waste treatment plants that discharged their effluents directly into the Houston, (TX) Ship Channel. Samples were extracted and tested in the Salmonella mutagenicity assay (TA1538 only). A sludge and effluent obtained from one of the two plants was not mutagenically active; however, an effluent from the second plant was unequivocally positive, demonstrating a very high mutagenic activity ratio. A method was then designed to isolate and identify the constituents in the effluent responsible for the activity. Fractionation of the sample showed that the neutral and acid portions were clearly mutagenic, whereas the alkaline fraction was only marginally active. Aliquots of the neutral and acid fractions were further partitioned by thin layer chromatography, and the separation of at least two different mutagens was achieved.

DISCUSSION

The findings presented in this chapter have demonstrated that short-term bioassays can reliably and expeditiously measure the genotoxic potential of hazardous industrial wastes and effluents. Research in this area began about a decade ago, and a relatively large body of scientific data has been generated since that time. Petrochemical wastes have been studied in some detail, especially discharges from chemical manufacturing plants (which are generally toxic) and textile and dye effluents (which, as a group, demonstrate substantial genotoxicity). To the contrary, there is a dearth of information on effluents from pesticide manufacturers. The most extensive and methodical evaluations have been conducted on effluents from pulp and paper mills. The research performed in this area is exemplary and illustrates the effectiveness of cooperative ventures between environmental chemists and toxicologists. The systematic efforts of these investigations have helped to determine which pulping plants generate the most genotoxic effluents, they have established which process wastes are most hazardous, they have isolated and identified the compounds responsible for the genotoxic activity, they have described the environmental fate of these compounds, they have evaluated the types of genetic damage likely to occur upon exposure to the effluents, and they have identified several treatment methods effective at reducing the genotoxicity of the effluents.

Many of the genotoxic effluents reported here were obtained just prior to release into receiving waters used for recreational or agricultural purposes or for human consumption. Thus, it is evident that current waste discharge/disposal regulations (in this country as well as other countries) are not adequately protecting human health and the environment from exposure to mutagens/carcinogens. Indeed, the criteria promulgated by the federal agencies in this country to identify hazardous wastes and effluents rely solely upon chemical characterizations of the waste or waste extract; biological measurements for endpoints such as carcinogenicity, mutagenicity, teratogenicity, etc. are not incorporated into the regulations. However, several states, including New Jersey, California, and Illinois, have taken it upon themselves to utilize information obtained from short-term bioassays. In an attempt to control the discharge of mutagenic, carcinogenic, and teratogenic effluents, these states are taking steps to incorporate results from short-term genotoxicity assays (specifically, the Salmonella assay) in the permitting process and, potentially, in other regulatory decisions.

It cannot be argued that chemical analyses are indispensible for identifying and quantitating the hazardous chemicals present in an effluent; however, only a toxicological assessment can establish which compounds or effluents pose the greatest risk. Unfortunately, the activities of the environmental chemist and toxicologist often remain parallel and distinct, and comprehensive investigative studies are uncommon. The importance of coupling both chemical and biological analyses in evaluations of potential hazards cannot be overstated.

NOTICE

Although the research described in this article has been supported by the U.S. Environmental Protection Agency, it has not been subjected to Agency review and therefore does not necessarily reflect the views of the Agency.

REFERENCES

1. Ander, P., Eriksson, K.-E., Kolar, M.-C., Kringstad, K., Rannug, U., and Ramel, C. "Studies on the Mutagenic Properties of Bleaching Effluents," *Sven. Papperstidn.* 80:454–459 (1977).
2. Andon, B., Jackson, M., Houk, V., and Claxton, L. "Evaluation of Chemical and Biological Methods for the Identification of Mutagenic and Cytotoxic Hazardous Waste Samples," in *Hazardous and Industrial Solid Waste Testing: Fourth Symposium, ASTM STP 886,* J. K. Petros, Jr., W. J. Lacy, and R. A. Conway, Eds., American Society for Testing and Materials, Philadelphia, PA, 1986, pp. 204–215.
3. Battelle Columbus Laboratories. "Collaborative Study for the Evaluation of a Selected Method for Hazardous Waste Analysis. Final Report," EPA Contract 68-02-3169, Columbus, OH (8 December 1981).
4. Bjørseth, A., Carlberg, G. E., and Møller, M. "Determination of Halogenated Organic Compounds and Mutagenicity Testing of Spent Bleach Liquors," *Sci. Total Environ.* 11:197–211 (1979).
5. Brem, H., Stein, A. B., and Rosenkranz, H. "The Mutagenicity and DNA-Modifying Effect of Haloalkanes," *Cancer Res.* 34:2576–2579 (1974).
6. Brown, K. W. and Donnelly, K. C. "Mutagenic Potential of Water Concentrates from the Effluent of a Waste Storage Pond," *Bull. Environ. Contam. Toxicol.* 28:424–429 (1982).
7. Brown, K. W. and Donnelly, K. C. "Mutagenic Activity of the Liquid Waste from the Production of Acetonitrile," *Bull. Environ. Contam. Toxicol.* 32:742–748 (1984).
8. Carlberg, G. E., Johansen, S., Loras, V., Møller, M., Soteland, N., and Tveten, G. "Bleaching of Sulfite Pulps Using Conventional and Unconventional Sequences," *Das Papier* 36:270–278 (1980a).
9. Carlberg, G. E., Gjøs, N., Møller, M., Gustavsen, K. O., Tveten, G., and Renberg, L. "Chemical Characterization and Mutagenicity Testing of Chlorinated Trihydroxybenzenes Identified in Spent Bleach Liquors from a Sulfite Plant," *Sci. Total Environ.* 15:3–15 (1980b).
10. Carlberg, G. E., Drangsholt, H., and Gjøs, N. "Identification of Chlorinated Compounds in the Spent Chlorination Liquor from Differently Treated Sulfite Pulps with Special Emphasis on Mutagenic Compounds," *Sci. Total Environ.* 48:157–167 (1986).
11. Castegnaro, M., Malaveille, C., Brouet, I., Michelon, J., and Barek, J. "Destruction of Aromatic Amines in Laboratory Wastes through Oxidation with Potassium Permanganate/Sulfuric Acid into Non-Mutagenic Derivatives," *Am. Ind. Hyg. Assoc.* 46:187–191 (1985).
12. Cerna, M. and Kypenova, N. "Mutagenic Activity of Chloroethylenes Analyzed by Screening System Tests," *Mutat. Res.* 46:214–215 (1977).
13. Chen, D. J., Deaven, L. L., Meyne, J., Okinaka, R. T., and Strniste, G. F.

"Determination of Direct-Acting Mutagens and Clastogens in Oil Shale Retort Process Waster," in *Short-term Bioassays in the Analysis of Complex Environmental Mixtures III*, M.D. Waters et al., Eds., Plenum Press, New York, 1983, pp. 269–275.

14. Commoner, B. "Chemical Carcinogens in the Environment," in *Identification and Analysis of Organic Pollutants in Water,* L. H. Keith, Ed., Ann Arbor Science, Ann Arbor, MI, 1977, pp. 49–71.
15. Commoner, B., Vithayathil, A. J., and Dolara, P. "Mutagenic Analysis of Complex Samples of Aqueous Effluents, Air Particulates, and Foods," in *Short-Term Bioassays of Complex Environmental Mixtures I,* M.D. Waters et al., Eds., Plenum Press, New York, 1978, pp. 49–71.
16. Cunningham, M. L., Haugen, D. A., Kirchner, F. R., and Reilly, C. A. "Toxicologic Responses to a Complex Coal Conversion By-Product: Mammalian Cell Mutagenicity and Dermal Carcinogenicity," in *Short-Term Bioassays in the Analysis of Complex Environmental Mixtures IV,* M.D. Waters et al., Eds., Plenum Press, New York, 1985, pp. 113–123.
17. DeMarini, D. M., Brusick, D. J., and Lewtas, J. "Use of Limited Protocols to Evaluate the Genotoxicity of Hazardous Waste in Mammalian Cell Assays: Comparison to Salmonella," *J. Toxicol. Environ. Health* 22:225–239 (1987a).
18. DeMarini, D. M., Inmon, J. P., Simmons, J. E., Berman, E., Pasley, T. C., Warren, S. H., and Williams, R. W. "Mutagenicity in Salmonella of Hazardous Wastes and Urine from Rats Fed These Wastes," *Mutat. Res.* 189:205–216 (1987b).
19. DeMarini, D. M. and Houk, V. S. "Assessment of Hazardous Wastes for Genotoxicity," in *Hazardous Waste: Detection, Control, Treatment,* R. Abbou, Ed., Elsevier, Amsterdam, 1988, pp. 1107–1115.
20. DeMarini, D. M., Gallagher, J. E., Houk, V. S., and Simmons, J. E. "Toxicological Evaluation of Complex Industrial Wastes: Implications for Exposure Assessment," *Toxicol. Lett.* 49:199–214 (1989).
21. de Raat, W. K., Hanstveit, A. O., and de Kreuk, J. F. "The Role of Mutagenicity Testing in the Ecotoxicological Evaluation of Industrial Discharges into the Aquatic Environment," *Fd. Chem. Toxicol.* 23:33–41 (1985).
22. Donnelly, K. C. and Brown, K. W. "The Development of Laboratory and Field Studies to Determine the Fate of Mutagenic Compounds from Land Applied Hazardous Waste," in *Land Disposal of Hazardous Waste; Proc. Ann. Res. Symp., 7th,* EPA 600/9-81-002b, Cincinnati, OH, 1981, pp. 224–239.
23. Donnelly, K. C., Brown, K. W., and Scott, B. R. "The Use of Short-Term Bioassays to Monitor the Environmental Impact of Land Treatment of Hazardous Wastes," in *Short-term Bioassays in the Analysis of Complex Environmental Mixtures III,* M.D. Waters et al., Eds., Plenum Press, New York, 1983, pp. 59–78.
24. Donnelly, K. C., Brown, K. W., and Kampbell, D. "Chemical and Biological Characterization of Hazardous Industrial Waste. I. Prokaryotic Bioassays and Chemical Analysis of a Wood-Preserving Bottom-Sediment Waste," *Mutat. Res.* 180:31–42 (1987a).
25. Donnelly, K. C., Brown, K. W., and Scott, B. R. "Chemical and Biological Characterization of Hazardous Industrial Waste. II. Eukaryotic Bioassay of a Wood-Preserving Bottom Sediment," *Mutat. Res.* 180:43–53 (1987b).
26. Donnelly, K. C., Davol, P., Brown, K. W., Estiri, M., and Thomas, J. C. "Muta-

genic Activity of Two Soils Amended With a Wood-Preserving Waste," *Environ. Sci. Technol.* 21:57–64 (1987c).

27. Donnini, G. P. "Reduction of Toxicity and Mutagenicity of Chlorination Effluents with Sulfur Dioxide," *Pulp Pap. Can.* 82:106–111 (1981).
28. Donnini, G. P. "The Effect of Chlorine Dioxide Substitution on Bleaching Effluent Toxicity and Mutagenicity," *Pulp Pap. Can.* 84:74–80 (1983).
29. Douglas, G. R. et al. "Mutagenic Activity in Pulp Mill Effluents," in *Water Chlorination: Environmental Impact and Health Effects,* Vol. 3, R. L. Jolley, W. A. Brungs, and R. B. Cumming, Eds., Ann Arbor Science, Ann Arbor, MI, 1980, pp. 865–880.
30. Douglas, G. R. et al. "Mutagenicity of Pulp and Paper Mill Effluent: A Comprehensive Study of Complex Mixtures," in *Short-Term Bioassays in the Analysis of Complex Environmental Mixtures III,* M. D. Waters et al., Eds., Plenum Press, New York, 1983, pp. 431–459.
31. Douglas, G. R. et al. "Determination of Potential Hazard from Pulp and Paper Mills: Mutagenicity and Chemical Analysis," in *Carcinogens and Mutagens in the Environment,* Vol. 5, H.F. Stich, Ed., CRC Press, Inc., Boca Raton, FL, 1985, pp. 151–164.
32. Ellenton, J. A., Douglas, G. R., and Nestmann, E. R. "Mutagenic Evaluation of 1,1,2,3-tetrachloro-2-propene, a Contaminant in Pulp Mill Effluents, Using a Battery of *In Vitro* Mammalian and Microbial Tests," *Can. J. Genet. Cytol.* 23:17–25 (1981).
33. Ellis, D. D., Jones, C. M., Larson, R. A., and Schaeffer, D. J. "Organic Constituents of Mutagenic Secondary Effluents from Wastewater Treatment Plants," *Arch. Environ. Contam. Toxicol.* 11:373–382 (1982).
34. Epler, J. L., Larimer, F. W., Rao, T. K., Nix, C. E., and Ho, T. "Energy-Related Pollutants in the Environment: Use of Short-Term Tests for Mutagenicity in the Isolation and Identification of Biohazards," *Environ. Health Perspec.* 27:11–20 (1978).
35. Eriksson, K.-E., Kolar, M. C., and Kringstad, K. "Studies on the Mutagenic Properties of Bleaching Effluents, II," *Svensk Papperstidn.* 82:95–104 (1979).
36. Eriksson, K.-E., Kringstad, K., de Sousa, F., and Stromberg, L. "Studies on the Mutagenic Properties of Spent Bleaching Liquors," *Svensk Papperstidn.* 85:R73-R76 (1982).
37. "Hazardous Waste Proposed Guidelines and Regulations and Proposal on Identification and Listing," *Fed. Regis.* 43(243):58946–58994 (1978).
38. "Hazardous Waste and Consolidated Permit Regulations," *Fed. Regis.* 45(98):33066–33133 (1980).
39. Goettems, E. M. P., Teixeira, R. L., and Malabarba, L. R. "Biological Aspects in the Evaluation of Tertiary Lagoons and Efficiency in the Removal of Organic Pollutants," *Water Sci. Tech.* 19:1259–1261 (1987).
40. Hoglund, C., Allard, A.-S., Neilson, A. H., and Landner, L. "Is the Mutagenic Activity of Bleach Plant Effluents Persistent in the Environment?" *Svensk Papperstidn.* 82:447–449 (1979).
41. Holmbom, B. R., Voss, R. H., Mortimer, R. D., and Wong, A. "Isolation and Identification of an Ames-Mutagenic Compound Present in Kraft Chlorination Effluents," *Tappi* 64:172–174 (1981).
42. Holmbom, B. R., Voss, R. H., Mortimer, R. D., and Wong, A. "Fractionation,

Isolation, and Characterization of Ames Mutagenic Compounds in Kraft Chlorination Effluents," *Environ. Sci. Technol.* 18:333–337 (1984).

43. Houk, V. S. "Screening Complex Hazardous Waste for Mutagenic Activity Using the TLC/Ames Assay," Master's Thesis, Department of Environmental Sciences and Engineering, University of North Carolina, Chapel Hill, NC (1984).
44. Houk, V. S. and Claxton, L. D. "Screening Complex Hazardous Wastes for Mutagenic Activity Using a Modified Version of the TLC/Salmonella Assay," *Mutat. Res.* 169:81–92 (1986).
45. Houk, V. S. and DeMarini, D. M. "Use of the Microscreen Phage-Induction Assay to Assess the Genotoxicity of 14 Hazardous Industrial Wastes," *Environ. Mol. Mutagen.* 11:13–29 (1988).
46. Ishiguro, Y., LaLonde, R. T., Dence, C. W., Santodonato, J. "Mutagenicity of Chlorine-Substituted Furanones and their Inactivation by Reaction with Nucleophiles," *Environ. Toxicol. Chem.* 6:935–946 (1987).
47. Jungclaus, G. A., Lopez-Avila, V., and Hites, R. A. "Organic Compounds in an Industrial Wastewater: A Case Study of their Environmental Impact," *Environ. Sci. Technol.* 12:88–96 (1978).
48. Kamra, O. P., Nestmann, E. R., Douglas, G. R., Kowbel, D. J., and Harrington, T. R. "Genotoxic Activity of Pulp Mill Effluent in Salmonella and *Saccharomyces cerevisiae* Assays," *Mutat. Res.* 118:269–276 (1983).
49. Keleti, G., Bern, J., Shapiro, M. A., Gulledge, W. P., and Moore, G. T. "Mutagenicity of SRC-II Coal Liquefaction Wastewater Treatment Residues," *Environ. Sci. Technol.* 16:826–830 (1982).
50. Kringstad, K. P., Ljungquist, P. O., de Sousa, F., and Stromberg, L. M. "Identification and Mutagenic Properties of Some Chlorinated Aliphatic Compounds in the Spent Liquor from Kraft Pulp Chlorination," *Environ. Sci. Technol.* 15:562–566 (1981).
51. Kringstad, K. P., Ljungquist, P. O., de Sousa, F., and Stromberg, L. M. "Contributions of Some Chlorinated Aliphatic Compounds to the Mutagenicity of Spent Kraft Pulp Chlorination Liquors," in *Water Chlorination: Environmental Impact and Health Effects,* Vol. 4, R. L. Jolley, W. A. Brungs, and R. B. Cumming, Eds., Ann Arbor Science, Ann Arbor, MI, 1983, pp. 1311–1323.
52. Kringstad, K. P., de Sousa, F., and Stromberg, L. M. "Evaluation of Lipophilic Properties of Mutagens Present in the Spent Chlorination Liquor from Pulp Bleaching," *Environ. Sci. Technol.* 18:200–203 (1984).
53. Kringstad, K. P. and Lindstrom, K. "Spent Liquors from Pulp Bleaching," *Environ. Sci. Technol.* 18:236A-248A (1984).
54. Lee, E. G.-H., Mueller, J. C., and Walden, C. C. "Biological Characteristics of Pulp Mill Effluents (Part 1)," CPAR Report 678–1, Environmental Protection Service, Department of the Environment, Ottawa, Ontario (1978).
55. Lee, E. G.-H., Mueller, J. C., Walden, C. C., and Stich, H. "Mutagenic Properties of Pulp Mill Effluents," *Pulp Pap. Can.* 82:69–77 (1981).
56. Lindgaard-Jorgensen, P. and Folke, J. "Organics in Wheat and Rye Straw Pulp Bleaching and Combined Mill Effluents. II. Ecotoxicological Screening," *Toxicol. Environ. Chem.* 10:25–40 (1985).
57. Lunn, G. and Sansone, E. B. "Reductive Destruction of Dacarbazine, Procarbazine Hydrochloride, Isoniazid, and Iproniazid," *Am. J. Hosp. Pharm.* 44:2519–2524 (1987).
58. McCann, J., Choi, E., Yamasaki, E., Ames, B. N. "Detection of Carcinogens as

Mutagens in the Salmonella/Microsome Test: Assay of 300 Chemicals," *Proc. Natl. Acad. Sci.* U.S.A 72:5135–5139 (1975).

59. McGeorge, L. J., Louis J. B., Atherholt T. B., and McGarrity, G. J. "Mutagenicity Analyses of Industrial Effluents: Results and Considerations for Integration into Water Pollution Control Programs," in *Short-term Bioassays in the Analysis of Complex Environmental Mixtures IV,* M.D. Waters et al., Eds., Plenum Press, New York, 1985, pp. 247–268.
60. McKague, A. B., Lee, E. G.-H., and Douglas, G. R. "Chloroacetones: Mutagenic Constituents of Bleached Kraft Chlorination Effluent," *Mutat. Res.* 91:301–306 (1981).
61. Møller, M., Landmark L. H., Bjørseth, A., and Renberg, L. "Characterization of Industrial Aqueous Discharges by the TLC/Ames Assay," *Chemosphere* 13:873–879 (1984).
62. Møller, M., Bjørseth A., and Houk, V. S. "Chemical Separation and *in situ* Mutagenicity Testing," in *Mutagenicity Testing in Environmental Pollution Control,* F. K. Zimmermann and R. E. Taylor-Mayer, Eds., J. Wiley & Sons, Inc., New York, 1985, pp. 47–67.
63. Møller, M. A., Carlberg, G. E., and Soteland, N. "Mutagenic Properties of Spent Bleaching Liquors from Sulfite Pulps and a Comparison with Kraft Pulp Bleaching Liquors," *Mutat. Res.* 172:89–96 (1986).
64. Monarca, S., Hongslo, J. K., Kringstad, A., and Carlberg, G. E. "Mutagenicity and Organic Halogen Determination in Body Fluids and Tissues of Rats Treated with Drinking Water and Pulp Mill Bleachery Effluent Concentrates," *Chemosphere* 13:1271–1281 (1984).
65. Nazar, M. A. and Rapson, W. H. "Elimination of the Mutagenicity of Bleach Plant Effluents," *Pulp Pap. Can.* 81:75–82 (1980).
66. Nazar, M. A. and Rapson, W. H. "pH Stability of Some Mutagens Produced by Aqueous Chlorination of Organic Compounds," *Environ. Mutagen.* 4:435–444 (1982).
67. Nestmann, E. R., Lee, E. G.-H., Mueller, J. C., and Douglas, G. R. "Mutagenicity of Resin Acids Identified in Pulp and Paper Mill Effluents Using the Salmonella/Mammalian-Microsome Assay," *Environ. Mutagen.* 1:361–369 (1979).
68. Nestmann, E. R., Lee, E. G.-H., Matula, T., Douglas, G. R., and Mueller, J. C. "Mutagenicity of Constituents Identified in Pulp and Paper Mill Effluents Using the Salmonella/Mammalian-Microsome Assay," *Mutat. Res.* 79:203–212 (1980).
69. Nestmann, E. R. and Lee, E. G.-H. "Mutagenicity of Constituents of Pulp and Paper Mill Effluent in Growing Cells of *Saccharomyces cerevisiae, Mutat. Res.* 119:273–280 (1983).
70. Nestmann, E. R., Kowbel, D. J., Kamra, O. P., and Douglas, G. R. "Reduction of Mutagenicity of Pulp and Paper Mill Effluent by Secondary Treatment in an Aerated Lagoon," *Haz. Waste* 1:67–72 (1984).
71. Nestmann, E. R. "Detection of Genetic Activity in Effluents from Pulp and Paper Mills: Mutagenicity in *Saccharomyces cerevisiae,"* in *Mutagenicity Testing in Environmental Pollution Control,* F. K. Zimmermann and R. E. Taylor-Mayer, Eds., J. Wiley & Sons, Inc., New York, 1985, pp. 118–124.
72. Nestmann, E. R. and Lee, E. G.-H. "Genetic Activity in *Saccharomyces cerevisiae* of Compounds Found in Effluents of Pulp and Paper Mills," *Mutat. Res.* 155:53–60 (1985).

73. Rannug, U. and Ramel, C. "Mutagenicity of Waste Products from Vinyl Chloride Industries," *J. Toxicol. Environ. Health* 2:1019–1029 (1977).
74. Rannug, U. "Mutagenicity of Effluents from Chlorine Bleaching in the Pulp and Paper Industry," in *Water Chlorination: Environmental Impact and Health Effects,* Vol. 3, R. L. Jolley, W. A. Brungs, and R. B. Cumming, Eds., Ann Arbor Science, Ann Arbor, MI, 1980, pp. 851–863.
75. Rannug, U., Jenssen, D., Ramel, C., Eriksson, K.-E., and Kringstad, K. "Mutagenic Effects of Effluents from Chlorine Bleaching of Pulp," *J. Toxicol. Environ. Health* 7:33–47 (1981).
76. Rapson, W. H., Nazar, M. A., and Butsky, V. V. "Mutagenicity Produced by Aqueous Chlorination of Organic Compounds," *Bull. Environ. Contam. Toxicol.* 24:590–596 (1980).
77. Ravindran, P. N. and Ravindran, S. "Cytological Irregularities Induced by Water Polluted with Factory Effluents: A Preliminary Report," *Cytologia* 43:565–568 (1978).
78. Sanchez, P. S., Sato, M. I. Z., Paschoal, C. M. R. B., Alves, M. N., Furlan, E. V., and Martins, M. T. "Toxicity Assessment of Industrial Effluents from S. Paulo State, Brazil, Using Short-Term Microbial Assays," *Toxicity Assessment* 3:55–80 (1988).
79. Schaeffer, D. J. and Kerster, H. W. "Estimating the Mass of Mutagens in Indeterminate Mixtures," *Ecotoxicol. Environ. Safety* 10:190–196 (1985).
80. Simmon, V. F., Kauhanen, K., and Tardiff, R. G. "Mutagenic Activity of Chemicals Identified in Drinking Water," in *Progress in Genetic Toxicology, Developments in Toxicology and Environmental Science,* Vol. 2, D. Scott, B. A. Bridges, and F. H. Sobel, Eds., Elsevier, Amsterdam, 1977, pp. 249–258.
81. Simmons, J. E., DeMarini, D. M., and Berman, E. "Lethality and Hepatotoxicity of Complex Waste Mixtures," *Environ. Res.* 46:74–85 (1988).
82. Somani, S. M., Teece, R. G., and Schaeffer, D. J. "Identification of Cocarcinogens and Promoters in Industrial Discharges into and in the Illinois River," *J. Toxicol. Environ. Health* 6:315–331 (1980).
83. Somashekar, R. K. and Arekal, G. D. "Chromosomal Aberrations Induced by Electroplating Waste Water," *Cytologia* 48:621–625 (1983).
84. Somashekar, R. K., Gurudev, M. R., and Ramiah, S. "Somatic Cell Abnormalities Induced by Dye Manufacturing Industry Waste Water," *Cytologia* 50:129–134 (1985).
85. Somashekar, R. K. "Meiotic Abnormalities Induced by Dye Industry Waste Water in *Chlorophytum amaniense*," *Engler. Cytologia* 52:647–652 (1987).
86. Spanggord, R. J., Mortelmans, K. E., Griffin, A. F., and Simmon, V. F. "Mutagenicity in *Salmonella typhimurium* and Structure-Activity Relationships of Wastewater Components Emanating from the Manufacture of Trinitrotoluene," *Environ. Mutagen.* 4:163–179 (1982).
87. Stockman, L., Stromberg, L., and de Sousa, F. "Mutagenic Properties of Bleach Plant Effluents: Present State of Knowledge," *Cellul. Chem. Technol.* 14:517–526 (1980).
88. Sundvall, A., Marklund, H., and Rannug, U. "The Mutagenicity on *Salmonella typhimurium* of Nitrobenzoic Acids and Other Wastewater Components Generated in the Production of Nitrobenzoic Acids and Nitrotoluenes," *Mutat. Res.* 137:71–78 (1984).

89. Van Hoof, F. and Verheyden, J. "Mutagenic Activity in the River Meuse in Belgium," *Sci. Total Environ.* 20:15–22 (1981).
90. Van Hoof, F. and Manteleers, G. "Chlorination of Mutagenic Fractions of Coke-Plant Effluents," in *Water Chlorination: Environmental Impact and Health Effects,* Vol. 4, R. L. Jolley, W. A. Brungs, and R. B. Cumming, Eds., Ann Arbor Science, Ann Arbor, MI, 1983, pp. 1325–1332.
91. Waters, M. D. et al. "Use of Computerized Data Listings and Activity Profiles of Genetic and Related Effects in the Review of 195 Compounds," *Mutat. Res.* 205:295–312 (1988).

CHAPTER 12

Detection of Genetic Hazards from Environmental Chemicals with Plant Test Systems

B. S. Gill and S. S. Sandhu

INTRODUCTION

As a result of advancements in modern technology, a large number of synthetic chemicals and industrial waste products have been introduced into our environment, some of which have been found to be carcinogens. Thus, there is a need to screen these chemicals for health safety especially for chronic toxic effects. Because testing for carcinogenicity and other genetic hazards in animals is expensive and time-consuming, the development of alternative test methods for screening chemicals for their potential health effects has been emphasized. The discovery in the mid-70s that a significant proportion of carcinogens were mutagenic in bacteria encouraged the use of a large number of short-term bioassays for evaluating the genotoxic potential of environmental chemicals. The test organisms employed in these short-term bioassays ranged from viruses to laboratory animals. Only rarely were plant test systems used for the toxicological evaluation of xenobiotics. However, plants have played an important role in genetics research from the elucidation of the principles of heredity by Gregor Mendel to the discovery of transposons by Barbara McClintock.

The utility of using plant test systems in toxicological studies was explored by two workshops organized by the National Institute of Environmental Health Sciences (de Serres and Shelby, 1978; Constantin et al., 1981) and summarized by a review of several plant bioassays under the Gene Tox Program of U.S. Environmental Protection Agency (*Mutation Research*, Vol. 99, 1982). A current collaborative study between the U.S. EPA and IPCS (International Program on Chemical Safety) is continuing to explore the use of plant assays for assessing environmental hazards.

RELEVANCE OF PLANT TEST SYSTEMS FOR GENETIC HAZARD IDENTIFICATION

Higher plants offer many advantages for the screening of environmental chemicals for their potential mutagenicity and for monitoring health hazards from uncontrolled industrial waste sites. Green plants, like animals, are eukaryotes, contain similar amounts of DNA per cell, and have a similar amount of repetitive DNA as do animals. In addition, the regulation of gene action relative to ontogeny and cell heredity are typical of higher organisms (Constantin, 1978; Freeling, 1978). Although animal cells lack a cell wall and plant cells lack centrioles, the chromosome organization of plants and animals, including the genome organization, is very much alike. Clive (1981) surveyed the literature and made comparisons of the response of chemical mutagens in various test systems (plants, mammals in vivo and in vitro, *Drosophila*, and bacteria) and reported that response of the plant systems appears to more closely resemble that of the mammalian systems than *Drosophila* and bacteria. These findings were supported by the analysis of data on assay systems compiled under the U.S. EPA Genetic Toxicology Program. Plant assays have certain advantages including the relative ease of handling, amenability to diverse growth and testing conditions, and relatively low cost compared to animal systems. Elaborate protocols for aseptic cultures are usually unnecessary, and the training of technicians is easily accomplished. Perhaps the greatest advantage that plants offer over other test systems is the ease with which genotoxicity in somatic and germ cells can be measured in the same organism. Furthermore, the data presented in a recent issue of *Mutation Research* (Plewa and Gentile, 1988) show that plants are capable of metabolizing a variety of environmental chemicals from promutagens into mutagens.

In this paper we briefly describe some model plant test systems selected on the basis of the extent of validation, ease of handling, their utility for screening individual chemicals, and their potential use in onsite environmental hazard assessment.

PLANT TEST SYSTEMS

The most commonly used endpoints in plant test systems are gene mutation, chromosome aberrations, and micronuclei. Cytogenetic test systems are usually employed in the laboratory to evaluate the clastogenic potential of environmental substances, whereas specific locus tests are used to evaluate the mutagenicity of the ambient environment. However, either of the test systems could be used for in situ monitoring or laboratory screening. Furthermore, the same indicator organism (e.g., *Tradescantia*, *Zea*) may be used for chromosome aberrations as well as for specific gene mutation.

The assays are organized into cytogenetic and specific locus test systems.

1. Cytogenetic test systems
   - *Vicia faba* root tip assay
   - *Tradescantia* micronuclei assay
   - Wheat seedling assay for aneuploidy
2. Specific-locus test systems
   - *Tradescantia* stamen hair assay
   - *Arabidopsis* embryo assay
   - Maize *waxy* locus assay

PREPARATION OF CHEMICAL SOLUTIONS

The chemicals to be tested, if not soluble in water, are first dissolved in appropriate solvent (e.g., DMSO) and then diluted with the required volume of buffer solution to maintain proper pH. Chemical solutions should be prepared fresh for each treatment because aqueous solutions of many chemical compounds are often unstable.

The information on the collection, preparation, and chemical characterization of samples of complex mixtures to be used for bioassays has been reported in a series of publications (Pellizzari, 1979; Kobler et al., 1981; Duke and Bean, 1983; Donnelly et al., 1983; Tabor et al., 1985, and Wang et al., 1987).

CYTOGENETIC TEST SYSTEMS

Vicia faba Root Tip Assay

Because roots come in direct contact with soil and leachate, cytogenetic changes in these may serve as useful indicators of the level of toxic chemicals in the environment. Seed germination and root elongation are considered to be useful indicators for toxicity of chemical or complex chemical mixtures (Brusick and Young, 1981). Most of the earlier work on the ability of radiation and chemicals to induce chromosome aberrations was performed on plant root cells and even the phenomenon of sister-chromatid exchange was discovered in plant root cells (Taylor, 1958). Root tips are relatively inexpensive to obtain in large numbers, can be handled easily, and the material is available throughout the year. However, root tip contains a heterogenous population of cells with only moderate synchrony. Therefore, different root tip cells may differ in their response to prescribed dose levels of various chemicals.

The different species generally used for root tip studies are broad beans (*Vicia faba*, Kihlman, 1975; Grant, 1982; Sandhu and Acedo, 1988), onions (*Allium cepa*, Fiskesjo, 1988; Das et al., 1988), barley (*Hordeum vulgare*, Constantin and Nilan, 1982), and spiderwort (*Tradescantia paludosa*, Ma, 1982a). These species have a small number of large-sized chromosomes. Although protocols for cytogenetic evaluation in each of these test systems vary because of differences in growth requirements, duration of the mitotic cycle, and the period of DNA synthesis, the principles upon which these test systems are based are similar.

Broad bean (*Vicia faba*) is the most commonly used plant species for studies on mutagenesis. Ma (1982b) reported cytogenetic evaluation of 76 chemicals in the *Vicia* root tip mitosis test. For evaluating the genotoxicity of hazardous wastes, *Vicia faba* seed can be grown at the test site and samples of root tips can be obtained a few days after seed germination. Alternatively, seed may be grown in soil samples in the laboratory. Most of the studies using the *Vicia* root tip assay for chromosome breaks have been performed on individual compounds. However, Degrassi and Rizzoni (1982) used micronuclei formation in *Vicia* root tips for detecting the genotoxicity of drinking water. Sandhu and Acedo (1988) also employed this assay for evaluating the ability of certain compounds to induce aneuploidy.

A modification of the general procedure proposed by Kihlman (1975) for the root tip assay for chromosomal aberrations using chemicals or sample extracts dissolved in the appropriate solvents is described below.

Growing of Lateral Roots

1. Variety minor of *Vicia faba* is commonly used. Seeds are stored at low temperature.
2. The seeds can be disinfected before use by immersing in 0.5% sodium hypochlorite solution for 5 min. The seeds are then rinsed for 10 min under running tap water.
3. The disinfected seeds are soaked for 6 to 8 hr at room temperature in distilled water or in Hoaglands salt solution.
4. The soaked seeds are germinated in moist Perlite or Vermiculite at 20°C for approximately 4 days to obtain primary roots of 3- to 4-cm length.
5. The tips (5 mm) of the primary roots are cut to stimulate growth of the lateral roots and the seedlings are transferred to a water tank at constant temperature of 20°C during the growth period by using thermo-regulated running tap water. At the time of transfer, the shoots are also cut off. If the seedlings are grown in vials or tubes, water should be changed daily while maintaining a constant temperature at 20°C.
6. The seedlings are maintained in tap water or nutrient solution for about 4 days or until the lateral roots are of suitable length (1 to 2 cm) to be used in the test.

Treatment

1. Treatment is carried out by suspending the roots in different concentrations of freshly prepared solutions of chemicals to be tested.
2. Usually, each chemical or complex mixture sample is evaluated at four doses. A concurrent positive control and solvent control are maintained for each experiment.
3. Treatment is carried out for 1 hr in the dark. The temperature during the treatment and recovery periods is maintained at 20°C. Normally three seedlings are used for each treatment.

4. Since chemicals differ in producing chromosomal aberrations at different stages of the cell cycle, different recovery periods (8, 18, 22 and 26 hr) are used.
5. The roots are treated with 0.05% colchicine solution during the last 3 hr of the recovery period.
6. The roots are thoroughly washed with tap water after chemical treatment or colchicine treatment to remove the chemicals on the surface of the roots.
7. The roots are fixed at the end of the recovery period in a freshly prepared fixative containing three parts absolute alcohol and one part glacial acetic acid. The roots are stored in the refrigerator for 24 hr and the fixative is then replaced with 70% alcohol for longer storage.

Data Collection

1. The roots are thoroughly washed with distilled water before hydrolysis in 1 N HCl for 6 to 7 minutes at 60°C. After hydrolysis, the roots are washed again and are transferred to Feulgen stain for 45 to 60 min at 20°C.
2. The stained root tips are cut and placed on a clean slide in a drop of 45% acetic acid. Three root tips per slide give sufficient cells for chromosome analysis. A cover slip is put on the slide and squash preparations are produced by tapping gently on the cover slip. A slight heating over a flame and gentle tapping is recommended for adequate spreading of the cells.
3. Slides are examined under a microscope and chromosomal aberrations at metaphase are scored. Three hundred metaphases per treatment are considered adequate.
4. Major types of chromosomal aberrations are shown in Figure 1.

***Tradescantia* Micronucleus Assay**

The *Tradescantia* micronucleus test (Trad-MCN), where the plant cuttings are exposed to the test substance, is the simplest and perhaps the most sensitive of all the plant cytogenetic assays currently used. This test has been reported to be suitable for both liquid and gaseous mutagens (Ma, 1979) and has been used for in situ monitoring of air pollutants (Ma et al., 1983) and low level pollutants in drinking water (Ma et al., 1985). Results of 140 agents tested with this assay were compiled by Ma et al. in 1984. A modification of the general procedure for this assay proposed by Ma (1979) is presented below.

Growing Tradescantia Plants

1. Clone 4430 of *Tradescantia paludosa* is suitable for both micronucleus (MCN) and stamen hair mutation assays. This clone is a hybrid between *T. hirsutiflora* and *T. subcaulis* and is incapable of sexual reproduction and thus ensures genetic homogeneity. Crowded pots can be subdivided and started into new clones, thereby ensuring a constant supply of plants.
2. *Tradescantia* plants should be grown in a greenhouse under a temperature range of 21 to 25°C during the daytime and at 16°C during the nighttime with a relative humidity around 60 to 80%. Plants should be grown in a nonpolluted atmosphere using relatively clean soil. The flowering can be promoted

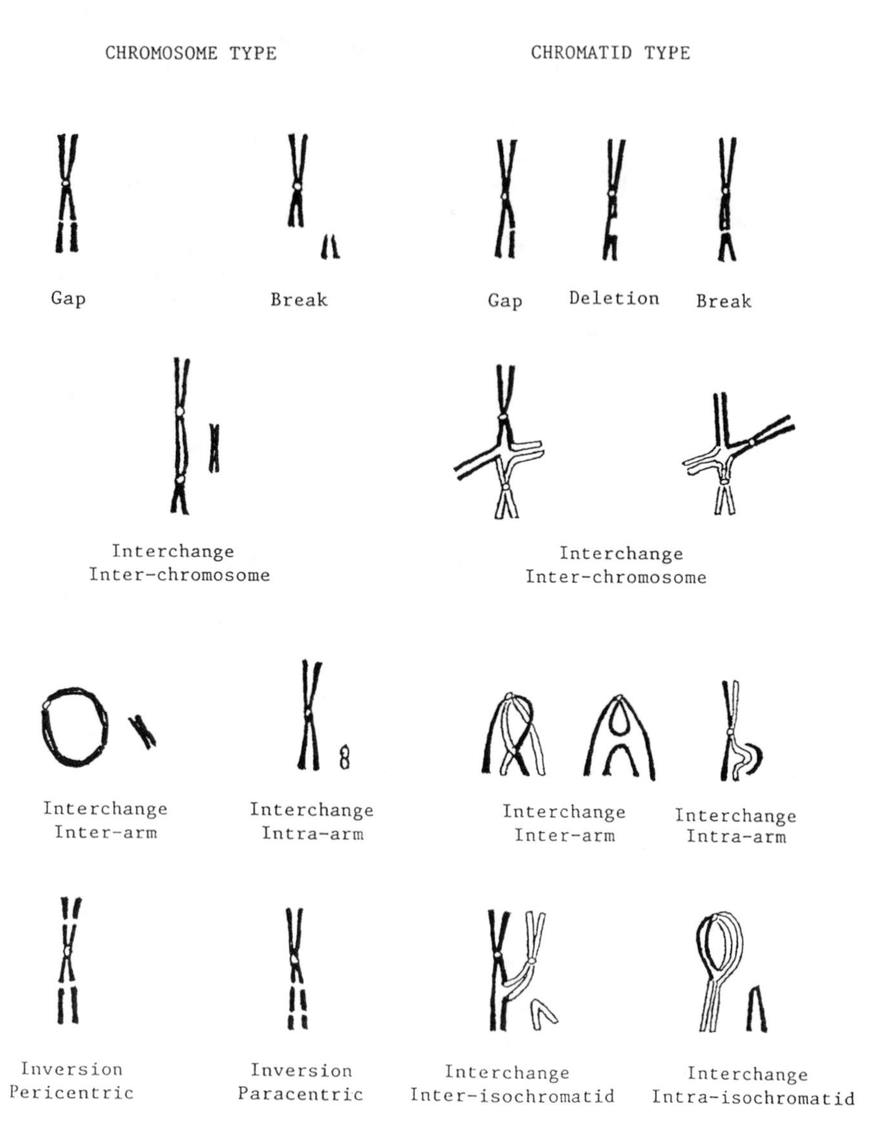

Figure 1. Major types of chromosomal aberrations.

using a 16/8-hr (light/dark) photoperiod. Plants can be grown in growth chambers for low MCN background or in the field for large populations and economy. Incandescent light is adequate if grown in the greenhouse or in the field; however, for the growth chamber, the intensity of fluorescent light should be about 1800-ft candles and incandescent light should be about 180-ft candles measured at the tip of the plants.

3. To obtain X number of inflorescences per week for experimental work, about a 10X number of well-grown plants are needed.

4. Nutrients are provided by daily watering with Hoaglands solution.

Treatment

1. Only young inflorescences bearing at least nine exposed buds are selected for treatment. Generally, 15 cuttings having a 10 to 15-cm long stem per treatment are sufficient for making 5 to 7 slides of the proper early-stage tetrads. The cuttings can be maintained in tap water or Hoaglands solution (1/3 dilution) before treatment.
2. Cuttings are treated for 6 hr followed by a 24-hr recovery period in order to allow the early prophase pollen mother cells, which received the treatment, to arrive at the early tetrad stage. Generally, the duration of treatment should not exceed 12 hr except in testing extremely low concentrations. In exceptional cases, a continuous 30 (or more) hr of treatment without recovery time are recommended. If there is an indication of mitotic delay with any treatment, a longer than 24-hr recovery period is required.
3. The inflorescences after treatment are fixed in freshly prepared fixative containing three parts absolute alcohol and one part glacial acetic acid and then are transferred after 24 hr to 70% alcohol for storage.
4. Slides are prepared for the early tetrad stage using the aceto-carmine squash method. The early tetrad stage has the common wall structure around the four cells and usually does not separate easily under normal heating and pressing.
5. The temporary slides can be used for scoring the micronucleus frequencies. Slides can be made permanent if needed with the "Dry Ice" technique and mounted in Euparal.

Data Collection

1. Scoring of micronucleus frequencies is done under 400 × magnification. Normal tetrads and tetrads containing 1, 2, 3, 4, 5, or more micronuclei are scored under separate categories. At least 300 tetrads from each of the five slides of each treatment are scored. The frequency of micronuclei is expressed as the number of micronuclei per 100 tetrads. The background frequency of micronuclei for the control should not exceed 3 to 4 per 100 tetrads.
2. Sample means and standard deviations are calculated from five slides of each experimental group. The significance of difference between the treated and control groups is analyzed with Student's t-test or the Dunnett t-statistic as appropriate.
3. *Tradescantia* plant and a tetrad with a micronucleus are shown in Figures 2A and 2B.

Wheat Seedling Assay for Aneuploidy

Normal segregation of chromosomes at cell division is a very complex phenomenon that involves the proper functioning of a large number of cell organelles. Malfunctioning of any one of these organelles may lead to the unequal distribution of chromosomes in the cell progeny. In humans, aneuploidy may be the most significant factor in the human genetic disease burden. Recent conferences and publications are testimony to the significance of the role of

Figure 2A. *Tradescantia* plant.

aneuploidy in human morbidity and mortality (Bond and Chandley, 1983; Dellarco et al., 1985; Vig and Sandberg, 1987; and Zimmerman, 1988). Several assay systems have been proposed for detecting aneuploidy-inducing agents (Zimmerman, 1988). An assay using hexaploid wheat for detecting chemically induced aneuploidy has been described recently by Redei and Sandhu (1988). In this assay, chromosome loss or gain is detected by green or white stripes on virescent leaves, respectively.

For details on the cytogenetic basis of this assay, the reader is referred to

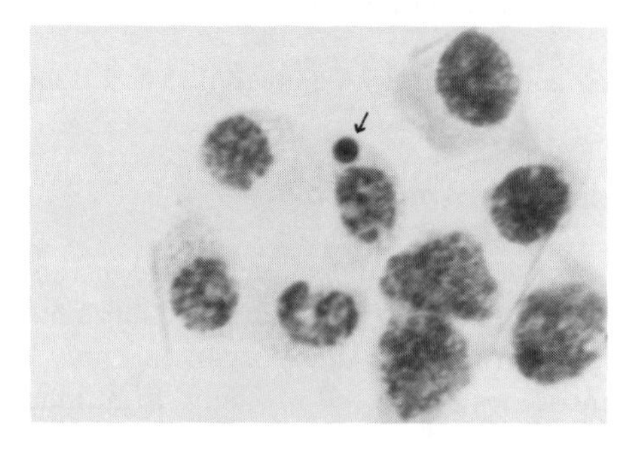

Figure 2B. A tetrad with a micronucleus.

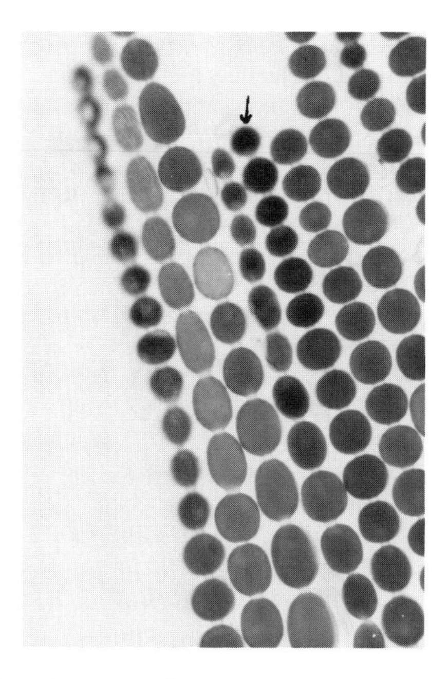

Figure 2C. Stamen Hairs with mutant cells.

Redei and Sandhu (1988). The procedure used in this assay involves the following steps:

Preparation of Seed and Growing Wheat Plants

1. Seeds of Neatby's Virescent Chinese Spring Wheat are used in this assay.
2. Seeds are soaked in distilled water for 8 hr at room temperature before treatment with test chemicals.
3. After treatment, germinated seeds are transplanted into pots (containing equal parts of soil, fine sand, and peat moss) and grown for 3 to 4 weeks in the greenhouse or growth chamber at 18°C under a 16-hr daily light regime at approximately 1000-ft candles (ca. 11,000 lux).
4. Nutrients are provided by daily watering with Hoaglands solution.

Treatment

1. The presoaked seeds are treated for 20 hr at 25°C on a rotary shaker.

Data Collection

1. Green and white sectors are scored on 2nd and 3rd leaves as they emerge within 3 to 4 weeks. Green sectors are easily identified on the virescent background, but white sectors are more difficult to observe. However, after scoring of green sectors, if the plants are subjected to temperatures above 26°C, the virescent plants develop chlorophyll and the white sectors (if any) become prominent.

2. The data are expressed as number of sectors per plant, calculated by dividing the number of sectors by the number of plants scored. The mean and standard error are calculated for statistical comparison of treatments using the Student *t*-test.
3. Wheat seedlings are shown in Figure 3A and green and white sectors as they appear in different forms are shown in Figures 3B thru 3D.

SPECIFIC-LOCUS TEST SYSTEMS

Tradescantia Stamen Hair Assay

The *Tradescantia* stamen hair assay was originally developed by Sparrow (Underbrink et al., 1973). In situ monitoring of ambient air using this assay was reported by Schairer et al. (1978). Based on the results of this study, it was possible to rank ambient air quality of several cities in the United States. Similarly, studies conducted with *Tradescantia* by Ichikawa (1981) around nuclear power plants in Japan showed significantly increased mutation frequencies. This assay has been used for monitoring the mutagenicity of leachate from sewage sludges, water pollution, bottom sediment from a drinking water reservoir, leachate from hazardous sediments, chemical landfills, diesel emission exposures, air pollution from petrochemical plants, soil and air pollution from a lead smelter, and assessing the hazards of peroxyacetyl nitrate and related photo-oxidants in ambient atmospheric smog and under laboratory conditions (for review see Sandhu and Lower, 1987). This assay is based on the

Figure 3A. Wheat seedlings.

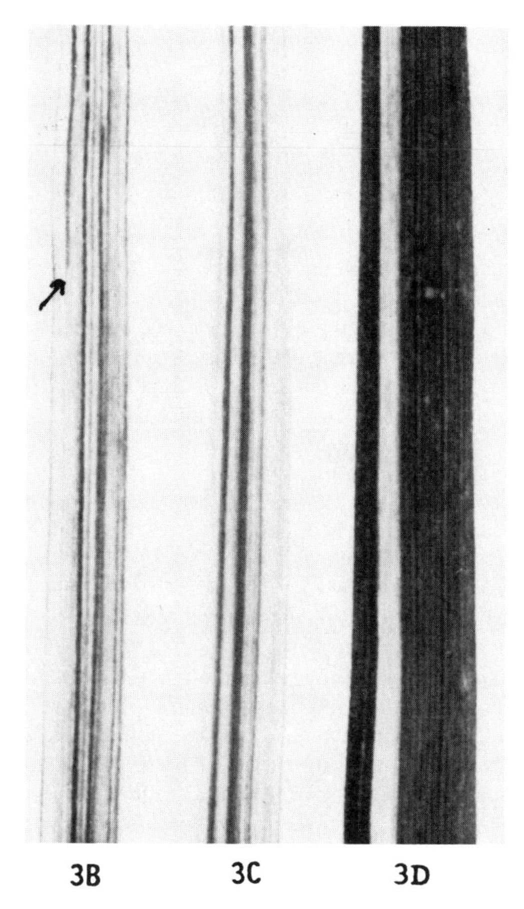

Figure 3. **B)** Appearance of a new sector on virescent leaf; **C)** Green sectors on virescent leaf; **D)** White sector on green leaf.

detection of changes in the color of stamen hair cells from blue (dominant) to pink (recessive). A modification of the general procedure for this assay proposed by Van't Hof and Schairer (1982) for chemicals in solution is listed below:

Growing Tradescantia Plants (See Tradescantia Micronucleus Assay)

Treatment

1. Fresh *Tradescantia* cuttings, each with a young inflorescence containing flower buds in a range of developmental stages, are used for exposure.
2. *Tradescantia* cuttings are exposed to liquid test chemicals by simple immersion of the inflorescences and to gaseous mutagens by fumigation in open air

or gas-light chambers. For details on exposure in gas chamber, see Van't Hof and Schairer (1982).
3. Cuttings after exposure are grown in aerated Hoaglands nutrient solution under controlled environmental conditions.
4. If prolonged observations are required, potted plants can also be used.

Data Collection

1. The flowers are analyzed each day as they bloom for approximately 2 weeks after treatment.
2. At maturity and under optimal growth conditions, untreated stamens have 40 to 75 hairs, each with an average of 24 cells. A mutant cell produced early in development is capable of having several mutant progeny that appear as a string of pink cells. A continuous string of pink cells is noted as a single event. The treatment may also produce a lethal lesion-stopping cell division and a stunted mature hair would result.
3. Induced pink-event rates are expressed as the mean of the rates for several consecutive peak response days. Usually 11 to 15 days are used for radiation and 7 to 12 days for chemical exposures.
4. Lethal doses are determined by counting the number of cells per hair. If there are less than 23 cells per hair, the dose is too high.
5. Generally 3 to 4 $\times$ 10^5 stamen hairs are counted for each treatment.
6. The data are expressed as the number of pink mutant sectors per 100 or 1000 hairs. The mean number of mutations per hair is calculated by summing the number of pink sectors from individual flowers blooming during the peak response period and dividing by the number of flowers multiplied by an estimate of the number of hairs per flower. The number of hairs per flower is calculated by counting the hairs of several stamens and is taken to be a constant for that treatment. A hair is counted as a single mutant event if it has one or more pink cells.
7. Stamen hair and mutated cells are shown in Figure 2C.

Arabidopsis **Multilocus Assay**

Arabidopsis thaliana is a short cruciferious plant that can complete its life cycle in 4 to 6 weeks. Thousands of plants can be grown in a small growth chamber in the laboratory or greenhouse. A single plant can produce as many as 5,000 seeds. This assay is capable of simultaneously detecting mutation at multiple loci (Redei, 1982; Redei et al., 1984a). The test can be performed by observing segregating M_2 embryos in the fruits of M_1 plants or by observing M_2 progeny. Observations of M_2 embryos are preferred because it is faster and more efficient. This method of embryo analysis was first used for assessment of mutagenicity by Muller (1963, 1965). This plant is capable of metabolically activating many promutagens to mutagens (Redei et al., 1984b).

Approximately 120 compounds belonging to various chemical classes have been tested in this assay. Redei et al. (1984b) have evaluated in *Arabidopsis* the mutagenicity of 37 compounds that had been tested extensively in multiple bioassays under an international program (de Serres and Ashby, 1981). Relative to other in vitro and in vivo bioassays employed in this study, *Arabidopsis* was the most predictable test system.

Availability of sufficient quantities of test samples needed for adequate testing is often a limiting factor in applying plant bioassays to the genetic analysis of hazardous wastes. Most often, the hazardous waste samples have to be extracted and fractionated, and only very small quantities of such fractions are available for testing. For most of the in vivo bioassays (including plants), this scarcity of test samples creates a problem. However, because of the high sensitivity and very small seeds of *Arabidopsis*, even small samples can be tested adequately in this assay. Recently, Sandhu (1987) tested a few industrial waste samples for mutagenicity in *Arabidopsis* and compared these results with those obtained in an Ames *Salmonella* assay. All samples that were positive for mutagenicity in *Arabidopsis* were also positive in *Salmonella*; however, a few samples that were negative in *Salmonella* yielded a positive response in *Arabidopsis*. This may be attributable to the fact that the routine *Salmonella* assay detects only gene mutation at the histidine locus, whereas the *Arabidopsis* assay is capable of simultaneously detecting mutations at multiple loci as well as chromosome aberrations. A modification of the general procedure proposed by Redei (1982) for this assay is given below:

Growing Arabidopsis plants

1. Strain Columbia Wild of *Arabidopsis thaliana* is used in this assay.
2. The treated seeds are planted on a "Promix" medium contained in petri plates previously pasteurized at 100°C for 1 hr. Alternatively, a medium consisting of equal parts of soil, fine sand, and peat moss contained in small pots can be used.
3. The petri plates are kept in the growth chamber for a 16-hr photoperiod at 24°C at approximately 1000-ft candles of light.
4. Nutrients are provided by daily watering with Hoaglands solution.
5. The fruits can be scored for embryo mutations approximately 4 weeks after planting.

Treatment

1. Approximately 300 to 500 seeds of *Arabidopsis thaliana* (strain Columbia Wild) are picked up using a long-tipped glass pipette. Seeds are wrapped in a cloth bag for ease of handling.
2. The cloth bags containing seeds are lowered into small containers containing 10 ml of the appropriate concentrations of treatment solution.
3. The seeds are treated for 15 hr and then washed in running tap water for approximately 1 hr.
4. The washed seeds are dispersed in water and planted in pots.

Data Collection

1. One hundred plants are scored per treatment using three fruits per plant.
2. The fruits are opened with sharp-pointed forceps and scored for any of the following genetic endpoints:
   - chlorophyll mutants as cream-colored, light green or white embryos
   - lethal as brown or shrivelled embryos
   - sterile as empty spaces in the fruit
4. The mutation frequency for each treatment is computed by dividing the number of fruits with mutations by the total number of fruits scored and expressed as a percentage.
5. The results are considered positive if a treatment causes a doubling in the frequency of mutations over the control or a dose response is observed with at least three doses.
6. An *Arabidopsis* plant and the fruit of *Arabidopsis* segregating for cotyledon color are shown in Figures 4A and 4B.

Maize *waxy* Locus Assay

This assay system was utilized by Plewa and his associates (1978) to show that maize plants grown under field conditions provide the needed metabolic activation to transform certain pesticides into mutagens (Gentile et al., 1978). The plants homozygous for *waxy* alleles (wx/wx) produce endosperm that contain only amylopectin; the plants homozygous (Wx/Wx) or heterozygous (Wx/wx) produce starch in the endosperm containing amylopectin and amylose. Because amylose stains a dark blue-black when it reacts with iodine, whereas amylopectin does not react with iodine (endosperm stains red), these two phenotypes could be easily identified.

This system has been used for in situ monitoring of the mutagenicity of chemical pesticides (Plewa 1978; Gentile et al., 1978), ambient air quality (Lower et al., 1978), and municipal sludges (Hopke et al., 1982), and it has also been employed with ionizing radiation (Erikson, 1963, 1971; Erikson and Tavrin, 1965).

A modification of the general procedure proposed by Plewa (1982) is as follows:

Preparation of Seed and Growing Maize Plants

1. Early-Early Synthetic variety of *Zea mays* is recommended for this assay. This variety tassels in less than 4 weeks and attains a height of approximately 50 cm.
2. The kernels are surface sterilized in 0.5% sodium hypochlorite solution for 5 min. The kernels are then rinsed for 10 min under running tapwater and are soaked in aerated water for 72 hr at 20°C.

Treatment

1. The presoaked kernels are usually treated for 4 to 12 hr using approximately 2 to 5 mL of test solution per kernel and aerated constantly.

Figure 4A. *Arabidopsis* plant.

2. After treatment, the kernels are rinsed 2 to 3 times with water with a final rinse under running tapwater for 0.5 to 4 hours to ensure that all the test chemical is removed.
3. After treatment, one kernel is planted per pot (10-cm diameter) containing equal parts of soil, fine sand, and peat moss.
4. The pots are placed in a plant growth chamber that is set for a 17-hr photoperiod at an illumination of 1000-ft candles keeping day and night temperature at 25°C and 20°C, respectively.

Data Collection

1. The tassels are harvested at early anthesis. Each tassel is labeled and placed in 70% ethanol for storage.
2. The tassel to be analyzed is removed from the storage jar and agitated in a 1-L jar of clean 70% ethanol.

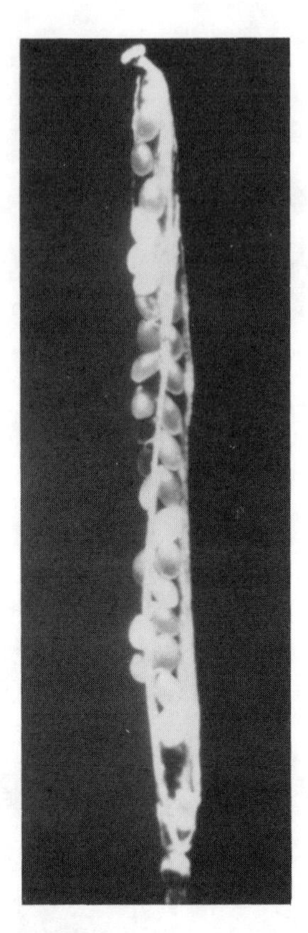

Figure 4B. Pod with segregating embryos.

3. The anthers are dissected from 10 to 12 unopened florets and are placed in a petri dish filled with 70% ethyl alcohol.
4. The anthers are placed into a stainless steel Virtis microhomogenation cup containing 0.6 mL of a gelatin-iodine stain and are minced with scissors. The anthers are homogenized for 30 sec.
5. The homogenate is filtered through a double thickness of cheesecloth on the surface of a large microscope slide (51 x 76 mm) and a large cover slip (40 x 50 mm) is placed over the pollen suspension.
6. The slide is examined under 40x magnification for forward mutation Wx → wx and pollen abortions.
7. The total number of viable pollen grains is estimated by counting pollen grains from 20 randomly chosen 1 mm^2 areas and multiplying this value by a factor to cover the entire area of the cover slip.
8. Approximately 2.5 X 10^4 pollen grains are analyzed per tassel.
9. Frequency of mutant pollen grains can be estimated by dividing the number of mutant pollen grains by the total estimated number of pollen grains. Generally 1 x 10^6 pollen grains per treatment group are analyzed.
10. The number of aborted pollen grains can also be recorded at the time of

estimating the number of viable pollen grains. The frequency of aborted pollen grains provides a measure of the gametophytic death.

11. The mean, standard error of the mean, and the analysis of variance may be used to test for significant differences among treatment groups and the control.
12. A Maize plant and a forward pollen mutant (Wx→wx) pollen are shown in Figures 5A and 5B.

Figure 5A. Maize plant.

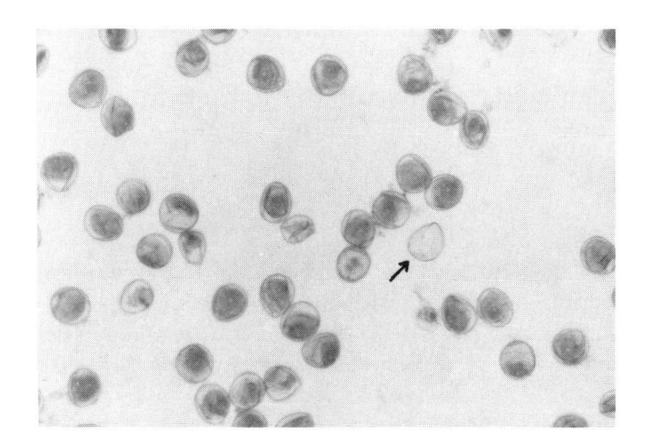

Figure 5B. Pollen showing forward mutation (Wx—wx).

CONCLUSION

In conclusion, plant assays are a simple and cost-effective means for evaluating the genetic hazards associated with environmental agents. Some plant species with small chromosome numbers, but large chromosomes, are specifically useful for evaluating the cytogenetic effects of xenobiotics. Leachates from hazardous wastes are one of our major concerns because of the possibility that these may seep into drinking water aquafers. Plant roots first come into contact with the soluble components of hazardous wastes and may provide a sensitive indicator of the clastogenic potential of such wastes. Plant species such as *Allium cepa* and *Vicia faba* have been used to some extent for this purpose. Similarly, *Tradescantia paludosa* stamen hair and micronucleus assays have been reported to be quite efficient in detecting mutagens and clastogens in the ambient atmosphere. The *Arabidopsis* embryo assay is a simple, sensitive, and reliable test system for mutagenicity evaluation of individual compounds and for hazardous waste samples. The utility of this assay for in situ environmental assessment should be evaluated.

The data from plant assays cannot, of course, serve as a substitute for data from animal assays or from human epidemiology surveys, but these assays may serve as a convenient tool in the initial stages of evaluation for potential human hazard.

ACKNOWLEDGMENTS

The authors are grateful to Dr. Bruce Casto for his assistance in preparing this manuscript and to Ms. Kathy Rouse for secretarial help.

NOTICE

This manuscript has been reviewed and approved for publication by the Health Research Laboratory, U.S. Environmental Protection Agency. Approval, however, does not necessarily reflect the views and policies of the Agency, and neither does mention of trade names constitute an official endorsement for use.

REFERENCES

1. Bond, D. J. and Chandley, A. C. *Oxford Monographs on Medical Genetics No. II, Aneuploidy,* Oxford University Press, New York, 1983, pp. 1–198.
2. Brusick, D. J. and Young, R. R. "IERL-RTP Procedures Manual: Level 1 Environmental Assessment Biological Tests," EPA Report 600/8-81-024 (1981), pp. 1–136.
3. Clive, D. "Comparative Mutagenicity," in *Comparative Mutagenesis: Can We Make Risk Estimates?,* F. J. de Serres and M. D. Shelby, Eds., Plenum Press, New York, 1981, pp. 1039–1066.

4. Constantin, M. J. "Utility of Specific Locus Systems in Higher Plants to Monitor for Mutagens," *Environ. Health Pers.* 27:69–75 (1978).
5. Constantin, M. J. and Nilan, R. A. "Chromosome Aberration Assays in Barley (*Hordeum vulgare*). A Report of the U.S. Environmental Protection Agency Gene-Tox Program," *Mutation Res.* 99:13–36 (1982).
6. Constantin, M. J., de Serres, F. J., Nilan, R. A., Sandhu, S. S., and Shelby, M. D. "Pollen Systems to Detect Biological Activity of Environmental Pollutants," *Environ. Health Pers.* 37:1–168 (1981).
7. Das, S., Panda, K. K., and Panda, B. B. "Biomonitoring of Low Levels of Mercurial Derivatives in Water and Soil by Allium Micronucleus Assay," *Mutation Res.* 203:11–21 (1988).
8. Degrassi, F. and Rizzoni, M. "Micronucleus Test in *Vicia faba* Root Tips to Detect Mutagen Damage in Fresh-Water Pollution," *Mutation Res.* 97:19–33 (1982).
9. Dellarco, V. L., Voytek, P. E., and Hollaender A., Eds., *Aneuploidy—Etiology and Mechanism,* Plenum Press, New York, 1985, pp. 1–562.
10. de Serres, F. J. and Shelby, M. D. "Higher Plant Systems as Monitors of Environmental Mutagens," *Environ. Health Pers.* 27:1 (1978).
11. de Serres, F. J. and Ashby, J., Eds. *Evaluation of Short-Term Tests for Carcinogens,* Elsevier North-Holland, New York, 1981, pp. 1–827.
12. Donnelly, K. C., Brown, K. W., and Scott, B. R. "The Use of Short-Term Bioassays to Monitor the Environmental Impact of Land Treatment of Hazardous Wastes," in *Short-Term Bioassays in the Analysis of Complex Environmental Mixtures III,* M. D. Waters, S. S. Sandhu, J. Lewtas, L. Claxton, N. Chernoff, and S. Nesnow, Eds., Plenum Press, New York, 1983, pp. 59–78.
13. Duke, K. M. and Bean, J. B. "Preparation of Hazardous and Complex Samples for Ecological Testing," in *Short-Term Bioassays in the Analysis of Complex Environmental Mixtures III,* M. D. Waters, S. S. Sandhu, J. Lewtas, L. Claxton, N. Chernoff, and S. Nesnow, Eds., Plenum Press, New York, 1983, pp. 27–38.
14. Eriksson, G. "Induction of *waxy* Mutants in Maize by Acute and Chronic Gamma Irradiation," *Hereditas* 50:161–178 (1963).
15. Eriksson, G. "Variation in Radiosensitivity and the Dose Effect Relationship in the Low Dose Region," *Hereditas* 68:101–114 (1971).
16. Eriksson, G. and Tavrin, E. "Variation in Radiosensitivity During Meiosis of Pollen Mother Cells in Maize," *Hereditas* 54:156–169 (1965).
17. Fiskesjo, G. "The Allium Test—an Alternative in Environmental Studies: The Relative Toxicity of Metal Ions," *Mutation Res.*197:243–260 (1988).
18. Freeling, M. "Maize Adh1 as a Monitor of Environmental Mutagens," *Environ. Health Pers.* 27:91–97 (1978).
19. Genitle, J. M., Wagner, E. D., and Plewa, M. J. "A Comprehensive Analysis of the Mutagenic Properties of Pesticides Used in Commercial Maize Production," *Mutation Res.* 53:112 (1978).
20. Grant, W. F. "Plant Mutagen Assays Based Upon Chromosome Mutations," in *Environmental Mutagenesis, Carcinogenesis and Plant Biology,* E. Klekowski, Ed., Praeger, New York, 1982, pp. 1–24.
21. Hopke, P. K., Plewa, M. J., Johnston, J. B., Weaver, D., Wood, S. G., Larson, R. A. and Hinesly, T. "Multitechnique Screening of Chicago Municipal Sludge for Mutagenic Activity," *Environ. Sci. Technol.* 16:140–147 (1982).
22. Ichikawa, S. "In Situ Monitoring with *Tradescantia* Around Nuclear Power Plants," *Environ. Health Pers.* 37:145–164 (1981).

23. Kihlman, B. A. "Root tips of *Vicia faba* for the Study of the Induction of Chromosomal Aberrations," *Mutation Res.* 31:401–412 (1975).
24. Kobler, A., Wolff, T., Hughes, T., Pellizzari, E., Sparacino, C., Waters, M., Huisingh, J. L., and Claxton, L. "Collection, Chemical Fractionation, and Mutagenicity Bioassays of Ambient Air Particulate," in *Short-Term Bioassays in the Analysis of Complex Environmental Mixtures II,* M. W. Waters, S. S. Sandhu, J. L. Hiusingh, L. Claxton, and S. Nesnow, Eds., Plenum Press, New York, 1981, pp. 21–44.
25. Lower, W. R., Rose, P. S., and Drobney, V. K. "*In Situ* Mutagenic and Other Effects Associated with Lead Smelting," *Mutation Res.* 54:83–95 (1978).
26. Ma, T. "Micronuclei Induced by X-Rays and Chemical Mutagens in Meiotic Pollen Mother Cells of Tradescantia, A Promising Mutagen Test System," *Mutation Res.* 64:307–313 (1979).
27. Ma, Te-Hsiu. "*Tradescantia* Cytogenetic Tests (Root-Tip Mitosis, Pollen Mitosis, Pollen Mother-Cell Meiosis). A Report of the U.S. Environmental Protection Agency Gene-Tox Program," *Mutation Res.* 99:293–302 (1982a).
28. Ma, Te-Hsiu. "*Vicia* Cytogenetic Tests for Environmental Mutagens. A Report of the U.S. Environmental Protection Agency Gene-Tox Program," *Mutation Res.* 99:257–271 (1982b).
29. Ma. T. H., Neas, R. E., Anderson, V. A., Harris, M. M., and Lee, T. S. "Mutagenicity of Drinking Water-Detected by Tradescantia-Micronucleus Test," *Can. J. Genet. Cytol.* 27:143–150 (1985).
30. Ma, T., Lower, W. R., Harris, F. D., Poku, J., Anderson, V. A., Harris, M. M, and Bare, J. L. "Evaluation by the Tradescantia-Micronucleus Test of The Mutagenicity of Internal Combustion Engine Exhaust Fumes from Diesel and Diesel-Soybean Oil mixed Fuels," in *Short-Term Bioassays in The Analysis of Complex Environmental Mixtures III,* M. D. Waters, S. S. Sandhu, J. Lewtas, L. Claxton, N. Chernoff and S. Nesnow, Eds., Plenum Press, New York, 1983, pp. 89–99.
31. Ma, Te-Hsiu, Harris, M. M., Anderson, V. A., Ahmed, I., Mohammad, K., Bare, J. L., and Lin, G. "Tradescantia-Micronucleus (Trad-MCN) Tests on 140 Health-Related Agents," *Mutation Res.* 138:157–167 (1984).
32. Muller, A. J., "Embroynentest zum Nachweis rezessiver Letalfaktoren bei *Arabidopsis Thaliana,*" *Biol. Zentralbl.* 83:133–163 (1963).
33. Muller, A. J. "The Chimerical Structure of M_1 Plants and its Bearing on the Determination of Mutation Frequencies in *Arabidopsis* Induction of Mutation and the Mutation Process," *Symp. Czech. Acad. Sci. Prague* (1965), pp. 46–52.
34. Pellizzari, E. D. "State-of-the-Art Analytical Techniques for Ambient Vapor Phase Organics and volatile Organics in Aqueous Samples form Energy-Related Activities," in *Application of Short-Term Bioassays in the Fractionation and Analysis of Complex Environmental Mixtures,* M. D. Waters, S. Nesnow, J. L. Huisingh, S. S. Sandhu, and L. Claxton, Eds., Plenum Press, New York, 1979, pp. 195–226.
35. Plewa, M. J. "Activation of Chemicals into Mutagens by Green Plants: A Preliminary Discussion," *Environ. Health Pers.* 27:45–50 (1978).
36. Plewa, M. J. "Specific-Locus Mutation Assay in *Zea mays.* A report of the U.S. Environmental Protection Agency Gene-Tox Program," *Mutation Res.* 99:317–337 (1982).
37. Plewa, M. J. and Gentile, J. M. "Activation of Promutagen by Plant Systems," *Mutation Res.* 197(Special Issue):173–336 (1988).

38. Redei G. P. "Mutagen Assay with Arabidopsis. A Report of the U.S. Environmental Protection Agency Gene-Tox Program," *Mutation Res.* 99:243–255 (1982).
39. Redei, G. P. and Sandhu, S. S. "Aneuploidy Detection With a Short-Term Hexaploid Wheat Assay," *Mutation Res.* 201:337–348 (1988).
40. Redei, G. P., Acedo, G. N., and Sandhu, S. S. "Mutation Induction and Detection in Arabidopsis," in *Mutation, Cancer, and Malformation,* Ernest H. Y. Chu and Walderico M. Generoso, Eds., Plenum Press, New York, 1984a, pp. 285–313.
41. Redei, G. P., Acedo, G. N. and Sandhu, S. S. "Sensitivity Specificity and Accuracy of The *Arabidopsis* Assay in The Identification of Carcinogen," in *Mutation, Cancer, and Malformation,* Ernest H. Y. Chu and Walderico M. Generoso, Eds., Plenum Press, New York, 1984b, pp. 689–708.
42. Sandhu, S. S. "Application of a simple Short-Term Bioassay for Identification of Genotoxins from Hazardous Wastes," *Proc. 3rd Ann. U.S. EPA Sym. Solid Waste Test. Qual. Assurance.* 1:2/63–2/78 (1987).
43. Sandhu, S. S., and Lower, W. R. "In situ Monitoring of Environmental Genotoxins," in *Short-Term Bioassays in the Analysis of Complex Environmental Mixtures V,* S. S. Sandhu, D. M. DeMarini, M. J. Mass, M. M. Moore, J. L. Mumford, Eds., Plenum Press, New York, 1987, pp. 145–160.
44. Sandhu, S. S. and Acedo, G. N. "Detection of Chemically Induced Aneuploidy by the *Vicia faba* Root-Tip Assay," *Toxicol. Ind. Health* 4:257–267 (1988).
45. Schairer, L. A., Van't Hof, J., Hayes, C. G., Burton, R. M., and de Serres, F. J. "Exploratory Monitoring of Air Pollutants for Mutagenicity Activity with the *Tradescantia* Stamen Hair System," *Environ. Health Pers.* 27:51–60 (1978).
46. Tabor, M. W., Loper, J. C., Myers, B. L., Rosenblum, L., and Daniel, F. B. "Isolation of Mutagenic Compounds form Sludges and Wastewaters," in *Short-Term Bioassays in the Analysis of Complex Environmental Mixtures III.,* M. D. Waters, S. S. Sandhu, J. Lewtas, L. Claxton, G. Strauss, and S. Nesnow, Eds., Plenum Press, New York, 1985, pp. 269–288.
47. Taylor, J. H. "Sister Chromatid Exchanges In Tritium-labeled Chromosomes," *Genetics* 43:515–529 (1958).
48. Underbrink, A. G., Schairer, L. A., and Sparrow, A. H. "Tradescantia Stamen Hairs: A Radiobiological Test System Applicable to Chemical Mutagenesis," in *Chemical Mutagens: Principles and Methods for Their Detection,* A. Hollaender Ed., Plenum Press, New York, 1973, pp. 171–207.
49. Van't Hof, J. and Schairer, L. A. "*Tradescantia* Assay System for Gaseous Mutagens. A Report of the U.S. Environmental Protection Agency Gene-Tox Program," *Mutation Res.* 99:303–315 (1982).
50. Vig, B. and Sandberg, A. A. "Aneuploidy Part 1: Incidence and Etiology," in *Program Topics in Cytogenetics,* Alan R. Liss, New York, 1987, pp. 1–434.
51. Wang, Y. Y., Flessel, C. P., Williams, L. R., Chang, K., DiBartolomeis, M. J., Simmons, B., Singer, H., and Sun, S. "Evaluation of Guidelines for Preparing Wastewater Samples for Ames Testing," in *Short-Term Bioassays in the Analysis of Complex Environmental Mixtures V,* S. S. Sandhu, D. M. DeMarini, M. J. Mass, M. M. Moore, and J. L. Mumford, Eds., Plenum Press, New York, 1987, pp. 67–88.
52. Zimmerman, F. K. "Chromosomal Malsegregation and Aneuploidy," *Mutation Res.* 201(2)(Special Issue):255–445 (1988).

SECTION V

QUALITY ASSURANCE/ QUALITY CONTROL

CHAPTER 13

Quality Assurance and Quality Control

Mark F. Marcus, Eugene J. Klesta, and John W. Nixon

The Resource Conservation Recovery Act (RCRA) program initiated a framework for the management of hazardous waste in the United States. In this framework, and inherent in the program, is the management of the risk and minimization of the liability associated with the proper disposal of hazardous waste.

A situation was identified in which the defensibility of waste management decisions was crucial. The key element in the defensibility was the in-house analytical chemistry performed by the laboratories. Making sound disposal decisions on new waste streams, monitoring of receiving loads, supporting safe onsite processing, and evaluating environmental media are all dependent on a comprehensive Analytical Quality Assurance / Quality Control Program.

The time demands created by the chemical processing of hazardous wastes and the costs resulting from delaying transportation equipment were critical in evaluating which quality assurance and quality control procedures could be included in the overall plan. The processing of hazardous wastes like most other industrial processes requires a short turnaround time for analysis. Therefore, the generation of the data must be done in a timely manner.

QUALITY ASSURANCE

Definition

The definition of quality assurance has been widely debated and is subject to many interpretations. A simple definition that incorporates the tenants of analytical chemistry, guidance from the Association of Official Analytical Chemists, and the regulatory mandates of the EPA's Federal Insecticide Fungicide, Rodenticide Act (FIFRA) and Toxic Substances Control Act (TSCA) is as follows:

> "Quality Assurance of Analytical Chemistry is a set of followed principles and procedures that provide valid, defensible data on a timely basis".

The principles and procedures are generally documented as a Quality Assurance Policy and Standard Operating Procedures (SOPs).

Standard Operating Procedures

Standard operating procedures (SOPs) are subject to as many definitions as quality assurance. The word procedure is basically incorrect because it implies an activity at the bench level of detail. SOPs are really at the policy level and are best carried over from good laboratory practices to include the principles which will be discussed later. These, in turn, are customized to a specific laboratory and applied at the bench level as site-specific practices. The SOPs serve as the template for these.

Standard operating procedures which apply to all laboratories have been written and distributed to the laboratory managers. The information found in these SOPs dictates the manner in which the test facilities (laboratories) should function. The SOP must include title, scope, procedure, corrective action, responsibility, and definitions. Each SOP has been written, reviewed, edited, and approved before being issued as a standard. The areas covered by SOPs are (1) organization, (2) internal assessment, (3) sampling, (4) sample control, (5) generation of analytical methods, (6) instrument maintenance and calibration, (7) data handling, (8) laboratory records, (9) quality assurance unit, (10) facility adequacy for laboratories, (11) corrective action, and (12) laboratory safety. Descriptions of some of these key principles follow.

Principles

Defensible Documentation

Documentation within the laboratory shall include two major components: (1) organization protocols, permanent staff records and company or regulatory agency required documents; and (2) SOPs for recording information regarding (a) the analytical methods, (b) quality control procedures, (c) chain-of-custody, and (d) instrumentation parameters. Accountable documents are used to provide a complete set of defensible records. These include, but are not limited to, a quality assurance plan, quality control procedures, training program, standard operating procedures, analytical methods, logbooks and raw data documents. The process of documentation (record keeping) should also be evaluated. Completeness of records, consistency of procedures through time and appropriate procedures for correction, archiving and security must be in place and practiced.

Computerization of data records must follow strict guidelines for security and defensibility and such guidelines must be described in the Quality Assurance Procedures Manual. A historical sequence of policies and procedures is also necessary to ensure recreation of the laboratory record at any point in time.

Analytical Methods

All analytical methods must be documented in a site-specific Analytical Methods Manual including those described in the site-specific Waste Analysis Plan. Unless statutes and regulations require otherwise, the Analytical Methods Manual should not be a part of the site permit, due to the dynamic nature of the methodology. However, it may be referred to in the site permits. Changes or modifications to methods cannot be made without significant proof of equivalency or improvement. A format for writing methods and a process for validation should exist. The process for validation should be followed diligently.

All modified standard methods should be submitted for approval. All non-standard methods must be written in the prescribed nine-point format and approved before using. The analytical methods in use must have the Method Detection Limit (MDL) and Limit of Quantitation (LOQ) determined and documented. The methods must be followed as written by the laboratory staff.

Copies of recognized standard methods must be on hand for reference by the analysts. Use of standard methods must be followed precisely to assure the defensibility of the data.

Sampling and Sample Control Methodology

Within the laboratory, security and integrity of samples are essential for the defensibility of the data generated for the samples. Standard operating procedures must be used to assure that appropriate sampling techniques and equipment are used and chain-of-custody of the samples remains inviolate (test samples are properly prepared, and test portions are appropriately taken from the test sample). The standard operating procedures shall include proper labeling techniques, the use of proper containers, and documentation requirements. Storage requirements regarding holding time, location, and safety are also delineated.

Facility Adequacy

The laboratory layout, furniture, floors, cabinets, hoods, and size should be evaluated. The isolation of incompatible preparation and analytical activities should be maintained. Design for the free flow of samples, people, and paperwork should be assessed. Proper equipment and essential instrumentation should be on hand to perform the analysis of interest.

Sufficient laboratory fume hood space with proper air handling equipment shall be provided to assure the safety of all employees in the laboratories. Housekeeping guidelines shall be used to maintain a proper working environment for performing analysis.

Equipment Maintenance and Calibration

All equipment maintenance and daily performance checks must be documented as permanent laboratory records. Outside service provided to the laboratory shall be properly noted and recorded. Instruments which require calibration shall be scheduled for service on a regular basis. Equipment may be sent out or have the proper service performed in the laboratory as required. All calibration curves and printouts must be properly documented and stored.

Evaluation of an instrument maintenance plan should be performed. The frequency of calibration and the use of service contracts should be assessed to determine the effectiveness of the instruments in use. Existence of calibration records and standardization practices should be reviewed during a system audit. Protocols for corrective action should be in place and the adherence to the protocols should be evaluated.

Personnel Training

Each site must have written training protocols, developed to initiate new employees into the use of current analytical and quality assurance/quality control procedures. Ongoing training shall continue periodically, according to written procedures, to keep all staff members aware of new methods and technologies.

A training plan should be drawn up and followed. This plan should include specific requirements for determining proficiency and special requirements for retraining and review. Job descriptions should be reviewed for levels of responsibility and experience. Personnel curricula vitae should be reviewed to determine adherence to the job description requirements. Increased responsibilities follow from individual educational improvements. Attendance at symposia, training courses, and further formal education is encouraged to develop well-trained personnel. Training which has been completed must be documented for each employee and kept as a part of the training records.

Quality Control Policy and Procedures

A quality control policy and quality control procedures are mandatory at all laboratories. Specific procedures are required and regular quality control audits are performed to assure compliance to company policy. Monthly quality control reports are required. The overall quality of analysis shall be compiled and summarized by the Quality Assurance/Quality Control unit. Acceptable levels of performance shall be presented and monitored to assess the quality of analysis.

QUALITY CONTROL

Definition

Quality control is a defined subset of quality assurance. The determination of an analyte's identification and the measurement of its concentration with known precision and accuracy is the essence of quality control. A set of principles and procedures are established to guide the analyst in the generation of data with known criteria. This is especially critical in hazardous waste analyses because of the variety of matrices encountered and the absence of standard methods that yield results of known precision and accuracy.

Generation

Most quality control programs require the analysis of duplicate samples and spiked samples. The use of instrument performance checks and quality control (QC) check samples may also be required in the program. Participation in external proficiency evaluations and split analysis with another laboratory are good ways to evaluate the laboratory's performance. The generation of this data is not the only requirement for a good program. The conversion of data to information, the interpretation of trends, and the problem solving of unacceptable performance are important activities which should be carried out after the generation of QC data.

The quality control program must establish some standard mechanisms for data generation and frequency. The calculation of percent error or percent difference or percent recovery must be standardized between laboratory operating groups or between laboratories when two or more laboratories are under the jurisdiction of the quality control program. After the generation of the QC data, a format for reporting the information must be established. The use of control charts, the calculation of control limits, and the generation of some basic statistics such as the mean and standard deviation must be standardized and used universally by the laboratory personnel.

Principles and Procedures

The elements necessary to produce data of known quality and a flow chart with decision points are shown in the following figure. Each of the principles is broken down to a detailed procedure for each of the analyses performed. Examples of procedures are included, but it must be emphasized that the procedures must be customized to each laboratories' needs. Each of the elements are defined as follows:

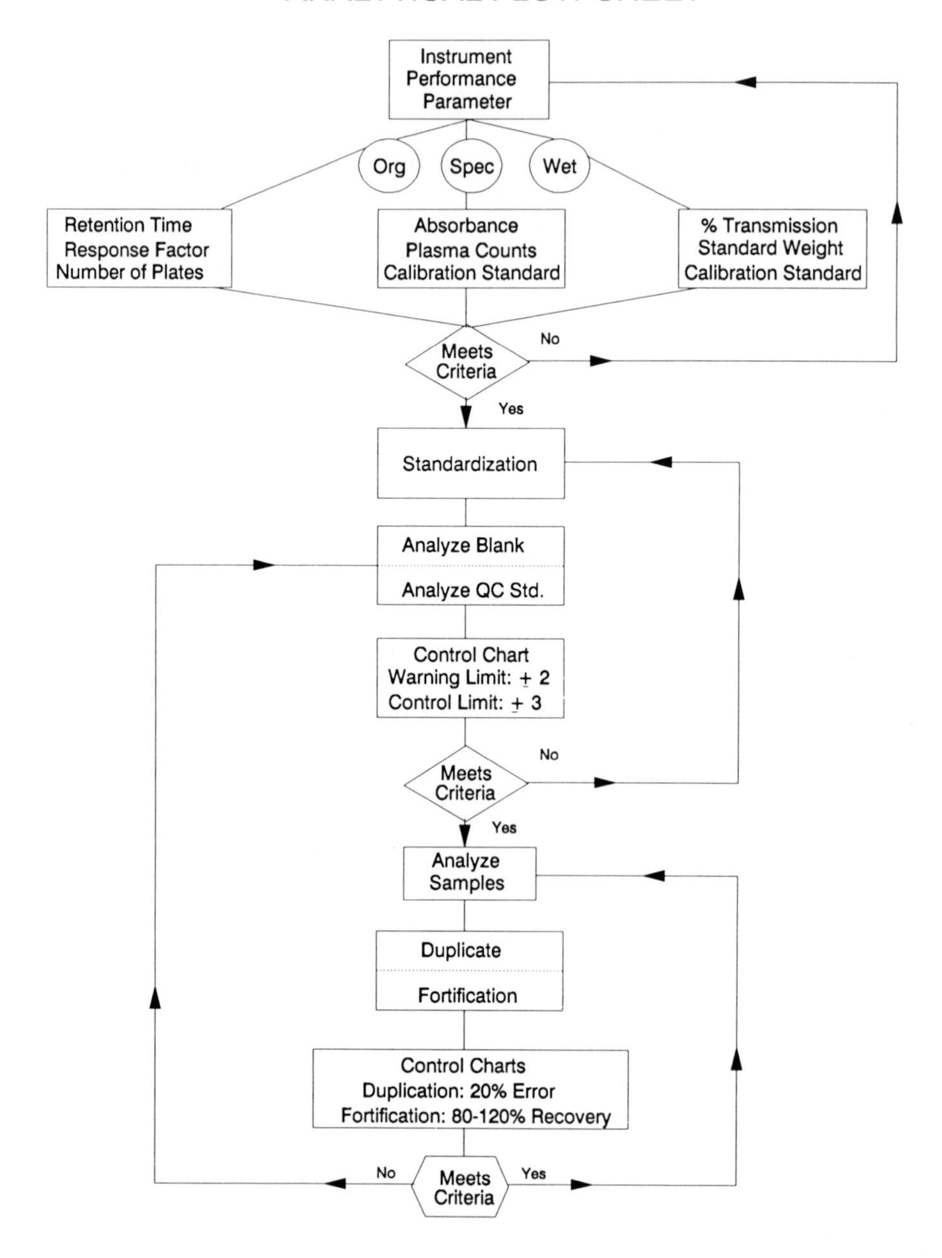
ANALYTICAL FLOW SHEET
Instrument Performance Parameter
Org
Spec
Wet
Retention Time
Response Factor
Number of Plates
Absorbance
Plasma Counts
Calibration Standard
% Transmission
Standard Weight
Calibration Standard
Meets Criteria
No
Yes
Standardization
Analyze Blank
Analyze QC Std.
Control Chart
Warning Limit: ± 2
Control Limit: ± 3
Meets Criteria
No
Yes
Analyze Samples
Duplicate
Fortification
Control Charts
Duplication: 20% Error
Fortification: 80-120% Recovery
No
Meets Criteria
Yes

Instrument Performance Parameters

Principle. All instrumentation must be evaluated through the use of an instrument check standard and calibration blank before standardization can be initiated. Divergence from acceptable benchmark criteria specified in the Quality Control Procedures Manual requires correction before analyses can be performed. Blank and instrument check standard results are recorded in a company-bound instrument log book which will also contain evaluation parameters, benchmark, and maintenance records.

Procedures. Performance parameters are checked each time the instrument is operated. Documentation of operating conditions must be found in the log book. Verification of conditions for each day of operation should be made. An instrument check standard and a calibration blank are run. The results must be recorded in bound instrument log books, one for each instrument with appropriate formats which are easily reviewed.

Organics. The number of plates is a measurement of the quality of the column. The precision of retention times monitors the flow rate and the possible occurrence of leaks at the inlet side of the column. The response evaluates the detector efficiency and the occurrence of leaks.

Contamination Evaluation

Principle. Reagent blanks must be prepared with each batch of samples and analyzed to insure that sample contamination has not occurred. Whenever possible, a field blank should be taken and analyzed accordingly. If blank analyses do not fall within acceptable limits, as specified in the Quality Control Procedures Manual, modification of reagents or modification of the analytical method must be implemented.

Procedure. Verification of contamination is made when the method blank yields a result which is greater than the detection limit for the method. A method blank should be analyzed with each batch of samples being run for a particular parameter and compared to the calibration blank.

Calibration Blank. A portion of the solvent equal in volume to the amount injected during an analysis to show background, if present. The results of which are subtracted from the instrument performance standard.

Method Detection Limits. MDLs can be established from several techniques:

1. From the analysis of 10 replicates of a method blank, a mean and standard deviation can be calculated. The method detection limit is equal to three times the standard deviation of the blank.
2. Serial dilutions are made until a standard at or near an approximation of the expected detection limit is found. Ten runs of the solution at or near the method detection limit are made through the entire method as written. The

results are used to calculate the mean and standard deviation. Three times the standard deviation will equal the MDL.

Limit Of Quantitation (LOQ). The limit of quantitation is recommended to be equal to 10 times the method detection limit. This limit is generally much higher than what is recommended by the American Chemical Society because of the complexity of matrices which occur in hazardous waste.

Organic. A method blank must be run with each set of samples. An amount of solvent equal to a sample is processed through all the steps of the procedure without any sample actually being present. Comparison is made to the calibration blank.

Acceptance Criteria: Less than 10 × Method Detection Limit.

Quality Control Check Sample

Principle. A quality control solution or sample material should be analyzed every day to show that calibration and standardization of instrumentation is within acceptable limits. This procedure informs the laboratory that prescribed precision and accuracy are being maintained. Results of these analyses will be reported monthly to the Manager of Quality Control.

Procedure. A quality control check sample is a standard separate from calibration standards which is processed and analyzed as a normal sample. The QC check sample is used for analyses control and should include a typical matrix. When unacceptable results are obtained, the analysis is out of control and must be stopped. The problem must be located and corrected before analysis can continue. All results taken between a QC result out of control and the previous one in control must be rerun.

Organic. PCB—An oil sample can be spiked with one or several Arochlors to be used as a check sample. This material is handled identically to sample procedures. Dilution, acid cleanup, and injection quantity must be standardized. The results for the QC check sample must be obtained daily.

Duplication and Fortification

Duplicate Analysis. Principle. A sample must be analyzed in duplicate for each 10 samples being analyzed for a particular parameter. A blind duplicate sample will be submitted to each laboratory on a weekly basis. Results of these duplicate analyses will be reviewed weekly and reported to the Manager of Quality Control monthly. Selection of duplicate samples should be specified in the Quality Control Procedures Manual.

Fortification of Samples. Principle. Fortifications are employed to monitor recoveries and maintain extraction and/or concentration techniques at acceptable levels. This procedure provides information about the effect of the sample matrix on the analyte in question. A ratio of one fortification for each 10 samples analyzed should be maintained. The same sample used for the duplicate analysis should be fortified according to the method prescribed in the procedure manual.

Procedure. Every 10th sample must be duplicated. The selection of the 10th sample must be random. A fortification must be done to the same sample that was duplicated. A known amount of analyte is added to the sample and analyzed in an identical manner.

When the analytical results for a particular parameter continually yield results below the limit of quantitation or are recorded as "none detected," then duplicate fortifications are strongly recommended for that parameter. Three weighings of the sample are made. A recommended amount of fortification material is added to two of the weighings. The percent error is calculated from these two fortified samples. The percent recovery is calculated from the mean result for the duplicates minus the result from the unfortified sample.

Organic. PCB—Typical quantities for analysis are as follows:

- 1 mL weighed (pipetable liquids)
- 3 g for sonication (viscous liquids-nonsoils)
- 10 g for soil (soxhlet)
- each final dilution is 1:100

Acceptance Criteria: ±20% error
80 to 120% recovery

In each case, a duplicated weighing must be made every 10th sample. The dilution with hexane, double acid cleanup, and Florisil cleanup should be done for both weighings as needed.

Blind Duplicate

Each week a sample should be arbitrarily chosen and logged in as a new sample without previous knowledge by analysts. All parameters that are tested by the laboratory in a month will have at least one blind duplicate done for that month. The Laboratory Manager should maintain a cross-reference record of samples submitted as "blind" duplicates. Evaluation of the laboratory procedures can be investigated from these results. The designated quality control person should make this evaluation. The "blind" duplicate results should be kept in a separate file so that they can be inspected during a quality control audit.

Reference Materials

Standard Reference Materials (SRMs) from the National Bureau of Standards or from the Environmental Protection Agency should be obtained and analyzed according to normal laboratory methodology to indicate accuracy of the methods. These materials should be analyzed at least quarterly, but preferably more frequently.

Round Robin Analyses

All site laboratories must participate in round robin sample analyses. These samples will be submitted by the Manager of Quality Control and will be for normal parameters associated with the hazardous waste industry. Results must be reported to the Manager of Quality Control before the due date. Digressions from the norms established by the majority of laboratories participating will be investigated and corrected by the Manager of Quality Control.

Reference Laboratory Evaluation

At least one sample per month for all parameters being analyzed in that laboratory for that month is to be sent to another laboratory for parallel analysis. These split sample results should be sent to the Quality Assurance (QA) Unit.

Reports

All laboratories must report monthly to the Quality Assurance Unit the following information on the specific Parameter Quality Control Chart:

1. number of samples analyzed
2. number of duplicates
3. number of fortifications
4. instrument used
5. frequency of occurrence of the quality control check sample within acceptable limits
6. mean and standard deviation for percent error and percent recovery for the analyte in question.

All laboratories must report intralab and round robin results as they occur and reference material analysis at least quarterly.

COMPUTERIZATION

The quality control program must establish a mechanism for laboratories to record, plot, and analyze quality control data through the use of personal computers. Our specialized program for the generation of monthly quality control reports was developed using Lotus Symphony®. A menu-driven system was established to enter quality control data such as duplicate analyses, fortified analyses, and quality control check sample analyses. All laboratories in the company use the same system for recording data. The control charts which are computer-generated are based on universal calculations at all laboratory facilities. The comparability of quality control data from one laboratory to another is performed easily.

The computerization of the actual data allows for the generation of important information regarding performance. The mean percent recovery and standard deviation for fortifications is readily calculated by the "macro" software. The quality control check sample data is converted into a mean and control limits for the succeeding month's data. The laboratory analyst can use these computer-generated limits to evaluate his performance on a daily basis. The duplicate analysis is summarized by the calculation of mean percent error and standard deviation. Outliers from established acceptance limits are readily noticed and can be addressed and corrected within a reasonable time frame by the use of computer programs.

PERFORMANCE AUDITS

General

Assessment of quality assurance practices and quality control procedures is essential in the generation of environmental data. Corrective action can only be effective when the areas in need of action are determined. Assessment can be internal and external and each of these aspects can concern itself with system evaluation and performance evaluation.

Internal Assessment

The laboratory should have a mechanism for internal review. The data generation process should be evaluated internally to be sure that all procedures are being followed. Internal systematic review should be done to determine compliance to the quality assurance plan. Corrective action should be done as soon as possible. The need for corrective action can only be determined when internal review is performed.

Internal assessment of the compliance to company QA/QC policies and procedures is conducted by the laboratory manager. Review of documentation, use of blind duplicates, and quarterly reference materials shall be used to evaluate individual performance within the laboratory.

External Assessment

External assessment of the adherence to the quality assurance policy and procedures and the quality control policy and procedures is the responsibility of the Quality Assurance/Quality Control unit. Quality Assurance and Quality Control Audit forms are developed and used on a periodic basis to evaluate performance of QA/QC procedures.

Centralization

Because of the complexity and versatility of hazardous waste, there is a need to develop corporate acceptance criteria based on real data from the laboratory facilities. The installation of personal computers in each of the laboratories allows for the extraction of critical information from each laboratory and the centralization of this information for further analysis and interpretation. Each laboratory is required to extract from each of its parameter files the key information that the corporate QA unit needs. The personal computers are equipped with modems and the corporate headquarters has established a 24-hr "bulletin board" to receive the information from the site laboratories. This information is compiled in a large database software system for interpretation and analysis.

Performance for particular parameters can be summarized for all sites which analyze the parameters to determine whether or not there are any sites which fall out of the limits of performance for that particular test. This information is useful to the corporate auditors to look for possible causes and establish corrective action at the facility. The data for a particular test for one site can be summarized through time to determine if a trend in either direction is occurring. Just as the monthly information for a particular test should be random, so should the statistical information from month to month be random. Pattern recognition and corrective action can also result from this type of summary.

The QA unit has the responsibility to assess and evaluate the performance of the individual laboratory, but it also has the responsibility of assessing and evaluating the performance of the analytical chemistry program. The generation, computerization, and centralization of the quality control data allows the QA unit to complete both of its responsibilities.

CHAPTER 14

Quality Assurance For Hazardous Waste Measurements

Dennis Hooton

THE CORNERSTONES OF QUALITY ASSURANCE

Quality assurance (QA) programs are developed to provide *systems* to help assure that information, products, or services are of proven and known quality. For hazardous waste management, policies and decisions are influenced by information provided through environmental measurements; actions that may affect entire communities or regions are based on the validity of those measurements.

From observation, the cornerstones of effective QA programs may be illustrated as the interrelationship of:

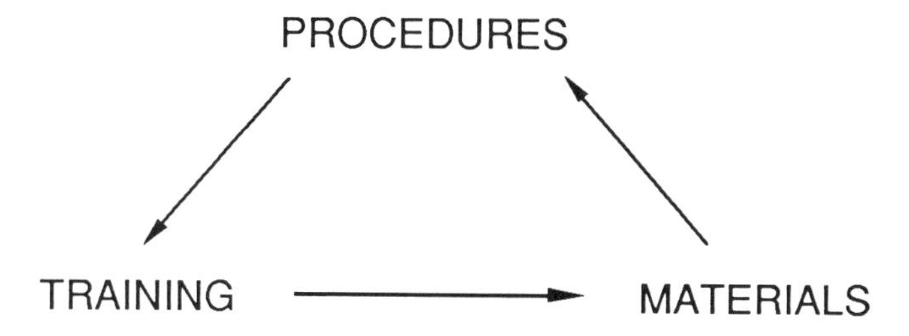

This simple diagram is meant to show that "known" quality is achieved when conscientious, qualified people perform their work using proven procedures with controlled materials and equipment. The elements of these "cornerstones" of quality as applied to hazardous waste management are presented in this chapter.

Usually there are two general misconceptions about QA: First, that the goal of all these systems is to always achieve the "best" quality, and second, that quality control (QC) is the same as quality assurance.

Quality assurance is affected by many factors such as the complexity of the project, sensitivity of information, basis of critical decisions, and costs. Also, there will always be "better mousetraps": new materials and better ways to accomplish tasks. The QA program must address the intended use of the data.

QC are deliberate and specific actions that are integrated into field, laboratory, or other research activities that will be used to collaborate and demonstrate the quality of the data.

Quality assurance functions as a system with the purpose of documenting QC activities and results, as well as assuring that such things as organizational accountability, qualifications, records management, data processing, and security are evident to the client or external agencies. Quality assurance may be thought of as protection of the environment and of an organization's reputation.

QUALITY TRAINING

The first cornerstone of QA is "quality training." The key elements of this concept are commitment, accountability, qualifications, certification, and performance. The sources of commitment must come from all levels within an organization beginning with a statement of organizational QA policy and objectives.

Policy and Objectives

For hazardous waste measurements, the purpose of an organizational QA program is to ensure that all environmental measurements and services performed, directed, or subcontracted to other laboratories by that organization result in scientifically valid and defensible data of known quality which are consistent with project requirements and the needs of the client. The specific policy of an organization's commitment to QA and the authority by which it is implemented should be clear and evident.

Administrative objectives may vary significantly, but certainly the following list suggests the scope of a QA program within an organization:

- method performance
- technical staff performance
- training needs
- instrument performance
- record keeping and report writing practices
- communication between scientists and management
- legal defensibility
- proposal preparation and marketing

Accountability

Accountability is also a key element of quality training. Organizational responsibilities must be understood and include the actual names and titles of key individuals and subcontractors. An accurate inventory of personnel and equipment resources, as well as indications of successfully completed projects are necessary for evaluating an organization's capabilities.

One of the first priorities of a viable program is to define the responsibilities regarding QA. Simple organizational charts help to show lines of authority and communication, as shown in the example below.

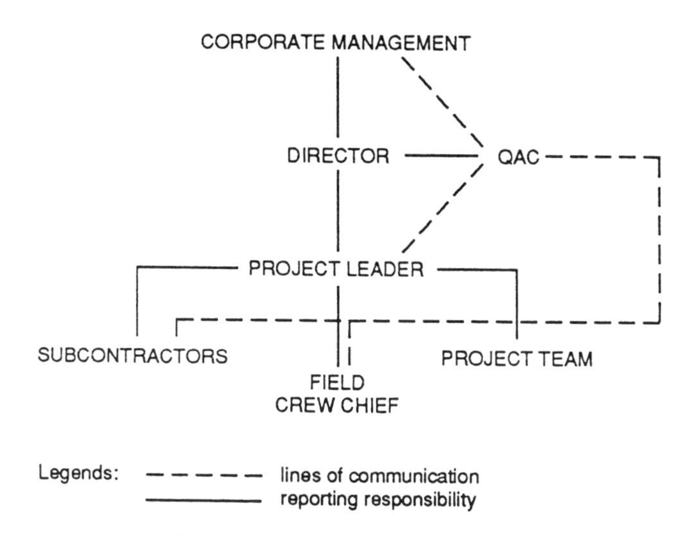

The following descriptions are generic and typical terms of key people in a QA program.

The QA organization usually starts at the director level. The director reports to corporate management and has ultimate authority and responsibility for all research and QA activities within the department. The director exercises general administrative control over each project undertaken by the department. The director ensures that the project teams have the talent and experience necessary to conduct project work most effectively and that all available resources are utilized to accomplish project objectives. Technical work and reports may be reviewed by the director to guarantee that the work is of the highest professional quality and that the research findings are presented to clients in a clear, concise, and useful manner.

A quality assurance coordinator (QAC) may report to a department director or directly to corporate management. This person would be responsible for the coordination and documentation of quality performance within the technical and administrative activities of the department or organization. To avoid conflict of interest, the QAC does not have direct responsibility for the generation of the analytical data, but would be technically qualified to review results of the work. Usually, the QAC has the unique advantage of reviewing and comparing many sources and types of work which may result in early detection of problems or specific recommendations for establishing QA policies. Traditional duties of the QAC include internal auditing, inspections of equipment or facilities, and evaluation of QC results and project deliverables in order to monitor the implementation and effectiveness of the QA program. Other functions of the QAC are to interact with external auditors, distribute audit reports, maintain QA files, and audit subcontractors or support laboratories.

The designated project leader is responsible for actually producing the data and assuring that all QA / QC measures have been performed correctly. Project leaders are usually the primary contact with clients. Their responsibilities include meeting a project's technical and fiscal goals and generating all project reports to the client. Project leaders plan the project work in detail on a daily basis, advise appropriate supervisors of any problem areas, and suggest suitable corrective action. Depending on the size and nature of a project, the project leader may personally perform most of the technical work or delegate it to key professionals, forming a project team.

For field activities, the crew chief would usually report to the project leader and be responsible for coordinating all activities during a field test, including technical and safety requirements. Pretest site visits are performed to identify test support and field equipment needs and also to identify potential and existing safety hazards associated with the facility or targets. The crew chief would assure the adequacy of technical training and safety precautions of the field crew members and that field test equipment is functional and appropriately maintained. The crew chief certifies the validity of field-generated data and the integrity of samples until these are relinquished to designated laboratory custodians.

The technical staff assigned to projects must have access to all information relevant to the technical methods, QC requirements, safety procedures, and security restrictions for the task at hand; this information needs to be reviewed and understood by the staff member prior to beginning the work. In other words, members must be trained and qualified. Nonfamiliarity with the assigned task requires that the work be performed under the supervision of a qualified staff member. The technical staff is required to fully document all tasks performed in a manner consistent with good laboratory practices, such as using the appropriate laboratory notebooks or project forms and recording unusual occurrences or deviations from stated test plans or methods. Major problems are usually avoided when the technical staff immediately notifies the project leader of any problem areas.

Complex projects sometime make it necessary for analyses to be subcontracted to other laboratories when the primary contractor does not routinely perform certain types of analyses or when workloads prohibit meeting deliverable schedules. Unless specified otherwise, the primary contractor is responsible for assuring the quality of the subcontracted work as well as its own. The quality of the subcontracted work must be reliable and consistent with project objectives. There are various ways to achieve this consistency, such as reserving the right to submit performance evaluation samples or blind duplicates, review QC results, audit supporting data, and perform inspections of subcontractor facilities and operations.

Qualifications, Certification, and Performance

Corporate commitment and accountability are only as good as the staff that performs the work. The QA program has a primary responsibility to develop a system that maintains records of qualification (e.g., diplomas, training courses, and resumes) for every staff member.

Certification and performance records of the project staff should be related to specific procedures they are asked to perform. For example, a "training program" may consist of the following steps:

- read and understand the written procedure
- perform the procedure under direct supervision
- perform the work independently with results inspected by a supervisor
- maintain records of performance

Continuing records of performance document the technical proficiency of personnel, help identify training needs, and provide valuable historical information on the accuracy and precision of the method.

QUALITY MATERIALS

The second cornerstone of quality assurance is "quality materials." The concept of "quality materials" for hazardous waste management includes assurances of proper maintenance and calibration of equipment, the certification of analytical reference standards and materials, and traceability of all components (equipment, standards, and staff) that are used to generate the data.

Measuring and Test Equipment

Tools, gauges, instruments, and other measuring and test equipment used for activities that affect data quality should be controlled through unique identification, use logs, and status tags (e.g., working, not working, etc.). Test personnel should have access to relevant standard operating procedures (SOPs) or operation manuals, which provide corrective action procedures in case of equipment failure.

Measurement equipment should have records on file that document all scheduled (preventive) or unscheduled (corrective action) maintenance performed. All calibrations and adjustments to maintain accuracy within prescribed limits should also be scheduled at specified periods and documented as performed. Most importantly, all generated data should be traceable to the instruments used. These records establish the reliability of the equipment during the time the data are generated.

Standards

The quality of the standards used for calibration, quantitation, and identification in hazardous waste measurements is critical in the final assessment and use of the data. Regulatory and enforcement actions are often based on legislated limits or emission standards, and significant consequences can occur if there are even minimum violation of these limits. Therefore, the reliability of the standard data helps assure that the regulations are appropriately applied.

Perhaps the most respected source of certification for calibration services or standard reference materials is the U.S. Department of Commerce, National Institute of Standards and Technology (NIST) formerly National Bureau of Standards (NBS). Calibration records are typically and purposefully made traceable to the NIST reference materials or test equipment originally calibrated by the NIST. Many commercially purchased standards may be prepared and tested under strict quality control requirements, but come without written guarantee or certification. An example of a NIST (NBS) certificate is presented in Figure 1.

Other excellent sources of reference materials are U.S. EPA reference materials and quality control samples. The U.S. EPA maintains histories of these samples and may establish the integrity of the reference materials through interlaboratory analyses.

Another option for an analytical laboratory is to prepare its own reference standards, which may then be verified through techniques such as using two independent sources or lots of reference material, making duplicate weighings for preparation of stock standards, or having a different analyst prepare a check standard. Characterization profiles on chemicals by chromatographic or spectrophotometric screening or simple physical parameter testing (e.g., melting point, etc.) may be used to establish identity and purity of questionable reference materials.

Many standards may degrade with time and require labeling preparations with expiration dates for disposal. These holding times may be established through controlled stability studies that simulate the matrix and storage conditions expected. Comparison of analytical performance to previous standards is also a good practice for establishing a history of acceptable criteria and detecting grossly deficient standards.

Even certified standards may be wrong at times, a general rule of thumb is to independently verify all standards used to generate critical data with a different method or source of material. Just as maintenance records establish the reliability of the equipment during the time the data are generated, the certification data and traceability of the standards help assure the accuracy of the test results.

QUALITY PROCEDURES

The final cornerstone of quality assurance may be defined in the general term "quality procedures." The concept of quality procedures for hazardous waste measurements encompasses key elements of both quality assurance and legal defensibility for generated data; these include reliability, representativeness, comparability, and data quality.

Reliability is the assurance that the procedure will give the same result on successive trials; the final results are of "known" quality. The EPA[1] defines representativeness as the degree to which the data accurately and precisely represent a characteristic of a population, parameter variations at a sampling point, or an environmental condition and comparability as a measure of the confidence with which one data set can be compared to another.

The choice of procedures and methods is reflective of the project objectives for data quality and the basis for determining the data quality indicators of accuracy, precision, and completeness. These quantitative measurements are discussed later.

Project and QA Plan

The basis for evaluating the "quality procedures" used for a particular project are the project plan and the QA plan. Sometimes these are separate documents; other times these may be incorporated into a single plan.

After sufficient background information about the project has been obtained and evaluated, a specific test plan is prepared based on the objectives and scope of the project. Simple projects may only require a memorandum which describes the project, identifies the objectives, and states how the work will be performed. More complex projects require that the project leader prepare a project plan detailing the project's scope, logistics, staffing, and schedules. Items that need to be addressed in the full-blown project plan are

- background information
- objectives
- scope of work
- pretest activities and planning
- data and reporting deliverables
- descriptions of sampling locations and monitoring methods
- analytical test parameters
- sampling procedures and analytical methodologies
- quality assurance / quality control requirements
- records management
- sample handling requirements
- project organization and key personnel
- equipment and resource needs
- safety requirements and safety plan
- schedules for field work, analyses, reports, etc.

National Bureau of Standards

Certificate of Analysis

Standard Reference Material 1581

Polychlorinated Biphenyls in Oils

This Standard Reference Material (SRM) is intended primarily for calibrating instrumentation and validating methodology used in the determination of polychlorinated biphenyl mixtures (PCB's) in motor and transformer oils. These PCB's are present as Aroclor 1242 and Aroclor 1260. Certified concentrations are shown in Table 1.

Table 1. Certified Concentrations of Aroclor 1242 and Aroclor 1260 in Motor and Transformer Oils.

Matrix	Aroclor Type	Concentration, μg/g*
Motor Oil	1242	100 ± 1
Motor Oil	1260	100 ± 2
Transformer Oil	1242	100 ± 1
Transformer Oil	1260	100 ± 3

*Uncertainty is expressed at the 95% confidence level.

NOTICE AND WARNING TO USERS

Handling: PCB-containing materials are reported to be toxic and should be handled with care. Contact your regional office of the U.S. Environmental Protection Agency for information regarding proper disposal.

Expiration of Certification: This certification is valid within the specified uncertainty limits for two years from the date of purchase. In the event that the certification should become invalid before then, purchasers will be notified by NBS.

Storage: Sealed ampoules, as received, should be stored in the dark at temperatures between 10 to 30° C.

Preparation and analytical determinations were performed at the Center for Analytical Chemistry, Organic Analytical Research Division, by S.N. Chesler, F.R. Guenther, and R.M. Parris.

Consultation on the statistical design and analysis of the experimental work was provided by K. Kafadar and K.R. Eberhardt of the Statistical Engineering Division.

The coordination of the technical measurements leading to certification was under the direction of S.N. Chesler and H.S. Hertz.

The technical and support aspects involved in the preparation, certification, and issuance of this Standard Reference Material were coordinated through the Office of Standard Reference Materials by R. Alvarez.

Washington, D.C. 20234
June 25, 1982

George A. Uriano, Chief
Office of Standard Reference Materials

Preparation and Analysis

The Aroclors were obtained from the U.S. Food and Drug Administration Industrial Chemical Repository (Aroclor 1242, FDA 399-DCT-71696; Aroclor 1260, FDA 371-DCT-71699). The virgin motor oil base stock and the mineral base transformer oil were obtained from commercial sources. The PCB solutions were prepared by weighing and mixing the individual Aroclors and oils. The four solutions were then dispensed into 5-mL amber ampoules and flame sealed. Samples representing early, middle, and final stages of ampouling were analyzed using gas chromatography. Prior to gas chromatography, the oil solutions were separated into fractions using a high performance liquid chromatographic (HPLC) procedure. In this HPLC procedure, an aminosilane semi-preparative scale HPLC column was used. The mobile phase was 100 percent hexane and the effluent was monitored at a wavelength of 254 nm with a UV detector.

The HPLC fractionated samples were analyzed with a gas chromatograph equipped with an injector splitter and a 30 m x 0.25 mm nonpolar, immobilized phase, wall coated, open-tubular column. A constant current Ni^{63} electron capture detector was used for these analyses. Quantitative results were obtained by using, as internal standards (IS), PCB isomers that were not detected in the analyte Aroclors. Calibration standards consisting of weighed amounts of the Aroclor and IS compounds in the PCB-free diluent transformer or motor oil were chromatographed to measure analyte response factors. The results of Aroclor 1260 are based on the areas of ten selected Aroclor PCB peaks and three IS peaks; for Aroclor 1242, on areas of nine Aroclor PCB and two IS peaks.

Table 2 lists the calculated and analytically determined concentrations of the Aroclors in the Oils.

Table 2. Calculated and Gas Chromatographic Results Used for Certification

Matrix	Aroclor Type	Concentration, µg/g	
		Calculated[a]	Gas Chromatography
Motor Oil	1242	100.00 ± 0.04[b]	100.1 ± 1.4[b]
Motor Oil	1260	100.00 ± 0.04	100.7 ± 2.0
Transformer Oil	1242	99.96 ± 0.04	100.4 ± 1.2
Transformer Oil	1260	99.98 ± 0.04	100.6 ± 2.7

[a] The calculated concentration is based on the total mass of the Aroclor added to the oil.
[b] Uncertainty is given as 95% confidence limits.

Figure 1. Example of NIST (NBS) Certificate of Analysis.

- project costs
- retention and storage of project files
- project records
- final disposition of samples

The project plan, when completed, represents an agreement between the requesting party and those individuals performing the work. There should be adequate time available for the agency or client to review and comment on the draft plan before any specific field, laboratory, or consultant activity is undertaken. The project leader is responsible for the distribution and control of copies of the draft project plans and maintains a record showing who received draft copies and when such copies were returned. Once all concerned parties agree to the project plan, it serves as a "working" document for the project. Any changes to the test plan, for whatever reason, should be explained and documented.

Aschwanden,[1a] an attorney with the National Enforcement Investigations Center of the EPA, recommends these general guidelines when obtaining "court-defensible" data for sensitive projects or planned litigation:

- It is of primary importance to obtain reliable and representative samples. Field surveys and sampling activities must be documented. Sufficient information is needed so that someone can reconstruct the sampling without reliance on the collector's memory. Failure to prove that a sample was properly taken could prevent subsequent data from being admitted into evidence.

- Sampling methods from the U.S. EPA "Test Methods for Evaluating Solid Waste: Physical/Chemical Methods," SW-846[2] are recommended, but may not be required. SW-846 is considered a "safe-harbor," i.e., reliable, representative, and comparable to similar type work. Alternative sampling methods can be used if equivalency to proven methods is demonstrated.
- The chain-of-custody record is a key element of court defensible data. A continuum from sampling to test results must be established. There must also be designated procedures that establish the regular course of business that adequately identifies what happens to the sample every step of the way through analysis.
- Testimony is critical to establishing custody, identity, and relevancy of data submitted as evidence. Excess paperwork and forms should be eliminated to minimize confusion. Documentation (records and procedures) should be used to support testimony about events that happened 2 or 3 years earlier.
- Documented training programs and Standard Operating Procedures (SOPs) allow the scientist to confidently testify that the right sample was analyzed in a controlled laboratory environment.
- Security measures are important to controlling access to samples and laboratory facilities. Security policies should be addressed in SOPs.
- Raw data are critical to court challenges regarding interpretation of test results, and therefore should be readily identified, traceable, and included in the project files.
- Modifications to analysis methods may be allowed if based on sound technical judgment; arbitrary changes to methods could cause data to be legally rejected. Some regulations require strict adherence to established methods—no exceptions. Technical justification for method modifications should be included with final test results.

QA Levels

Projects vary in QA requirements based on the complexity of the study, objectives, and costs. At a minimum, the EPA suggests[3] the QA project plan should include the following:

- project description
- data quality objectives for critical measurements

A complete and accurate description of the project's scope, activities, deliverables, and schedule is essential in avoiding misunderstandings with the client and maintaining control of the project.

Data quality objectives (DQOs) are qualitative or quantitative statements of the level of uncertainty that the client is willing to accept in results derived from data gathering activities. DQOs are usually defined in terms of accuracy, precision, completeness, comparability, and representativeness, and should always be based on the intended use of the data. DQOs may be derived from technical judgment (estimates), past performance, or contractual requirements; the basis for arriving at project DQOs needs to be specified in the project plan to avoid misunderstandings in the acceptability of the data.

In general, the Quality Assurance Project Plan (QAPP) should be consistent with the nature and scope of the project and the intended use of the data. For simple projects or for some research and feasibility-type projects, the description and objectives of the project can be incorporated into a letter proposal, test plan, or project memorandum. For other projects, client directives regarding QA/QC, work progression, or deliverables need to be formalized in writing and clearly understood by all parties. For projects that operate under a formal QA program, the QAPP becomes a contractually binding agreement on how the work is to be performed.

The EPA's policy[4] on quality assurance stipulates that every monitoring and measurement must have a written and approved QA project plan (QAPP). The minimum QA requirements that should be addressed in a formal QAPP are presented below. In addition, any additional client directives or method-specific QA/QC requirements should also be included or referenced in the QAPP:

- title page with approvals
- project description
- project and QA/QC organization
- data quality objectives for critical measurements
- sampling procedures
- sample custody
- calibration procedures and frequency
- analytical procedures
- data reduction, validation, and reporting
- internal QC checks
- plans for performance, systems, or data audits
- preventive maintenance
- calculation of data quality indicators
- corrective action
- QA/QC reports to management

The title page would identify the project, organizations involved, and distribution of the plan. It would also have the approval signatures and effective dates that authorize the agreement on how the work is to be performed. The project description should be a thorough, detailed summary of the scope and objectives of the project. Like a well-written newspaper article, the project description should summarize who, what, where, when, why, and how. Subsequent sections of the plan can elaborate on these topics.

All too frequently, QAPPs are not acceptable to an agency or client because of a vague, generic, or incomplete project description; it is unclear what the author plans to do, or why. In some cases, QAPPs are written prematurely, before solid project objectives are available. Consequently, other elements are also incomplete and numerous changes are needed before data collection can begin.

The QAPP should identify key personnel who will be responsible for the analyses and validation of the data, expected members of the project team, QA coordinators, and administrative management. If possible, statements of staff qualifications and relevant experience should be provided for key personnel. An organizational chart is useful for presenting this information.

Data quality objectives (DQOs) are primarily based on the intended use of the data. These objectives set the criteria for control of data collection or work progression and successful completion of the project. Critical versus noncritical data measurement activities should be identified for setting priorities and evaluating impact of any lost data.

Data quality objectives are generally expressed in terms of:

- precision
- accuracy
- completeness

Precision is the degree of mutual agreement among individual measurements made under prescribed conditions with a single test procedure. Examples of activities to assess precision are replicate samples, replicate injections, colocated monitors, and interlaboratory studies.

Accuracy is the degree of agreement of a measurement with an accepted reference or true value. Accuracy is expressed as (1) the difference between the two values, (2) a percentage of the reference or true value, or (3) a ratio of the two values. Examples of activities used to assess accuracy are analysis of independent performance samples, blind samples, spiked matrix samples, and spiked surrogate compounds.

Completeness is a measure of the amount of valid data obtained from a measurement system compared to the amount that was expected to be obtained under normal circumstances. Examples of activities to determine completeness would be analysis and verification of clean blank samples or a specified percentage of acceptable surrogate recoveries for validation of the data.

Collectively, these terms are sometimes called data quality indicators (DQIs). Precision, accuracy, and completeness are quantifiable values and are conveniently presented in tables to summarize QAPP target values. An example of such a table is presented in Figure 2.

Cautious judgment must be used in deciding how these criteria were derived, what means and frequency with which these will be determined, what should happen if these criteria are not met, and how this information supports the quality of decision data.

The numerical values must be justified; there is a danger of being too restrictive in setting criteria. Control limits should be reasonable and not

DATA QUALITY OBJECTIVES

Parameter	Matrix	Test condition	Precision (R %)	Accuracy: Mean recovery (%)	Completeness
Volatile POHCs					
	VOST traps	Instrument performance sample	NA	60–130	100%
	Waste-derived fuel	Duplicate analysis for 2 runs	< 30	NA	90%
	Fuel oil	Duplicate analysis for 2 runs	< 30	NA	90%
	Sludge	Duplicate analysis for 2 runs	< 30	NA	90%
	Organic liquid waste	Duplicate analysis for 2 runs	< 30	NA	90%
	Containerized waste	Duplicate analysis for 2 runs	< 30	NA	90%
	Low Btu liquid waste	Spiked water performance sample	NA	60–130	100%
PCDDs/PCDFs	Stack emission	Analysis of duplicate spiked blank filters and spiked XADs (method performance samples)	NA	50–150	100%
		Instrument performance sample	NA	60–120	100%
		Each sample spiked with ^{13}C-PCDD/PCDF method internal standards[c]	NA	50–150	90% of data obtained from all runs
Metals	All matrices being analyzed, except stack gas	Duplicate analysis of one run for each matrix	< 25	NA	90%
Chlorine, HV, ash, viscosity, specific gravity	Aqueous wastes	Duplicate analysis of each matrix for one run	< 20	NA	90%
Chlorine	Organic liquid wastes	Analyze three spikes at three different C1 levels	NA	80–120	100%
Hydrogen chloride	Stack gas impinger solutions	Analyze an independent check sample	NA	80–120	100%
		Duplicate analyses for 1 run	< 30	NA	100%

Figure 2. Example of data quality objectives.

always based on the optimal achievable values for a particular method or procedure.

Sampling procedures and custody of the samples should be specified in detail, including the following information:

- matrix
- location
- frequency of collection
- sample size
- types of containers and precleaning
- preservation of sample
- reference or inclusion of standard operating procedure
- QC requirements (number and types of blanks, duplicates, etc.)
- total number of samples to be collected
- total number of samples to be analyzed (e.g., after compositing)
- planned modifications to referenced methods
- means of transport
- designated field custodian

Much of this information can be presented in tables, such as in Fig. 3.

All analysis procedures and general laboratory procedures for sample handling should be specified, including the following:

- analysis parameters
- expected matrices
- reference or inclusion of preparation methods
- reference or inclusion of analysis methods
- reporting requirements
- QC requirements (number and types of blanks, duplicates, spikes, etc.)
- instrument calibration procedure
- identification of laboratory
- planned modifications to referenced methods
- laboratories involved
- sample custodian
- storage requirements
- holding times for extraction and/or analyses

Again, much of this information is easily presented in table form. Examples are presented in Figures 4 and 5.

Sample custody procedures and example forms should be provided in the QAPP to help ensure that adequate traceability and custody records will be produced. Calibration procedure, frequency, and acceptance criteria should be summarized and presented. Corrective actions may be developed ahead of time for common occurrences of specific problems.

Data reduction steps should be described with inclusion of example data entry forms. Forms should be traceable to the procedure used or may even include the procedure on the form for verification or noting deviations during sample processing. An example of this type of form is presented as Figure 6.

SAMPLING SCHEDULE FOR TEST CONDITION "X"

Sample	Location	Frequency	Method	Size	Analyses	Comment
Fuel oil	3	Each 30 min	Tap (S004)	100 mL	HV, A, VIS, SG, Cl, V, SV	Composite for analysis
	3	Each 30 min	Tap (S004)	100 mL	Metals	Composite for analysis
Dried solids	7	Once per run	Scoop (S002)	500 g	Metals	
Lime slurry	8	Once per run	Tap (S004)	500 g	Archive	
Stack gas	9	Continuous, 2 h	MM5-M	60–100 cu ft	Particulate, metals, moisture, temperature, velocity	
	9	Continuous, 2 h	MM5-0	60–100 cu ft	SV POHCs	
	9	Continuous, 2 h	MM5-0	60–100 cu ft	PCDD/PCDF	
	9	Continuous, 2 h	MM5-0	60–100 cu ft	SV, PCDD/PCDF, moisture, temperature, velocity, HCl	
	9	Each 30 min	VOST	20 L/trap pr	V POHCs, PICs	Four trap pairs at 30 ft per trap pair
	9	Continuous, 2 h	Integrated bag	15 L	V POHCs	
	9	Continuous, 2 h	EPA Ref. Method 3	20 L	Oxygen, carbon dioxide	

HV = Heating value
A = Ash
VIS = Viscosity
SG = Specific gravity
Cl = Chlorine
V = Volatiles
SV = Semivolatiles
PIC = Products of incomplete combustion
POHC = Principle organic hazardous constituent

Figure 3. Example of planned sampling schedule.

SUMMARY OF SCHEDULED ANALYSES

Analysis	Sample	Preparation	Method
Heating value (HV)	Waste derived fuel	NA	Calorimeter (D240-73)
	Low Btu liquid waste	NA	Calorimeter (D240-73)
	Fuel oil	NA	Calorimeter (D240-73)
	Sludge	NA	Calorimeter (D240-73)
	Organic liquid waste	NA	Calorimeter (D240-73)
	Bulk solids (soil)	NA	Calorimeter (D240-73)
	Bulk solids (spiked sand)	NA	Calorimeter (D240-73)
	Containerized waste	NA	Calorimeter (D240-73)
Ash (A)	Waste-derived fuel	NA	Ignition (D482-80)
	Low Btu liquid waste	NA	Ignition (D482-80)
	Fuel oil	NA	Ignition (D482-80)
	Sludge	NA	Ignition (D482-80)
	Organic liquid waste	NA	Ignition (D482-80)
	Bulk solids (soil)	NA	Ignition (D482-80)
	Bulk solids (spiked sand)	NA	Ignition (D482-80)
	Containerized waste	NA	Ignition (D482-80)
Viscosity (Vis)	Waste-derived fuel	NA	Viscometer (D88-81)
	Low Btu liquid waste	NA	Viscometer (D88-81)
	Fuel oil	NA	Viscometer (D88-81)
	Sludge	NA	Viscometer (D88-81)
	Organic liquid waste	NA	Viscometer (D88-81)
Specific gravity (SG)	Waste-derived fuel	NA	Pycnometer
	Low Btu liquid waste	NA	Pycnometer
	Fuel oil	NA	Pycnometer
	Sludge	NA	Pycnometer
	Organic liquid waste	NA	Pycnometer
Chlorine (Cl)	Waste-derived fuel	NA	Org. halide (D808-81, D4327-84, or E442-81)
	Low Btu liquid waste	NA	Org. halide (D808-81, D4327-84, or E442-81)
	Fuel oil	NA	Org. halide (D808-81, D4327-84, or E442-81)
	Sludge	NA	Org. halide (D808-81, D4327-84, or E442-81)
	Organic liquid waste	NA	Org. halide (D808-81, D4327-84, or E442-81)
	Bulk solids (soil)	NA	Org. halide (D808-81, D4327-84, or E442-81)
	Bulk solids (spiked sand)	NA	Org. halide (D808-81, D4327-84, or E442-81)
	Containerized waste	NA	Org. halide (D808-81, D4327-84, or E442-81)

Figure 4. Example of analysis schedule.

CONTAINERS, PRESERVATION, AND HOLDING TIMES

Measurement	Matrix	Containers[a]	Preservative	Recommended maximum holding time before extraction (days)	Holding time from extraction to analysis (days)
Volatile POHCs	Tenax	VOST cartridge	Chill with ice	14	NS
	Liquid organic waste	VOA vial	Chill with ice	14	NS
	Aqueous waste	VOA vial	Chill with ice	14	NS
Chlorine	Liquid waste	G	None	NS	30
Hydrogen chloride	KOH solution	G or P	None	NS	30
Stack gas particulate	Glass filter	Standard petri dish	None	NS	NS
PCDD/PCDF	MM5 components, organic, and aqueous waste feeds	XAD petri dish and G Teflon-lined caps	Chill with ice	NS	NS
Metals	MM5 components, organic, and aqueous waste feeds	Petri dish and G Teflon-lined caps	Chill with ice	NS	180[b]

KOH = Potassium hydroxide.
NS = Not specified.
[a] = Polyethylene (P) or glass (G).
[b] = Exception: mercury, 28 days.

Figure 5. Example of sample handling requirements.

PROJECT NO.:____________

DATE:____________

ROTARY EVAPORATION

Sample No.	Funnel/R-B Flask	Container Rinse Solvent	Container Rinse Lot No.	Hexane Lot No.	(Option C Only) PCDD/F Fraction I.D.	(Option C Only) PCB Fraction I.D.

COMMENTS OR MODIFICATIONS TO PROTOCOL:

ANALYSIS OPTIONS: A (PCBs) B (PCDD/PCDFs) C (PCBs & PCDD/Fs)
(circle one)

7.1 Rotary Evaporation

7.1.1 A sample to be concentrated using a rotary evaporation apparatus is placed in a standard $ 22 40 500-mL round-bottom flask. The flask should be no more than three-fourths full.

7.1.2 The apparatus is prepared for use.

7.1.2.1 Two hundred fifty milliliters of clean solvent (same as sample solvent) is placed in a clean round-bottom flask.

7.1.2.2 The vacuum is started (turn on water aspirator or vacuum pump). The vacuum line must have an in-line trap to prevent moisture from backing into the apparatus due to pressure or flow variations, and to prevent sample solvent from contaminating the vacuum source.

7.1.2.3 Turn on the water bath. Heat to the temperature necessary to evaporate the solvent used.

7.1.2.4 Turn on the condenser water.

7.1.2.5 Place the round-bottom flask with clean solvent on the apparatus. Clamp the round-bottom flask to the apparatus.

7.1.2.6 Close the pressure release valve and make sure that vacuum is achieved.

7.1.2.7 Turn on the motor and start the flask rotating slowly.

7.1.2.8 Lower the apparatus so that the solvent flask is in the water bath.

7.1.2.9 When the solvent begins to collect in the collection reservoir, adjust the water bath temperature so that the solvent is condensing at a rate of about 5 mL/min.

7.1.2.10 Allow the clean solvent flask to evaporate to almost dryness.

7.1.2.11 Raise the apparatus out of the water bath.

7.1.2.12 Turn off the motor.

7.1.2.13 Grasp the flask and open the pressure relief valve to slowly release the vacuum.

7.1.2.14 Remove the flask from the apparatus. The apparatus is ready for use.

7.1.3 Place the round-bottom flask with the sample on the apparatus. Proceed as in steps 7.1.2.6 through 7.1.2.14. If a water layer appears during concentration, immediately stop concentration and separate water as in step 6.1.3. Note on data sheet.

7.1.4 Do not allow samples to evaporate completely. Stop when the sample is reduced to 1 to 2 mL. If further concentration of the sample is necessary, nitrogen concentration will be performed.

7.1.5 The apparatus must be thoroughly rinsed with clean solvent in between samples to prevent cross contamination. Archive glassware rinses.

Chemists:

____

____

____

Reviewed by/Date:

____

Figure 6. Example of data entry form.

Example calculations or formulas should be presented; computer collection or manipulation steps and manual data verification points also need to be described.

Validation criteria should be described, listing the actual steps or checklist that comprise the data validation process. Figure 7 presents an example of the form used to define "valid" data. Technical research should include sources of literature, expressed qualifications of reported data, and the basis of any assumptions made in deriving information or conclusions.

The QAPP should clearly state what results, data, or narratives are to be reported. Reporting units, format of presentation, and deliverable dates should also be specified for evaluating the completeness of the final report.

Internal QC checks should be planned with specific purposes in mind. Strategic applications of QC can help determine the source of problems. For example, trip blanks are used to check for contamination in the transport and storage of samples and replicate samples taken in the field check the representativeness of the sampling. Control charting of specific and continuous test parameters are used to document system performance and trigger corrective action.

Audits should be coordinated with the sensitivity of data. Different types of audits are used in hazardous waste measurements, including systems audits, performance evaluation samples, and data quality audits.

Systems audits are conducted prior to, or early in, a project to evaluate that a viable system exists to complete the work in a manner consistent with the test plan or QAPP. Systems audits may include observation of actual field sampling or laboratory operations and inspection of equipment maintenance and calibration records.

Performance evaluation samples are usually introduced into the analytical flow as a check on analytical accuracy and reporting.

Data quality audits consist of reviewing all QC data, verifying the traceability of information, and independently recalculating representative results from randomly selected data points. The purpose of such a data audit is to find systematic errors; data audits are no guarantee of accuracy for every data point or reported value.

When predefined data quality objectives are not met or when there are significant losses of data, corrective action must be taken. At the very least, the problem should be identified, investigated, and any resultant actions documented in the project records. Sometimes, the full impact upon the data is not evident until the project is completed. Documentation of corrective action also helps identify recurring or systematic problems with specific matrices or methodology. QA/QC audits, inspections, and corrective actions are reportable to management and serve to strengthen the credibility of the technical data.

Sample matrix:
Date of extraction:
Date of analysis:

Batch no.:
Calibration reference standard:

	Validation steps	YES	NO	Comment
1.	Were all appropriate QC samples included with batch?			
2.	Was routine maintenance and inspection of analytical instruments performed prior to sample analyses?			
3.	Were initial and continuing calibration within criteria?			
4.	Have control charts been updated for current instrument performance, accuracy and precision of control spikes, and accuracy and precision of matrix spikes?			
5.	Are all charted data points within control limits?			
6.	Does the method blank indicate or confirm no background contamination or interference?			
7.	Are surrogate recoveries within criteria for all samples, blanks, and spikes (for SW-846 Soils Method?)			
8.	Are retention times of major Aroclor peaks within criteria windows?			
9.	Do Aroclor patterns found in samples match reference standards?			
10.	Have chromatograms been reviewed for consistency with the computer quantitation report?			
11.	Have all corrective actions (reanalysis, redilutions, additional cleanup, alternative calculation, etc.) been documnted?			
12.	Has the accuracy of intermediate data manipulations and/or transcribed results been verified?			
13.	Have all data generated with QC data outside of established criteria been flagged as "estimated value"?			

Chemist/reviewer: ______

Data: ______

Figure 7. Example of data validation checklist.

Reports

The result of most projects is information. Project reports vary with the type and complexity of the project and the deliverable agreements in the project plan. All project participants must assure that the facts within their individual contributions to the report and other documents related to the reported results are accurate, fully supportable, and presented in a professional manner. Typically, all reports are prepared in draft form for internal review prior to distribution in either draft or final form outside of the organization.

The security of the derived information should also be covered by quality assurance. Public disclosure or distribution of reports or data are strictly prohibited for information that is privileged or confidential, unless authorized in writing by the organization's management and the clients involved. All reports containing confidential information must be controlled and logged out by an authorized Document Control Officer, using strict procedures such as those for handling Toxic Substances Control Act Confidential Business Information.

Following completion and approval of the report, project files are collected, collated, and inventoried for final archiving. Archiving of the project records should be completed as soon as possible for security and easier retrieval.

Following contractual arrangements made in the project plan, the project leader issues a memorandum to the project files and to the sample custodian which specifies retention schedules and final disposition of any samples generated or remaining after project completion.

COMMITMENT TO QUALITY ASSURANCE

The commitment to QA varies with each organization and project. It may be driven by cost control, contractual agreements, a true concern for our environment, or pride and satisfaction in performing quality work. Whatever the driving force, QA should force us to look at the way we perform our work and the level of credibility and confidence that is evident in the documentation long after the project is completed.

REFERENCES

1. "National Enforcement Investigations Center Policies and Procedures Manual," U.S. EPA, Denver, CO, EPA-330/9-78-001-R (1986), p. II-53.
1a. Aschwanden, C. Personal communication (1986).
2. "Test Methods for Evaluating Solid Waste: Physical/Chemical Methods," (SW-846), U.S. EPA Office of Solid Waste and Emergency Response (November, 1986), Chapter 1.
3. "Air and Energy Engineering Research Laboratory Quality Assurance Procedures

Manual for Contractors and Financial Assistance Recipients," U.S. EPA, Section 3.4 (draft, July 20, 1987).

4. "Interim Guidelines and Specifications for Preparing Quality Assurance Project Plans," QAMS-005/80, U.S. EPA Office of Monitoring Systems and Quality Assurance (December 1980), Section 1.

LIST OF AUTHORS

Marilew Bartling, Martin Marietta Energy Systems, Inc., Oak Ridge, Tennessee 37831-7440

J. V. Burton, Woodward-Clyde Consultants, Wayne, New Jersey 07470

Kevin Cappo, State Water Survey Division, Savoy, Illinois 61874

John G. Cleland, Center for Technology Applications, Research Triangle Institute, Research Triangle Park, North Carolina 27709

William H. Cole, III, Advanced Monitoring Systems, Lockheed Engineering and Sciences Company, Las Vegas, Nevada 89119

David M. DeMarini, Genetic Bioassay Branch, Genetic Toxicology Division, Health Effects Research Laboratory, U.S. Environmental Protection Agency, Research Triangle Park, North Carolina 27711

Larry Eccles, Aquatic and Subsurface Monitoring Branch, U.S. Environmental Protection Agency, Environmental Monitoring Systems Laboratory, Office of Research and Development, Las Vegas, Nevada 89119

Robert E. Enwall, Advanced Monitoring Systems, Lockheed Engineering and Sciences Company, Las Vegas, Nevada 89119

Marianne L. Faber, Advanced Monitoring Systems, Lockheed Engineering and Sciences Company, Las Vegas, Nevada 89119

Aida Fuentes, Department of Environmental and Industrial Health, The University of Michigan, Ann Arbor, Michigan 48109

B. S. Gill, Environmental Health Research and Testing, Inc., Research Triangle Park, North Carolina 27711

Dennis Hooton, Midwest Research Institute, Kansas City, Missouri 64110

Virginia Stewart Houk, Genetic Bioassay Branch, Genetic Toxicology Division, Health Effects Research Laboratory, U.S. Environmental Protection Agency, Research Triangle Park, North Carolina 27711

Douglas S. Kendall, National Enforcement Investigations Center, U.S. Environmental Protection Agency, Denver Federal Center, Denver, Colorado 80225

Eugene J. Klesta, Quality Control, Chemical Waste Management, Inc., Technical Center, Riverdale, Illinois 60627

Steven P. Levine, Department of Environmental and Industrial Health, The University of Michigan, Ann Arbor, Michigan 48109-2029

Mark F. Marcus, Chemical Waste Management, Inc., Technical Center, Riverdale, Illinois 60627

James E. Martin, School of Public Health, The University of Michigan, Ann Arbor, Michigan 48109-2029

John W. Nixon, Waste Management of North America, Walker, Louisiana, 70785

M. L. Papp, Lockheed Engineering and Sciences Company, Las Vegas, Nevada 89119

Gregory A. Raab, Advanced Monitoring Systems, Lockheed Engineering and Sciences Company, Las Vegas, Nevada 89119

Brian A. Schumacher, Soils Quality Assurance Group, Lockheed Engineering and Sciences Company, Las Vegas, Nevada 89119

K. C. Shines, Lockheed Engineering and Sciences Company, Las Vegas, Nevada 89119

Milagros S. Simmons, Department of Environmental and Industrial Health, The University of Michigan, Ann Arbor, Michigan 48109

Martin A. Stapanian, Chemometrics Section, Lockheed Engineering and Sciences Company, Las Vegas, Nevada 89119

Robert L. Tidwell, Jr., Soils Quality Assurance Group, Lockheed Engineering and Sciences Company, Las Vegas, Nevada 89119

Dennis Wesolowski, U.S. Environmental Protection Agency, Region V, Chicago, Illinois 60605

Index